HADRONIC PHYSICS
AT INTERMEDIATE ENERGY

HADRONIC PHYSICS AT INTERMEDIATE ENERGY

Winter School held at Folgaria, Italy,
First Course, February 17-22, 1986

Edited by

Tullio BRESSANI

Istituto di Fisica Superiore,
Università di Torino,
I.N.F.N.,
Sezione di Torino,
Italy

and

Renato A. RICCI

Dipartimento di Fisica,
Università di Padova,
I.N.F.N.,
Laboratori Nazionali di Legnaro,
Italy

1986

NORTH-HOLLAND
AMSTERDAM · OXFORD · NEW YORK · TOKYO

© Elsevier Science Publishers B.V., 1986

All rights reserved. No part of this publication may be reproduced, stored in a retrieval system, or transmitted, in any form or by any means, electronic, mechanical, photocopying, recording or otherwise, without the prior permission of the publisher, Elsevier Science Publishers B.V. (North-Holland Physics Publishing Division), P.O. Box 103, 1000 AC Amsterdam, The Netherlands.

Special regulations for readers in the USA: This publication has been registered with the Copyright Clearance Center Inc. (CCC), Salem, Massachusetts. Information can be obtained from the CCC about conditions under which photocopies of parts of this publication may be made in the USA.

All other copyright questions, including photocopying outside of the USA, should be referred to the publisher.

ISBN: 0 444 87023 7

Published by:

North-Holland Physics Publishing
a division of
Elsevier Science Publishers B.V.
P.O. Box 103
1000 AC Amsterdam
The Netherlands

Sole distributors for the U.S.A. and Canada:

Elsevier Science Publishing Company, Inc.
52 Vanderbilt Avenue
New York, N.Y. 10017
U.S.A.

Library of Congress Cataloging in Publication Data

```
Hadronic physics at intermediate energy.

   Sponsored by the Italian Physical Society and the
University of Verona.
   Includes bibliographies and indexes.
   1. Hadrons--Congresses.  2. Hadron interactions--
Congresses.  I. Bressani, Tullio, 1940-    .  II. Ricci,
R. A.  III. Societá italiana di fisica.  IV. Università
di Verona.
QC793.5.H322H32 1986     539.7'216       86-23566
ISBN 0-444-87023-7 (U.S.)
```

PRINTED IN THE NETHERLANDS

PREFACE

The Winter School on "Hadronic Physics at Intermediate Energy", First Course, was held at Folgaria (Trento), on February 17–22, 1986, and was attended by 63 participants from 18 institutions. The School presented at least two original aspects:

– It was orientated towards students and young physicists of both Particle and Nuclear Physics with lecturers belonging equally to the two fields. It was a clear demonstration that, at least in the Intermediate Energy domain, there is a substantial unification between the two disciplines giving mutual benefits at a time in which specialization and fragmentation seem to be the general tendencies in physics. The School was organized following the results of the Workshop on Nuclear and Particle Physics at Intermediate Energies with Hadrons, held in Trieste, April 1–3, 1985 and, in preparation, at a tutorial level, to the Conference on the European Hadron Facility (E.H.F.) held at Mainz, March 10–14, 1986.
– The lectures were held in Italian with two exceptions. The reason was not a return to an out of date provincialism, which would certainly not be in the tradition of the community of Italian physicists, but due to the fact that the School was organized within a very short time bringing with it the utility of informal contacts between lecturers and students on a national level.

As a consequence no edition of the proceedings was planned and the lecturers were aware of that fact. However, at the end of the School, and following a round-table discussion between all participants, it appeared that the level and the completeness of the lectures were such that it was worthwhile to publish them in English in order to allow an international diffusion. As a result of a meeting in Mainz with Drs. A.P. de Ruiter, the North-Holland Physics agreed to publish the proceedings which now appear to be a useful complement, at an elementary level, to the Proceedings of the Conference on the E.H.F. Nearly all the lecturers agreed to prepare, within a very short time, the written version of their lecture.

The School was organized into four Sections :—

1. Theoretical Aspects of the Hadronic Interactions, *coordinator* G. Preparata;
2. Subnuclear Physics with Hadrons at Intermediate Energy, *coordinator* P. Dalpiaz;
3. Nuclear Physics with Hadrons at Intermediate Energy, *coordinator* T. Bressani;
4. Instrumental Aspects, *coordinator F. Bradamante.*

all touching on the neighbouring fields of heavy ion and muon physics.

The Istituto Nazionale di Fisica Nucleare (I.N.F.N.) supplied a generous financial grant for the organization of the School and we are grateful to the President, Professor N. Cabibbo, for his kindness and interest. We acknowledge the support of the Tourist Agency of Folgaria and, in particular, Dr. C. Taddei and Mr. E. Cappelletti for their help with the organization and the subvention of the social activities. The sponsorship of the Italian Physical Society and the University of Verona are gratefully acknowledged. Our thanks also go to Professor S. Galassini and Professor G. Moschini who acted as Scientific Secretaries.

T. BRESSANI
R.A. RICCI
Directors of the School and
Editors of the Proceedings

TABLE OF CONTENTS

Preface   v

I. **THEORETICAL ASPECTS OF THE HADRONIC INTERACTIONS**

    The Standard Model: Its Foundations and its Problems
        G. PREPARATA   3

    Hadron Spectroscopy -
    A Relativistic Quark-Diquark Model
        E. PREDAZZI   49

    Quark and Lepton Flavor Mixing:
    Role of Intermediate Energies
        N. PAVER   65

    Spin Effects at Short Distances
        J. SOFFER   85

    On the EMC Effect
        E. PREDAZZI   103

II. **SUBNUCLEAR PHYSICS WITH HADRONS AT INTERMEDIATE ENERGY**

    Hadron Spectroscopy from $\bar{p}p$ Annihilation
        P. DALPIAZ   119

    Rare Decays of K Mesons
        R. CESTER   129

    Charm and Beauty Physics at $\sqrt{s} = 20$ GeV with
    a $p\bar{p}$ Collider of High Luminosity
        P. PISTILLI   141

Introduction to Spin Phenomena in
High Energy Particle Physics
    P. SCHIAVON   149

Coherent Production
    G. BELLINI and L. MORONI   163

Use of the European Hadron Facility Combined with
an Underground Detector to Study
Neutrino Oscillations
    M. DE VINCENZI and P. PISTILLI   181

Present Knowledge of the Axial-Vector Weak
Interaction Coupling Constant
    A. BERTIN and A. VITALE   189

Muon Capture in Hydrogen and Deuterium:
Next Generation Experiments
    A. BERTIN and A. VITALE   201

Physics with Jet Targets at the SPS $\bar{p}p$ Collider
    L. DICK and W. KUBISCHTA
    (UA-6 Collaboration)   209

III. NUCLEAR PHYSICS WITH HADRONS AT INTERMEDIATE ENERGY

Perspectives of the Nuclear Physics:
The Role of the Hadronic Probes
    T. BRESSANI   223

Relevant Aspects and Perspectives of Low-Energy
Heavy-Ion Reactions
    R.A. RICCI   235

Color Degrees of Freedom in Nuclear Physics
    F. CANNATA   249

Hypernuclear Physics
    T. BRESSANI   259

Hadron Scattering on Nuclei
    S. COSTA   279

Antiproton-Nucleus Interaction:
Review of the Experimental Situation
    G. PIRAGINO   293

Exploring Quark-Gluon Degrees of Freedom in Nuclei
    E. CHIAVASSA   311

High-Energy Nucleus-Nucleus Collisions
and Nuclear Matter
    R.A. RICCI   321

IV. INSTRUMENTAL ASPECTS

High Intensity Proton Synchrotrons
and the EHF Project
    F. BRADMANTE   341

Use of Semiconductors in Experimental Physics
    F. FORTI, M.A. GIORGI and G. TRIGGIANI   363

The Radiofrequency Quadrupole Linear Accelerator
    M. PUGLISI   387

Antiprotons Trapping and Cooling
    N. BEVERINI, F. SCURI and G. TORELLI   427

An Introduction to Beam Cooling Techniques
    R. CALABRESE   435

Electron Cooling for $\bar{p}$-p Machines at
Intermediate Energies
    L. TECCHIO   447

Cherenkov Pick-Ups in the Microwave Band
    G. DI MASSA and V.G. VACCARO   455

List of Participants 465

Author Index 469

Subject Index 471

I

THEORETICAL ASPECTS OF THE HADRONIC INTERACTIONS

THE STANDARD MODEL: ITS FOUNDATIONS AND ITS PROBLEMS.

Giuliano PREPARATA(*)

INFN-Laboratori Nazionali di Frascati(Italy)

INTRODUCTION

These lectures have been prepared for the "1st Italian Winter School on Hadronic Physics", whose main aim was to bring together the communities of the "intermediate energy" nuclear physicists and of the "low energy" particle physicists, in the light of a convergence of interests in the development of a European Hadron Facility (EHF).

I had the not so easy task to introduce the Standard Model to an audience who knew either very little about it or too much, so that I stood a good chance to be useless to everybody, by entering into too many details for the nuclear physicists or too few for the particle physicists. I thus decided to divide my lectures into two parts, the first devoted to the foundations of the Standard Model, i.e. its physics, theoretical background, and phenomenology, the second focussed on its outstanding problems both in the electroweak sector (the Higgs problem) and in the colour sector (the problem of confinement and perturbative QCD), and on the very recent progress towards understanding colour confinement.

In this way I believe I could satisfy the need of nuclear physicists for having a general idea about the Standard Model without boring the particle physicists too much, for many elementary observations, especially those relating to the phenomenology of hadronic physics and the problems of perturbative QCD, are usually not found in the existing specialized literature, and reflect much of my own approach to the subject.

For all the above reasons the reader will not find in these lectures a systematic account neither of the foundations of Standard Model, nor of the developments of the alternative view of the Standard Model that I have been associated with: for this there exist several easily accessible reviews in the literature. The emphasis has been rather on giving a birdeye's view of this wide and complex problematics, that could be stimulating enough to induce the readers to wishing to learn about it in more depth and width and above all with a more critical eye.

Judging from the enthusiasm which students and teachers alike expressed

---
(*) On leave of absence from: Dipartimento di Fisica, Universita' di Bari, Via Amendola 173, BARI(Italy).

for seeing these lectures in print, I have the feeling that my effort, with all its limitations, was worth it and that the publication of these lectures might eventually have some small positive influence on the growth of a community of physicists which appears to have just been born: the hadronic physics community.

PART I: THE FOUNDATIONS OF THE STANDARD MODEL

I.1. PHYSICAL MOTIVATIONS

The Standard Model, as we know it today, is the final product of a long path of research both experimental and theoretical to understand the fundamental forces, which has occupied the physicists for the better part of this century.

Two are the key notions that are at the basis of any modern analysis of the fundamental forces of Nature:

a) the Symmetries,
b) the Dynamics.

The symmetries deal, so to say, with the <u>kinematics</u>, while the <u>dynamics</u> focusses on the time evolution of the systems we are interested in: the subnuclear particles and fields.

From the purely historical point of view it is natural that in a first, explorative phase the analysis concentrates on the symmetry aspect of the basic interactions, in search of meaningful patterns, that could unify a large number of physical phenomena which appear uncorrelated. It is the very need of the human mind for simplicity that makes the search of symmetries so basic and important.

Theoretically the analysis of symmetry patterns makes use of the powerful mathematical methods of Group Theory, which have proved very useful in the analysis of many fundamental physical problems, in atomic and solid state physics. There is however a danger in the somewhat automatized application of the mathematical instruments, that seems to be prevailing today. Indeed, the results one gets risk to be rather misleading if one does not realizes very clearly that there is an intimate and subtle relationship between the symmetries and the dynamics. An illustration particularly meaningful of such a relationship is the phenomenon known as "spontaneous" symmetry breaking.

Let me explain it with an example taken from scholastic philosophy: the Buridan's ass. (See FIG.1), which can be schematized in the following way. A pointlike ass is at the center of a circular arrangement of hay. The system ass plus hay possesses perfect rotational symmetry, and it would remain that way if the ass, having become hungry, would not head toward the hay to satisfy its hunger. The rotational symmetry is thus lost, or "spontaneously" broken.

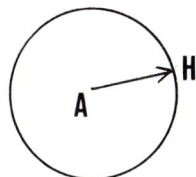

FIGURE 1
The schematized Buridan's ass A that breaks "spontaneously" the rotational symmetry by heading towards the hay H in a random direction (arrow).

This, as we shall see later on is typically what happens in gauge theories, where the equations of motion often possess much more symmetry than their solutions.

Thus when we study the symmetries, without having a clear grasp of how the dynamics works, we should not be surprised if in real life the systems under consideration have very little resemblance with what we are led to expect from the basic structure of the theory. We shall see that this is just what happens to non-abelian gauge theories in their perturbative realization.

After this brief methodological discussion, let me quickly review the main steps that have been made in search of the symmetries of the subnuclear particles and their interactions.

We shall begin by analyzing the hadronic symmetries. As is well known the first step in this direction was taken in the thirties, with the discovery by W.Heisenberg of the isospin symmetry $SU(2)_I$. It was the attempt to generalize isospin to include the strange particles, discovered in the fifties, that led M. Gell-Mann and Y.Neeman to propose in 1961 the unitary symmetry $SU(3)_f$ (f denotes the word: FLAVOUR, to distinguish it from the more fundamental colour symmetry)[1]. One of the main achievements of $SU(3)_f$ was its ability to classify according to simple irreducible representations (1,8 and 10 dimensional multiplets) all mesons and baryons (See FIG.2)

It must be stressed that from a fundamental point of view $SU(3)_f$, which is broken by mass differences, does not present today a significant interest, for its relevance appears to be related to a rather fortuitous accident: the

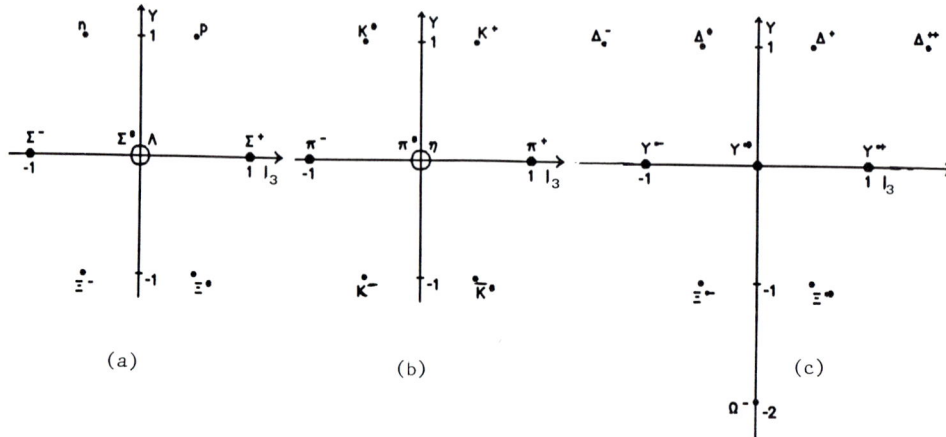

FIGURE 2
The meson (a) and baryon (b) octets, with the decuplet (c) of baryonic resonances, were the first SU(3)$_f$ families of hadrons to be discovered.

smallness of the mass differences among the so-called light quarks u,d and s. From a historical standpoint, however, it is difficult to underestimate the importance of SU(3)$_f$, for it led to the emergence of the all-important notion of quark.

Indeed a curious aspect of SU(3)$_f$, as compared with isotopic spin, is that the fundamental 3-dimensional representation (See FIG.3) does not seem to be realized in the hadronic world. Nevertheless by introducing this representation (and its complex conjugate) and calling it quark (antiquark), Gell-Mann and Zweig [2] were able in 1964 to show that the known mesons could be thought as composed of a $q\bar{q}$-pair while the baryons would consist of three quarks. In this way one achieves a <u>natural</u> explanation why mesons can be organized into one- and eight- dimensional multiplets, ($3\otimes 3=1\oplus 8$) and baryons span one-, eight- and ten-dimensional representations ($3\otimes 3\otimes 3 = 1\oplus 8\oplus 8\oplus 10$).

Needless to say that in all these developments, that took place in the better part of the sixties, no explanation is provided why the fundamental SU(3)$_f$ representation (the quark) is not realized in the real world, nor why representations of non zero <u>triality</u>(*) are similarly excluded from the physical spectrum.

It is to give a natural basis for describing eventually these very strange facts that another type of SU(3) group was invented, the group SU(3)$_c$ (c stands for colour). If, following O.W. Greenberg [3], we assume that each

---

(*) Triality is defined as the quark number N , modulo three, i.e. t = N (mod.3). Thus mesons and baryons have both t = 0.

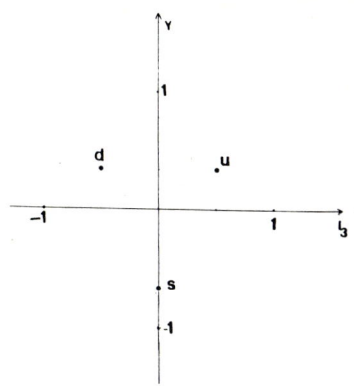

FIGURE 3
Unlike Isotopic Spin, $SU(3)_f$ does not have its fundamental representation among the physical states.

of the quarks u,d,s comes in <u>three colours</u> (say red, blue and green), then the zero-triality rule for physical hadronic states can be phrased as the requirement that physical states must be colour neutral, or colour singlets. This means that if we apply a $SU(3)_c$ transformation to any physical state, it must remain unaffected. Thus states such as q, qq, qq$\bar{q}$, ... cannot be physically realized, because they <u>cannot</u> be invariant under colour, while q$\bar{q}$, qqq, qq$\bar{q}\bar{q}$, ... can exist because one can project out of them well defined colour singlets.

It goes without saying that the colour hypothesis does not <u>explain</u> why only colour singlets are to be found in nature, rather it gives a simple group theoretical basis to <u>describe</u> the physical spectrum. We shall in fact see later that in order to explain such peculiarity of the hadronic spectrum, which we call <u>colour(or quark) confinement</u>, it will be necessary to get hold of the dynamics of the colour interaction; symmetry alone is obviously not sufficient.

Colour confinement is not the only motivation for $SU(3)_c$, though in my mind certainly the most important one. Let me simply mention other motivations, which historically have played an important role:

i.   the symmetry of the baryonic wave-functions;

ii. the value of the amplitude A ($\pi^0 \to \gamma\gamma$);

iii. the asymptotic value of the ratio R = $\sigma(e^+e^- \to hadrons)/\sigma(e^+e^- \to \mu^+\mu^-)$.

As for the flavour properties of subnuclear interactions (+) our present

---
(+) Among which the number of generations of quarks and leptons and their masses.

understanding is essentially nonexistent; the general impression being that it will only come from a deeper assessment of the nature of the other aspect of the Standard Model: the electroweak sector $SU(2)_L \otimes U(1)_Y$.

The idea that there could be an intimate link between weak and electromagnetic interactions, goes back to the very beginning of weak interaction theory, the famous 1934 paper of Enrico Fermi [4]. By proposing a current-current weak interaction Fermi clearly implied that the mechanism of transmission of the weak forces should have some fundamental point in common with the electromagnetic interaction. The idea got further corroborated by the discovery of the V-A theory in 1958, and became the present electroweak theory in 1967, when S.Weinberg ad A.Salam independently completed[5], through a well defined mechanism for the generation of the mass of the intermediate vector bosons (*), a theoretical proposal of S.Glashow[6].

Let us briefly describe the basic points of the electroweak theory. From the symmetry standpoint the well known charged weak currents $J^{\pm}_{\mu}(x)=(V-A)^{\pm}_{\mu}(x)$ determine, through the associated charges $Q^{\pm}_L = \int d^3x\, J^{\pm}_o(\vec{x},t)$, the simplest non trivial Lie algebra, of $SU(2)_L$ (L stands for left, V-A), which besides $Q^{\pm}$ contains another charge Q_3, which is then associated with a neutral current. It was in fact the discovery of a neutral weak current in the early seventies that lent very strong support to this particular group structure of the weak interactions and to the Glashow-Salam-Weinberg (GSW) theory, which is based upon it.

If to the charges $Q^{\pm}$, Q_3 we add the electric charge Q, and by a relation of the Gell-Mann-Nishijima type we introduce the weak hypercharge Y_W :

$$Q = Q_{3L} + Y_W/2, \qquad (I.1.1)$$

it turns out that this operator possesses a rather complex structure. It can in fact be decomposed into a left-handed, Y_L, and a right-handed, Y_R, component:

$$Y_W = Y_L + Y_R, \qquad (I.1.2)$$

and the relative eigenvalues of the six known leptons and of the six (including the still to be discovered top-quark) quarks can be readoff in TABLE I.

The four charges $Q^{\pm}_L$, Q_3 and Y_W are the generators of the electroweak group $SU(2)_L \otimes U(1)_Y$, which describes the (badly broken) symmetry of the weak and electromagnetic interactions. As far as we know all weak and electromagnetic phenomena have their foundation on the simple structure of

---

(*) We shall see that this mechanism - the Higgs mechanism - has however serious problems.

LEPTON	Q_{3L}	Y_L	Y_R	Q	QUARK	Q_{3L}	Y_L	Y_R	Q
ν_e	1/2	-1	0	0	u	1/2	2/3	4/3	2/3
e	-1/2	-1	-2	-1	d	-1/2	2/3	-2/3	-1/3
ν_μ	1/2	-1	0	0	c	1/2	2/3	4/3	2/3
μ	-1/2	-1	-2	-1	s	-1/2	2/3	-2/3	-1/3
ν_τ	1/2	-1	0	0	t	1/2	2/3	4/3	2/3
τ	-1/2	-1	-2	-1	b	-1/2	2/3	-2/3	-1/3

TABLE I
The Q_{3L}, Y_L, Y_R and electric charge assignments of leptons and quarks.

the electroweak symmetry. However a look at TABLE I, showing the complicated charge and hypercharge assignments of leptons and quarks, strongly suggests that the Standard Model:

$$SU(3)_C \otimes SU(2)_L \otimes U(1)_Y$$

cannot be the final answer to the problem of subnuclear forces. Many attempts at unifying and simplifying the various pieces of the standard model have recently been made, but so far with little or no success. It is my feeling that in order to go any further some crucial experimental input is needed from the still inaccessible TeV region.

I.2. GAUGE THEORIES: THE DYNAMICS OF SYMMETRIES

The experimental consolidation of the Standard Model, that took place at the end of the sixties and at the beginning of the seventies, has been guided in a decisive way by the development of the deep and very elegant theoretical framework of the Gauge Theories. The most distinctive features of this class of theories is to provide a natural link between the concept of symmetry and of local field theories, of which QED has been for more than four decades, so to speak, the unfailing "paradigm".

The central idea of Gauge Theories is the notion of "eichinvarianz" introduced by H.Weyl in 1917 [7], in an unsuccessful attempt to generalize Einstein's General Relativity to include the electromagnetic phenomena. As is well known a crucial role in the theory of curved spece-time (Riemann geometry) which is at the basis of Einstein theory of gravitation is played by the concept of "parallel transport of a vector", which can be illustrated as follows. (FIG.4)

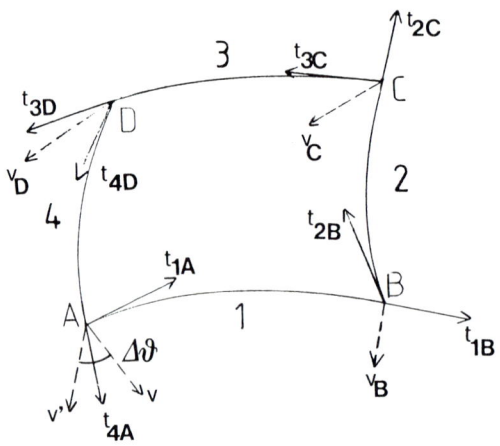

FIGURE 4
The parallel transport of a vector around a closed circuit on a curved surface.

Starting from the point A we can form a closed circuit by connecting four arcs of geodetics (AB,BC,CD and DA) which join the four points A,B,C and D by paths of minimum length. One distinctive feature of any geodetic line is that it is generated by "parallel transport" of its tangent vector: in addition the operation of "parallel transport" preserves the angles between any two vectors. Keeping all this in mind we can now easily understand why the parallel transport depicted in Fig.4 transforms the vector $\vec{v}$ at A, along successive steps, into $\vec{v}_B$, $\vec{v}_C$, $\vec{v}_D$ to end up in the vector $\vec{v}'$ at an angle $\Delta\theta$ with respect to the starting vector $\vec{v}$. The non-coincidence of $\vec{v}$ and $\vec{v}'$ is a typical signal for the curvature of space; indeed taking an infinitesimal circuit one can write

$$\Delta\theta \sim R\, \Delta\sigma \, , \qquad (I.2.1)$$

where R is related to the local curvature of space and $\Delta\sigma$ is the area of our circuit. Thus, if and only if space is locally flat (R=0) the parallel transported vector coincides with the initial one.

In a Riemannian geometry parallel transport changes only the direction of the vector, but not its length nor the scalar product (relative angle) with another vector. Weyl then thought that in the real space-time also the length ℓ of the vector $\vec{v}$ changes by parallel transport and he realized that, if this happened, in a way completely analogous to (I.2.1) the change could be written as

$$\Delta \ell \sim F_{\mu\nu}^{(x)} \Delta\sigma^{\mu\nu} \tag{I.2.2}$$

$$F_{\mu\nu}(x) = \partial_\mu A_\nu(x) - \partial_\nu A_\mu(x) \tag{I.2.3}$$

where $F_{\mu\nu}$ is an antisymmetric tensor, whose properties are precisely those of the Maxwell electromagnetic field strengths. In this way both gravity and electromagnetism would reflect an inherent property of the geometry of the space-time continuum, thus achieving a far reaching unification of the two basic forces of nature. In the same way as Einstein's gravity can be described by a Lagrangian that is invariant under a local redefinition of the coordinate system, Weyl's gauge-invariance requires that the Lagrangian be invariant under a local redefinition of the length of a vector, i.e. under the simultaneous transformations [$g_{\mu\nu}(x)$ is the metrical tensor]

$$\begin{aligned} g_{\mu\nu}(x) &\to \lambda(x)\, g_{\mu\nu}(x)\,, \\ A_\mu(x) &\to A_\mu(x) - \partial_\mu \lambda(x)\,; \end{aligned} \tag{I.2.4}$$

where $\lambda(x)$ is an arbitrary function of the space-time point x_μ.

However it was Einstein himself to show in 1920 that Weyl's theory could not describe nature for it would imply that the characteristic frequencies of an atom -its spectral lines- would change after a parallel transport, a phenomenon that has never been observed.

After the discovery of Quantum Mechanics Weyl soon realized [8] that his old idea of generating Maxwell theory by the requirement of gauge invariance had to do with the quantum <u>phase</u> $\phi(x)$ of charged field $\psi(x)$ rather than the <u>length</u> $\lambda(x)$ of a physical event.

Thus, according to Weyl, the QED Lagrangian should be invariant under the following transformations:

$$\psi'(x) = U(x)\psi(x), \tag{I.2.5}$$

$$A'_\mu(x) = U(x)A_\mu(x)U^{-1}(x) - \frac{i}{e}\partial_\mu U(x)\, U^{-1}(x), \tag{I.2.6}$$

where

$$U(x) = e^{ie\lambda(x)} \tag{I.2.7}$$

$\psi(x)$ is the charged electron field, and e is the electromagnetic charge. From (I.2.7) it is clear that the gauge-invariance associated with QED involves the transformations of an abelian U(1) group at each space-time point x. If we now write the QED Lagrangian as

$$L_{QED} = L_{gauge} + L_{matter} + L_{interaction}, \qquad (I.2.8)$$

with $[F_{\mu\nu}(x) = \partial_\nu A_\mu(x) - \partial_\nu A_\mu(x)]$

$$L_{gauge} = -1/4 \, F_{\mu\nu}(x) F^{\mu\nu}(x), \qquad (I.2.9)$$

and

$$L_{matter} = \bar{\psi}(x)(i\slashed{\partial}-m)\psi(x), \qquad (I.2.10)$$

the free Dirac lagrangian. Invariance under the gauge transformations (I.2.5) and (I.2.6) determines uniquely the structure of $L_{interaction}$ as:

$$L_{interaction}(x) = e\bar{\psi}(x)\gamma_\mu\psi(x)A^\mu(x), \qquad (I.2.11)$$

just the minimal interaction!

The procedure we have just followed can be immediately generalized to any non-abelian group G, and in particular to SU(N) [9]. Let us denote by T^α [$\alpha = 1,....,N-1$] the hermitian generators of the group SU(N), they form a Lie algebra:

$$[T^\alpha, T^\beta] = i f^{\alpha\beta\gamma} T^\gamma \qquad (I.2.12)$$

where the totally antisymmetric symbols $f^{\alpha\beta\gamma}$ are the "structure constants" of SU(N) and the T^α's are normalized as $Tr(T^\alpha T^\beta) = 2\delta^{\alpha\beta}$. The generic finite SU(N) transformation U(x) can be then written as (repeated indices are summed over)

$$U(x) = \exp i g \omega^\alpha(x) T^\alpha; \qquad (I.2.13)$$

and defining the gauge field operator $A_\mu(x)$ as:

$$A_\mu(x) = A_\mu^\alpha(x) T^\alpha, \qquad (I.2.14)$$

it is easy to check that the non-abelian field tensor:

$$F_{\mu\nu}(x) = \partial_\mu A_\nu(x) - \partial_\nu A_\mu(x) - ig\,[A_\mu(x), A_\nu(x)]\,, \qquad (I.2.15)$$

transforms covariantly under the gauge transformation (I.2.6) [the charge e is now replaced by the coupling constant g], i.e.

$$F'_{\mu\nu}(x) = U(x) F_{\mu\nu}(x) U^{-1}(x). \qquad (I.2.16)$$

So that we can immediately write down the gauge-invariant Lagrangian for the gauge field:

$$L_{gauge} = -1/2\,\mathrm{Tr}\,(F_{\mu\nu}(x) F^{\mu\nu}(x))\,; \qquad (I.2.17)$$

Note that the gauge invariance of L_{gauge} is an immediate consequence of the gauge covariance (I.2.16) of $F_{\mu\nu}(x)$, which requires the additional commutator term appearing in (I.2.15), as first noted in 1954 by C.N.Yang and R.Mills.

Introducing a Dirac matter field, transforming according to the fundamental SU(N)-representation, i.e. a N-component vector in internal space, its Dirac Lagrangian is:

$$L_{matter} = \bar{\psi}(i\slashed{\partial} - M)\psi \qquad (I.2.18)$$

where M is a hermitian NxN matrix, the mass matrix.

The requirement of gauge invariance under both (I.2.5) and (I.2.6) fixes again uniquely the interaction term as:

$$L_{int} = g\bar{\psi}(x)\,\slashed{A}(x)\,\psi(x)\,. \qquad (I.2.19)$$

Adding the three pieces (I.2.17), (I.2.18) and (I.2.19) for the relevant symmetry group, -in our case $SU(3)_c \otimes SU(2)_L \otimes SU(1)_Y$,- defines the gauge theory associated with the given symmetry.

The reason why Gauge Theories occupy such a focal position in the theoretical analysis of the symmetries of fundamental interactions is, to my mind, twofold:

i. the very notion of internal symmetry, defined at each space-time point, leads naturally to the existence of uniquely determined vector fields (gauge bosons);

ii. the coupling between gauge-fields and matter is unambiguously fixed.

The theory is so neat and simple that the "kit" for constructing the appropriate gauge description of a specified field of natural phenomena can be summarised in a very "easy" set of instructions:

a. Identify the relevant symmetry group G;
   In our case: $SU(3)_c \otimes SU(2)_L \otimes U(1)_y$.

b. Introduce as many gauge-fields as the number of generators of the group G;

   In our case:
   $$SU(3)_c \rightarrow 8 \text{ gluons}$$
   $$SU(2)_L \otimes U(1)_Y \rightarrow 3+1 \text{ vector bosons}(\gamma, W^{\pm}, Z^0).$$

c. Introduce the fermionic matter by assigning the fermions to the appropriate representation of G.

   In our case:

	QUARKS	LEPTONS
$SU(3)_c$	3	1
$SU(2)_L$	2	2
$U(1)_Y$	1	1

   ← Quark lepton universality

d. Write down the gauge-invariant Lagrangian:
   $$L = L_{YM} + L_F + L_I$$
   $$L_{YM} = -1/2 \, Tr(F_{\mu\nu} F^{\mu\nu})$$
   $$L_F = \bar{\psi}(i\slashed{\partial} - M)\psi$$
   $$L_I = g \, \bar{\psi} \slashed{A} \psi \quad .$$

e. Solve the dynamical problem defined by L.

Up to instruction (d) things have really proceeded smoothly, it is only when instruction (e) needs implementing that one realizes that the problems one must solve are rather non-trivial, like:

P_1 Find the ground state;
P_2 Quantize the theory around the ground state;
P_3 Find the spectrum, bound states, etc;
P_4 Calculate the Green's functions, the scattering amplitudes, etc;

So far we have been able to solve these problems only in perturbation theory, where the solutions S_i's to the above problem P_i's are:

S_1 The ground state is the classical vacuum (zero-field);
S_2 Free-field quantization;
S_3 The spectrum is what one "reads" in the Lagrangian (Fock space);
S_4 The systematic approach of Schwinger, Feynman and Dyson produces Green's functions, scattering amplitudes etc. to any order in the coupling constant.

Note however, that the perturbative strategy has produced completely satisfactory results only for QED. We shall see later what the problems are for the rest of the Standard Model.

I.3 A GENERAL PHENOMENOLOGICAL APPROACH TO HADRONIC PHYSICS

Our understanding of the dynamics of a non-abelian gauge theory such as QCD is still at a preliminary stage, even though, as I shall try to discuss at the end of these lectures, we are on the good road towards solving our problems. If it is true that much of this road has been "paved" only recently it should be recalled however that its original direction and scope is due to a broad attempt to get hold of the crucial phenomenological aspects of hadrons and their interactions [10]. This is what shall be discussed next.

Much of our understanding of what happens in the hadronic world depends on two key notions:

i. freedom at short (light-cone) distances;

ii. linear confining potentials at large distances.

According to (i) at short distances the physics is well described by (almost) free partons, which we obviously take to have the properties of quarks, spin 1/2 fermions with the internal quantum numbers (colour, flavour, charge etc.) dictated by the Standard Model. The basis of this very simple behaviour is a long sequence of experiments that were pioneered by the famous SLAC observations of deep inelastic electron-proton scattering, which began in 1967. Let me remind you that the surprising scaling behaviour observed there, later confirmed by countless experiments of high precision, was what gave respectability to the idea that hadrons consisted of (almost) free quarks, and impressed a great impulse to the building of the Standard Model.

And if the quarks at short distances are (almost) free, their confinement must reflect a simple property of their interaction at large distances. This is the real meaning of (ii), which stipulates that what keeps the quarks inside a hadron is a rather smooth interaction, whose associated energy increases linearly with the distance among quarks, without tampering with the (almost) free quark behaviour at short distances.

Both (i) and (ii) lie at the basis of a research programme in which I have been involved since 1971, and which through the successive theoretical steps of the Massive Quark Model (MQM, 1972 [11]), Quark GeometroDynamics (QGD, 1975 [12]) and Anisotropic Chromo Dynamics (ACD, 1980 [13]) has led to a precise and very successful phenomenological description of a large number of facts of hadronic physics. I believe that, no matter in what way the problems of QCD are going to be solved in the future, the phenomenological strength of this approach will constitute a precious guidance for making theory and experiments meet.

Another reason why I thought it useful to present a (necesarrily compressed) review of these development is that, as we shall see in Part II of these Lectures, the very popular approach to short distance physics based on perturbative QCD (PQCD) can be shown to be internally inconsistent (*), and as such cannot supply a viable description of nature. Thus, as far as I know, what I am going to present is the only comprehensive theoretical description of hadronic phenomena, in terms of the two key notions discussed above, that does not suffer from fundamental objections.

My discussion will focus on two subjects: highly inelastic hadron physics, and mesons spectroscopy in ACD. Those of you who want to know more about the MQM-QGD-ACD programme and of its achievements can consult Refs. 10-14.

I.3.a. HIGHLY INELASTIC HADRON PHYSICS

The application of the MQM-QGD-ACD theoretical framework to the physics of highly inelastic hadron collisions rests upon three basic points:

i. Hadrons are made of a small number of "almost" free quarks ($q\bar{q}, qqq, \ldots$) combined in a colour singlets, confined in typical space-time domains, whose size increases <u>linearly</u> with their mass.

ii. There is a basic <u>perturbative</u> <u>structure</u> in the number of quarks that can take part in a given hadronic process;

iii. Hadrons emerge in the final state from the (perturbative) evolution (decay) of the primitive ($q\bar{q}, qqq, \ldots$) states.

(*) i.e. PQCD does not represent a dynamically stable realization of the basic QCD theory. (See later)

Point (i) stems from the fact that the confining forces become strong only at large distances, where the physics is dominated by <u>linearly</u> increasing potentials. Point (ii) is related to the well known fact that hadronic interactions do not find it very easy to excite from the vacuum $q\bar{q}$-pairs (how could we understand otherwise the Zweig rule which is operative even at the light quark level?). The remarkable slowness of the quark pair creation process is thus at the basis of the assumed perturbative structure. Finally point (iii) is just the consequence of the other two, that is particularly relevant for the determination of the structure of hadronic final states in high energy collisions.

From (i) one can easily construct [12] a simple approximation to the spectra of the $q\bar{q}$-and qqq-states, (essentially all that is needed in practice) in terms of a "minimal" set of inputs:

m_q, the quark masses
R^2, the rate of increase of the hadronic size with mass.

As a result one obtains for the $q\bar{q}$-system a set of Regge trajectories described by the simple formula

$$M^2_{n\ell} \approx \pi/R^2 \, (2n+\ell), \qquad (\text{I.3.1})$$

where ℓ is the angular momentum and n is the radial quantum number (FIG.5).

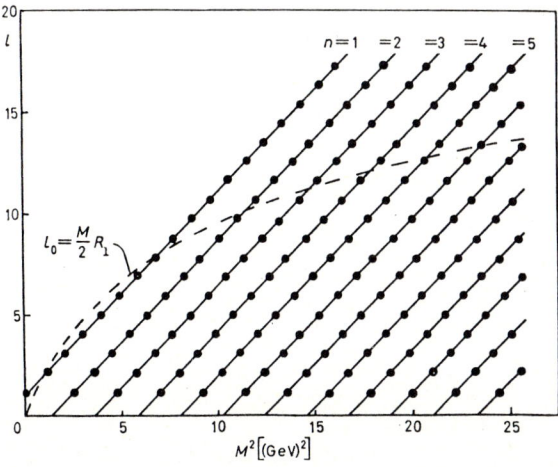

FIGURE 5
The Chew-Frautschi plot of the $q\bar{q}$-states according to QGD.

One can also derive the asymptotic meson w.f.'s (FIG.6), whose structure is given by:

$$\psi_{n\ell}(p,k) \xrightarrow[M_{n\ell} \text{large}]{} F(p_1^2, p_2^2) Y_\ell^m(\vec{\Omega}_k), \qquad (I.3.2)$$

where $\vec{\Omega}_k$ is the direction of the relative $q\bar{q}$-momentum k in the meson rest frame, and

$$F(p_1^2, p_2^2) \sim G(p_1^2) G(p_2^2), \qquad (I.3.3)$$

with the "quark" propagation function $G(p^2)$, having the following behaviours

$$G(p^2) \xrightarrow[p^2 \simeq m_q^2]{} \frac{\sin R^2(p^2-m_q^2)}{(p^2-m_q^2)}, \qquad (I.3.4)$$

$$\xrightarrow[|p^2| \gg m_q^2]{} \left(\frac{1}{p^2}\right). \qquad (I.3.5)$$

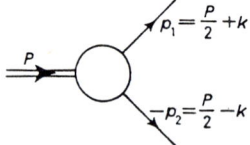

FIGURE 6
The Bethe-Salpeter w.f. describing a $q\bar{q}$-state of 4-momentum p.

Note that while (I.3.4) is a consequence of confinement (absence of the "mass-shell" quark singularity), (I.3.5) reproduces the free field behaviour at short distances (high p^2).

The quasi-degeneracy in the mass M of states with different angular momenta (See FIG.5), allows us to construct a new set of states whose w.f's are ($\ell_0 = MR_\perp /2$)

$$\psi(p,k;\Omega_o) = \sum_{\ell=0}^{\ell_o} \sum_{\ell=-\ell}^{\ell} \psi_{n\ell m}(p,k) Y_\ell^m(\vec{\Omega}_k) \bigg|_{2n+\ell \, fixed} . \qquad (I.3.6)$$

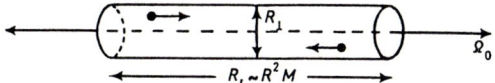

FIGURE 7
The spatial structure of the state (I.3.6).

In configuration space the structure of this state is reported in FIG.7, whereupon it is evident why we can call such a state a Fire-String (FS), oriented along Ω_o. A FS describes a qq-system in which the quarks move (approximately) in the Ω_o-direction with momenta $|\vec{p}| \simeq M/2$ (M is the mass of the FS). In a confined world, such as envisaged in our theoretical framework, the FS's are the closest approximation to the states of the parton model, whose phenomenological relevance is well known. Thus the FS represents a far-reaching link between two previously unrelated languages: Regge Theory and the Parton Model. In high energy hadronic behaviour the FS plays the same role as the coherent states of the radiation field in the semiclassical approximation of QED. We shall see later in fact that quantum interference effects in inelastic hadronic collisions are highly suppressed so that a semiclassical picture emerges in a most natural way.

Following the points (i) and (ii) above, we can envisage the highly inelastic interactions as comprising the following two steps

I. The production of a minimal set of FS's.

II. The decay of each FS into stable (lowest lying) states.

Let us discuss II) first.

The FS decay process consists in the break-up of the initial FS into two smaller FS's, through qq-pair creation in the colour field existing between the two quarks. The initial decay process:

$$FS \rightarrow FS_1 + FS_2 \qquad (I.3.7)$$

is then followed by similar processes involving FS and FS, and so on until the produced FS's have masses of the order of the hadronic low-lying states [$\pi, \rho, A_2, \ldots$]. The evolution of the FS-decay process is described in FIG.8, where there are also reported the diagrams that are used to calculate the amplitude of the decay (I.3.7).

As discussed thoroughly in Ref.15,

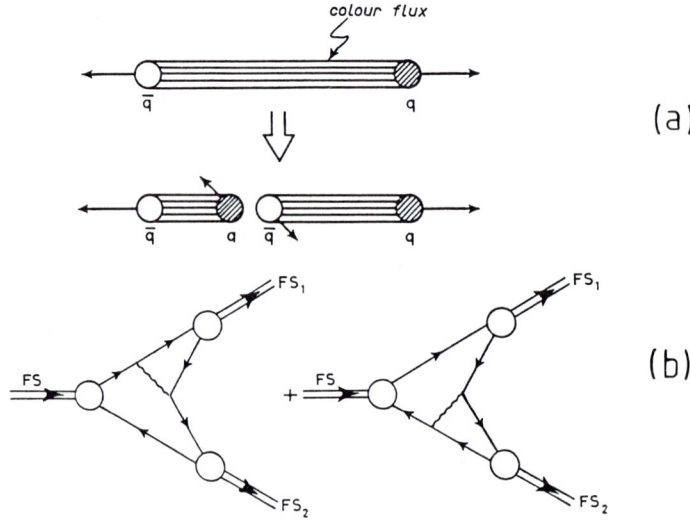

FIGURE 8
The evolution of the FS-decay process (a), and (b) the diagrams describing the basic FS break-up (I.3.7).

the "gluon" responsible for the quark pair creation, is <u>not</u> a dynamical gluon but rather the longitudinal field that is envisaged in ACD. This important difference with the perturbative QCD calculation insures that the calculated amplitude is free from the nasty collinear divergences that plague perturbative QCD.

Without entering into the details of the theoretical analysis which can be found in Ref. 15, I would like to make a few relevant comments:

a. The FS-break-up, unlike the most popular fragmentation models, is a <u>realistic</u> description of a physical process, which leads from a colourless confined state to two other such states which, according to confinement, are the only states that belong to the physical Hilbert space.

b. One can demonstrate the suppression of quantum interference effects, thus making it possible to calculate the FS decay process in terms of probabilities rather than quantum amplitudes. In order to get an idea of the kind of physics which is at work in FS-decay, in FIG.9 the probability $P(x_1,x_2)$ ($x_1 = (M_1/M)^2$, $x_2 = (M_2/M)^2$) is reported for one of the two decay configurations [the "gluon" emitted by the quark line].

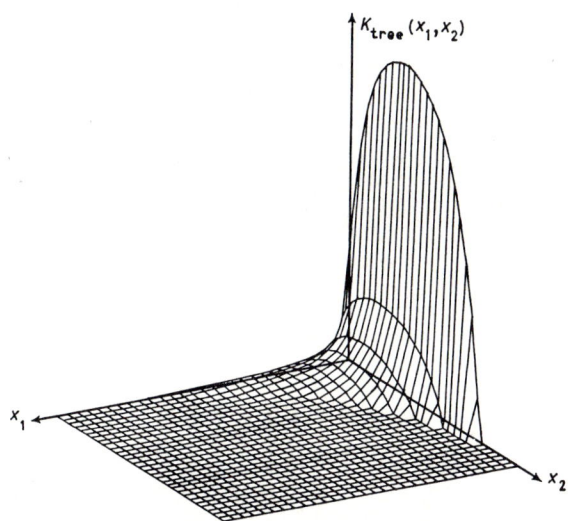

FIGURE 9
The $x_1 x_2$-distribution in the decay process of a FS.

One thing is immediately clear, the mass distributions of the two FS's is very asymmetrical, M_1 being almost always much less than M_2, except for a small probability tail, where also the transverse momentum gets large. This is the so called "3 jet-region".

c. So far we have only described the mesonic decay of the FS, but MQM-QGD allows us also to compute the main features of baryon-antibaryon production [16]. $B\bar{B}$ production arises in fact from the process

$$FS \rightarrow B(\vec{p}) + \bar{B}(-\vec{p}) \qquad (I.3.8)$$

where $B(\bar{B})$ are the lowest lying baryons (antibaryons) belonging to the 56(56) representation of SU(6). The diagram corresponding to (I.3.8) is reported in FIG.10 and it has been computed in Ref. 16.

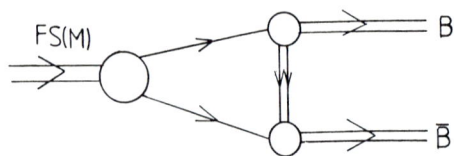

FIGURE 10
The diagram describing the decay process (I.3.8).

The results of this computation are shown in FIG.11, where the branching ratios for total BB production as well as for a few particular channels are plotted as a function of the FS mass M. Please note the strong suppression at high mass ($\sim 1/M$), which implies that BB production will show clear short-range correlations in rapidity.

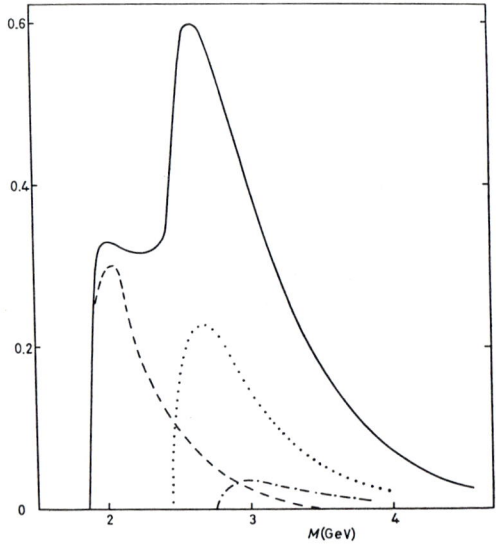

FIGURE 11
The unnormalized branching ratio $\Gamma_{B\bar{B}}(M)/\Gamma_{TOT}$ as a function of M_{FS} for a number of relevant channels —— all channels, --- $p\bar{p}$, ... $\Delta^{++} \bar{\Delta}^{++}$ -·-·- $\Sigma^{*+} \bar{\Sigma}^{*+}$.

Let us now briefly outline the physics of the main production mechanisms of FS's [17]. We shall consider two important classes of processes:

a. e^+e^--annihilation into hadrons;
b. deep-inelastic lepton-hadron scattering;

In e^+e^- annihilation we see at work the simplest mechanism of

FS-production. The process to consider is

$$e^+e^- \xrightarrow[\sqrt{s}]{} \gamma^* \to FS(\sqrt{s}) \qquad (I.3.9)$$

in which the virtual photon γ^* of energy $\sqrt{s}$, produced by the annihilation of the electron-positron pair, couples through the diagram in FIG.12 to a FS of mass $\sqrt{s}$.

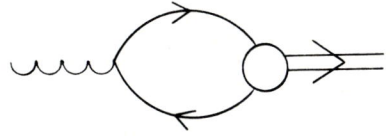

FIGURE 12
The conversion of a highly virtual timelike photon into a FS.

In ACD one can compute the total hadron annihilation cross section by evaluating the transition rates of a virtual photon to the $J^{PC} = 1^{--}$ excited states of the $q\bar{q}$-system, the result is the canonical [18]:

$$R = \frac{(e^+e^- \to \text{hadrons})}{(e^+e^- \to \mu^+\mu^-)} \xrightarrow[\text{large} \sqrt{s}]{} 3\Sigma Q_i^2, \qquad (I.3.10)$$

thus providing the first <u>realistic</u> derivation of this important confirmation of the existence of <u>colour</u> and <u>freedom</u> at short distances. A further visualization of the process (I.3.9) is afforded by FIG.13, where we see that the e^+e^- collision generates a FS in a direction , with respect to the collision axis, distributed according to the well known law:

$$\frac{d\sigma}{d\cos\theta} \simeq (1 + \cos^2\theta) , \qquad (I.3.10')$$

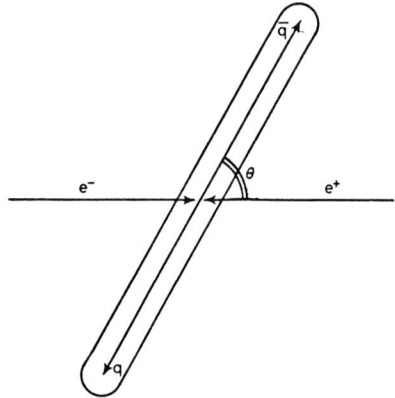

FIGURE 13
The production kinematics of the initial FS in e^+e^--annihilation.

a clear manifestation of the spin 1/2 nature of the quarks.

Now the theoretical description of e^+e^- annihilation is complete, and in order to extract predictions to compare with experiment we need only consider the decay process of the FS produced with the rate (I.3.10) and the angular distribution (I.3.10'), according to the theoretical ideas developed above. This has been done at length in Ref.15, the results, which I cannot discuss here, are extremely satisfactory.

As for deep inelastic scattering, in order to understand this fascinating class of processes we must first describe the quantum mechanical behaviour of the hadron present in the initial state. Let's consider a proton moving at high speed in the lab. If we take a snapshot of it which is quick enough we are likely to see it dissociated (See FIG.14) in a low-lying baryon and a qq-pair. Such a dissociation is the consequence of pair creation in the colour field holding the quarks of the proton together. [According to point (ii) above higher numbers of pairs are suppressed].
This is the picture of the proton that is seen by the incoming virtual photon (or $W^{\pm}$, or Z^0), its scattering however will most of the time take place on the qq-pair rather than on the virtual baryon. For in QGD-ACD the excitation of a baryonic FS is much more difficult due to the peculiar alignment that the three quarks must respect in order to give rise to a physical baryonic state. Thus one basic production mechanism in deep inelastic scattering is (See FIG.15)

$$J + N \rightarrow B^* + FS \qquad (I.3.11)$$

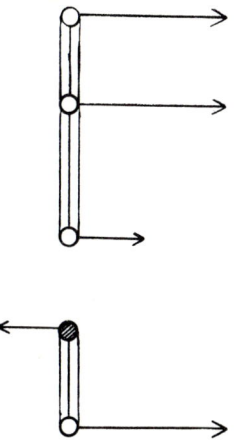

FIGURE 14
The incoming photon as seen by the initial electroweak current.

where the final state comprises an excited baryon (mostly a member of the SU(6) representation 56) and a FS. In order to calculate the diagram in FIG.15

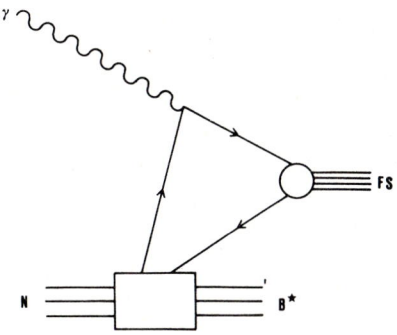

FIGURE 15
The diagram corresponding to (I.3.10).

we must evaluate the proton dissociation amplitude. QGD, providing explicit baryonic w.f's, allows us to perform the calculation [19]. But (I.3.11) is not the only deep inelastic mechanism, even though it dominates for $x = Q^2/2M_\nu \geq .2$, i.e. in the valence region. It may happen that the initial electroweak current J is itself virtually dissociated in a $q\bar{q}$-pair, in this case the physical process that takes place is (See FIG.16)

$$J + N \rightarrow B^* + FS_1 + FS_2 \qquad (I.3.11')$$

with the final state comprising 2FS's and an excited baryon B^*.

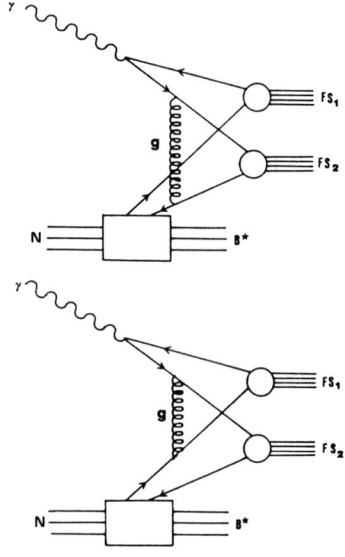

FIGURE 16
The diagrams corresponding to the DIS process (I.3.11).

The 2FS's -production mechanism has also been evaluated [19] and found to dominate the deep inelastic cross-section for small values of x [$x \lesssim 1$], i.e. in the "sea" region. One common feature of the two processes (I.3.11) and (I.3.11') is the presence in the final state of a "leading" baryon, a well known characteristic of high energy hadron-baryon scattering. The detailed evaluation of the dissociation process:

$$N \rightarrow B^* + q\bar{q} \qquad (I.3.12)$$

made possible by QGD shows that the leading baryon carries in the average about one half of the initial nucleon momentum. It is then clear that the momentum carried away by such a well defined physical system is lost to the inelastic collision that excites the FS's. This is a perfectly natural, and realistic answer to the famous problems of the "missing momentum" which is generally believed to imply the existence of dynamical gluons. Our analysis shows that this belief has no really compelling support, unless one wishes to consider the colour field "imprisoned" inside the leading baryon as a "gluon". But then the dynamics of these gluons is the dynamics of the quarks inside the leading baryon, as envisaged by the MQM-QGD-ACD approach.

Again I cannot describe in detail the calculations that have lead to the very good results of ref.19, reported concisely in FIG.17

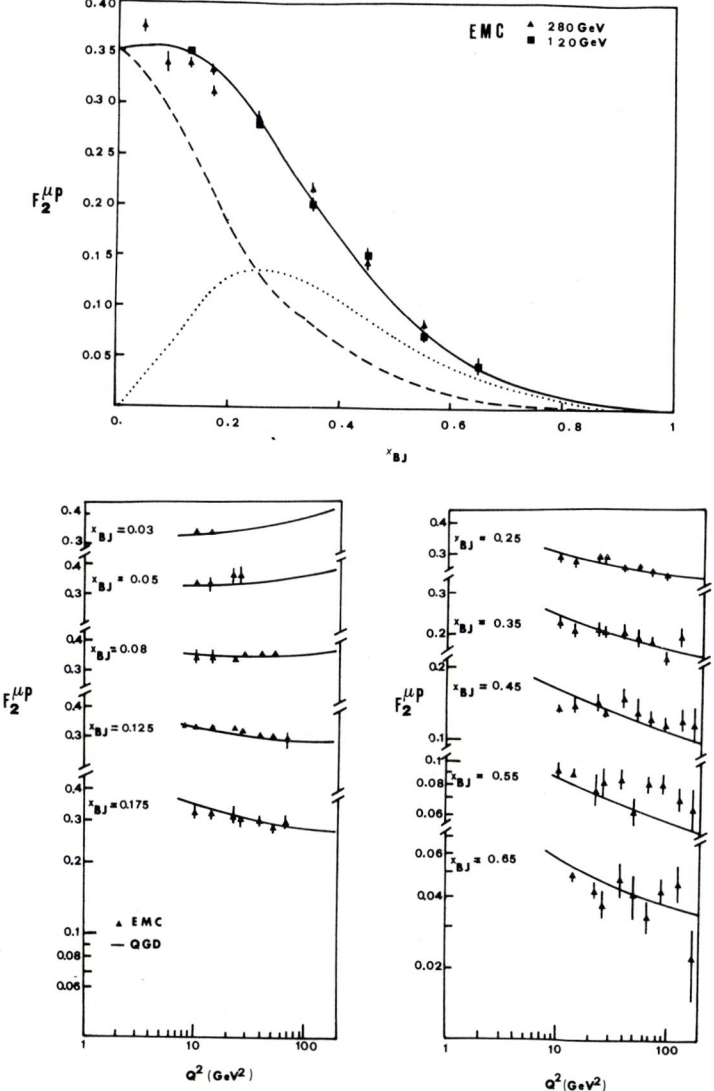

FIGURE 17
(a) The calculated F_2 (x) structure function at $Q^2=20$ GeV^2 (full line), compared with experimental data. The 1FS (dashed line) and 2FS (dotted line) contributions are also shown.
(b) Predictions for scaling violations at different values of x_{B_j} as a function of Q^2.

I.3.b. ACD AND THE MESON SPECTRUM

Before outlining the application of ACD to the calculation of the qq-spectrum, let me briefly describe the basic ideas of Anisotropic Chromo Dynamics [13].

ACD is a gauge theory of the colour degrees of freedom, whose associated fields are defined not on the usual space-time - the Minkowskian continuum M_4 - but on the extended manifold $M_4 \times \bar{S}_3$, which besides the usual space-time, contains a pseudosphere $\bar{S}_3$. I have called this manifold the Anisotropic Space-Time. Its "points" can be expressed by the pair of 4-vectors (x_μ, n_μ) $x_\mu \in M_4$, $n_\mu \in \bar{S}_3$ ($n^2 = -1$), and one can visualize them as "spacelike" arrows of unit length leaving the generic space-time point x_μ. The reason why these extra dimensions have been introduced is very simple. One wishes to be able to impose to the colour charge Q, centered at the space-time point $P \equiv (0, x)$, to be the source of a colour electric field $\vec{D}$ that, rather than being isotropically directed around P, has its lines of force in the direction $\vec{u}$ [$n \equiv (0, \vec{u})$]. (See FIG.18)

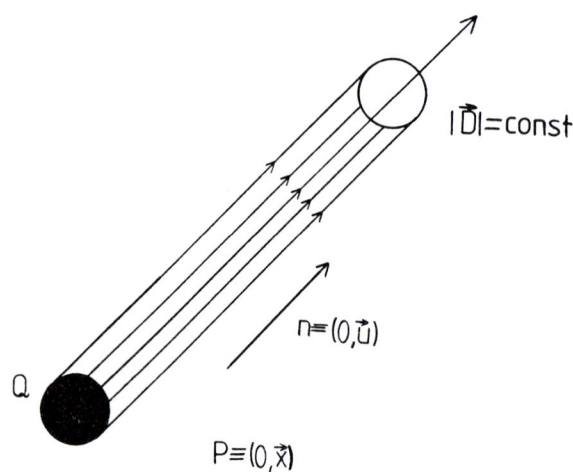

FIGURE 18
In the Anisotropic Space-Time a charge Q at (P,n) is the source of a colour electric field flux tube along the direction $\vec{u}$. The electric field is constant along the tube.

In this situation colour is trivially confined, for the energy necessary to isolate the charge Q is infinite due to the constancy of the electric field. How can this be realized is explained at length in Ref.13 and 20. Here I would like to stress that having a fundamental anisotropy in the dynamics of colour is necessary in order to produce the field configuration depicted in FIG.18. It is a recent exciting development that such anisotropic behaviour gets dynamically generated within QCD itself, as we shall discuss later on.

Even though the potentialities of ACD encompass all of hadronic physics, so far the field that has been more thoroughly studied is the qq̄-spectroscopy [20]. Let me briefly outline the main steps which lead to a calculation of the qq̄-spectrum.

The ACD Hamiltonian H, besides the free Dirac part, contains a basic colour current-current instantaneous interaction which, according to the previous discussion, <u>confines colour</u>. It is possible to separate in the Hamiltonian H two pieces:

$$H = H_0 + H' , \qquad (I.3.13)$$

where H contains the free Dirac term and the confining part, which has the peculiar property of conserving the number of quarks and antiquarks <u>separately</u>. Incidentally this interesting feature of ACD gives a natural explanation of the success of the quark model, for it shows that the phenomenon of confinement is not due to very strong forces among quarks within limited space-time domains, which would give rise to a copious qq̄-pair creation, but is associated to the peculiar polarization properties of the vacuum, responsible for field configurations of the type of FIG.18. Thus the process of qq̄-pair creation contained in H' may, and indeed will, be treated perturbatively.

By keeping H_0, in the Schrodinger picture we derive [20], without much ado, a rather simple equation for the wave function (FIG.19) $\psi_M(\vec{k})$, describing the momentum $\vec{k}$ distribution of the quarks inside the meson M:

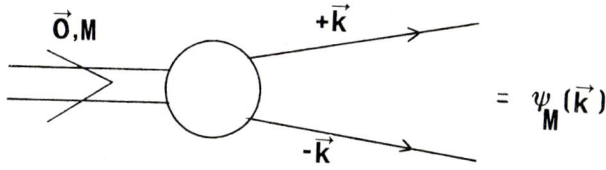

FIGURE 19
The Schrödinger wave-function in the meson rest frame.

$$[\sqrt{\vec{k}^2+m_q^2} + \sqrt{\vec{k}^2+m_{\bar{q}}^2} - M]\psi_M(\vec{k}) + \int \frac{d^3k'}{(2\pi)^3} V(\vec{k},\vec{k}') \psi_M(\vec{k}') = 0, \qquad (I.3.14)$$

where M is the meson mass, m_q ($m_{\bar{q}}$) is the quark (antiquark mass, and the non-local "potential" $V(\vec{k},\vec{k}')$ describes both confinement and spin effects. One can see that for small momentum transfers V is local (i.e. a function of $\vec{k}-\vec{k}'$ only) and its fourier transform $V(r) \sim \mu^2 r$, a "canonical" linear potential, μ^2 being the "string tension". In terms of a "minimal" set of inputs: the quark masses and μ^2 it is possible to compute the qq-spectrum for all flavours. In FIG.20 I report two examples of such calculations,

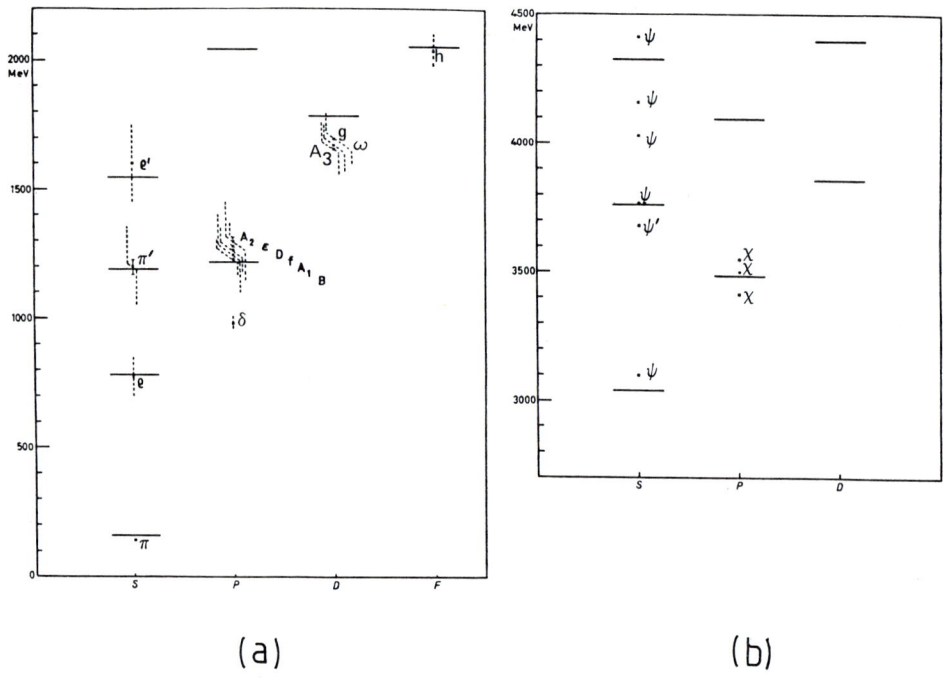

FIGURE 20
The calculated spectrum for the $\bar{u}u$ (a) and the $c\bar{c}$-mesons (b).

the uu- and the $c\bar{c}$-spectra, as one can see the agreement is good though not perfect. Indeed one has neglected a few effects such as the gluon-exchange part of the potential (the "Coulomb" term) and the (small) distortion introduced by H'; defects that can be remedied with not too much labour.

For a number of reasons, among which its heterodoxy, the exploration of ACD is still at the preliminary stages; let me summarize however its most important results, which make it a promising approach to hadrons and their interactions:

i.   the $q\bar{q}$-spectrum;

ii.   for $m_q \to 0$ the π-meson is the Goldstone boson of chiral symmetry [21];

iii. current-particles matrix elements [22];

iv.   calculation of the asymptotic value of R= ($e^+e^- \to$ hadrons) / ($e^+e^- \to \mu^+\mu^-$) = $3 \sum_i Q_i^2$ [18];

v.   weak and electromagnetic form factors of pseudoscalar mesons [23].

PART II: THE PROBLEMS OF THE STANDARD MODEL

II.1 THE PROBLEMS OF SU(2) XU(1).

The large masses of the intermediate vector bosons $W^\pm, Z^0$, as we know, indicate that $SU(2)_L \otimes U(1)_Y$ cannot be an exact symmetry of the electroweak interaction. It is thus absolutely necessary to understand how a symmetry breaking gets generated so substantial as to render what would be massless gauge bosons the heaviest of the known fundamental particles. We are then faced with a first problem: the electroweak symmetry SU(2)xU(1) cannot determine completely the theory, we must supplement it with a specific (ad hoc?) mechanism that is capable to generate the weak bosons' masses. In order to achieve this goal, the most popular mechanism is the one originally proposed by A. Salam and S. Weinberg [5], which makes use of an interesting observation made several years before by P.Higgs [24], in the context of QED. The reasoning goes as follows. Let's take a complex scalar field ϕ(x), which can be given the polar representation

$$\phi(x) = \rho(x) e^{ie\theta(x)} \qquad (II.1.1)$$

Its QED gauge-invariant lagrangian can be written as:

$$L(x) = -\frac{1}{2} (\partial_\mu - ieA_\mu)\phi(x) (\partial_\mu + ieA_\mu)\phi^*(x) - V(\phi^*\phi), \qquad (II.1.2)$$

where $A_\mu(x)$ is the e.m. vector potential, and V(z) is a quadratic polynomial in the variable z. The reason of this restriction is due to the requirement of renormalizability. Using the polar representation (II.1.1) the lagrangian L(x) can be rewritten:

$$\mathcal{L}(x) - \frac{1}{2}(\partial_\mu \rho - ie(A_\mu - \partial_\mu \theta)\rho)(\partial^\mu \rho + ie(A_\mu - \partial_\mu \theta)\rho) \qquad (\text{II}.1.3)$$
$$- V(\rho^2).$$

By using the gauge-invariance of the theory, by a gauge transformation we can replace $A_\mu(x)$ by $\tilde{A}_\mu(x) = A_\mu(x) - \partial_\mu \theta(x)$, and (II.1.3) becomes:

$$\mathcal{L}(x) = -\frac{1}{2}\partial_\mu \rho \partial^\mu \rho - \frac{e^2}{2}\rho^2 \tilde{A}_\mu \tilde{A}^\mu - V(\rho^2). \qquad (\text{II}.1.4)$$

It is at this point that $V(\rho^2)$ enters the picture, and with it the Buridan's ass. Let's suppose in fact that $V(\rho^2)$ has the shape drawn in FIG.21, so that

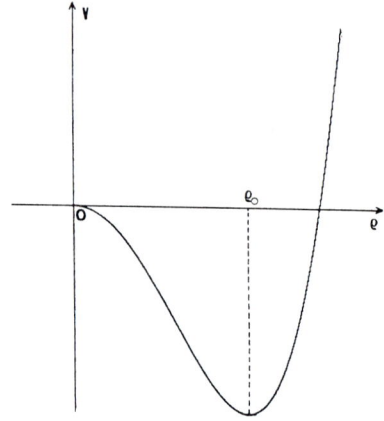

FIGURE 21
The assumed shape of the potential V.

the point $\rho = 0$ is not a local minimum but rather a maximum. Ordinary perturbation theory is clearly nonsense, for it corresponds to the dynamics of quantum fluctuations around the classical field $\rho = \rho(x) = 0$, which cannot be stable. A stationary point that is also stable is obviously $\rho(x) = \rho_0$ the quantum fluctuations around which, however, break the reflection symmetry $\phi \to -\phi$, possessed by the lagrangian (I.1.2). This is just what happens to the Buridan's ass, whose symmetrical configuration is not stable with respect to its hunger. In our case it is the energy to determine the preferred configurations, forcing us to quantize the theory by writing

$$\rho(x) = \rho_0 + \eta(x) \qquad (\text{II}.1.5)$$

$\eta(x)$ being the quantized field. By considering the "shift" (II.1.5) the lagrangian (II.1.4) plus the Maxwell term can be reduced to:

$$\mathcal{L}(x) = -\frac{1}{2}(\partial_\mu \eta(x))^2 - \frac{1}{2}V''(\rho_o)\eta(x)^2 - \frac{1}{4}F_{\mu\nu}(x)F^{\mu\nu}(x)$$
$$- \frac{e^2}{2}\rho_o^2 \tilde{A}_\mu(x)\tilde{A}^\mu(x) - \frac{e^2}{2}\eta(x)^2\tilde{A}_\mu(x)\tilde{A}^\mu(x) + \frac{\lambda}{4!}\eta^4(x) \qquad (II.1.6)$$

which exhibits several interesting features. The curvature $V''(\rho_o)$ of the potential around its mininum is the positive (unrenormalized) squared mass of the scalar quantum field $\eta(x)$ [the Higgs boson]; but the most important fact is that the photon also acquires a mass $m_\gamma^2 = e^2\rho_o^2$ as can be readily seen from the lagrangian (II.1.6).

Thus, as a result of the instability of the classical configuration $\rho = 0$, the theory has changed its structure completely. From a theory describing the interaction of a massless photon with a charged scalar particle, we end up with a quantum theory of a neutral massive boson (Higgs particle) interacting with a *massive* neutral vector field. Another interesting observation is that the Higgs phenomenon does not change the number of degress of freedom, it only leads to a rearrangement of them.

Indeed in the original symmetric theory at each space-time point we have 4 degrees of freedom: two associated to the complex field (real, and imaginary part) and two related to the transverse components of the photon. In the "spontaneously" broken theory the degrees of freedom are again 4, but distributed differently: 3 for the massive vector field (transverse plus longitudinal component) and 1 for the neutral scalar Higgs field.

One might say that the Higgs mechanism is a brilliant solution of the incompleteness problem of the electroweak sector, mentioned above. Not only does it provide a neat mechanism for the generation of the gauge-bosons mass but it predicts also the existence of a completely new type of elementary particle(s), thus providing a strong motivation for the ongoing quest for higher energies.

However, it must be pointed out that, as formulated, the theory is not internally consistent. The reason being that there has been in the last few years mounting evidence that field theories described by a Lagrangian (II.1.2) lead, after renormalization, to a trivial, non-interacting theory [25]. We have no idea, at present, whether in a theoretical context wider than the original GSW proposal, this serious problem of triviality might be overcome. For the time being the conclusion we must draw from all this is that the GSW model, a definite step forward from the original Fermi theory, cannot be considered yet as a fundamental theory, like QED. There is a sector of the theory -the Higgs sector- that can be at best considered as phenomenologically simulating the correct physics in the energy domain definitely below the TeV. We know that above a few TeV something must go wrong in our theoretical views: another good motivation for the quest for higher energies.

I should like to end this Subsection by recalling that in spite of a large number of attempts to overcome the Higgs problem that have been made in the last few years, so far no completely satisfactory proposal exists to cure the inadequacy of the GSW model. This should be a warning for the hunters of Higgs bosons.

II.2 THE PROBLEMS OF SU(3) AND THEIR POSSIBLE SOLUTION

So far the only theoretical framework that has produced a coherent picture of Quantum Chromo Dynamics -the dynamics of SU(3) - has been perturbation theory (PT).

The idea that PT must have something to do with any candidate theory of hadrons, such as QCD, is strongly motivated by the simple description that the parton model gives of physical processes at short light-cone distances, e.g. deep inelastic scattering. The discovery of Asymptotic Freedom (AF) [26] in perturbative QCD more than ten years ago has convinced the great majority of particle physicists that a deep theoretical reason had been found for the observed simple partonic behaviour, and that QCD represents the only believable candidate for a fundamental theory of hadrons and their interactions.

I believe that the overall success(*) of the research programme based on perturbative QCD has played a considerable role in diverting the proper attention from what is and remains the crucial problem of QCD, colour confinement. The favourable preliminary indications from the numerical simulations of Lattice Gauge Theories (LGT) [27] have again convinced the community that the difficult problem of confinement most likely could only be attacked through the development of appropriate computational tools for the numerical simulation of LGT's, such as the new, dedicated vector computers [28].

Thus the general assessment of the status of $SU(3)_c$ and of QCD today is

---

(*)Not without some notable exceptions like in spin physics at short distances, see J. Soffer's contribution to this school.

that short distance physics is simple and can be systematically analysed through PT, while long distance physics -the physics of confined quarks and gluons- is rather messy and can be (approximately) conquered through the massive computations required by LGT's on lattices of increasing size.

It is the purpose of the final part of these lectures to show that the generally accepted view of $SU(3)_c$ and QCD has serious problems, and to indicate how these problems can be remedied.

II.2.a THE PERTURBATIVE GROUND STATE IS "ESSENTIALLY" UNSTABLE

In order to understand the meaning of the analysis of Yang-Mills theories [29,30] that I shall discuss below , and its bearing upon the generally accepted view of QCD, it is necessary to look more closely at the dynamical problem of perturbation theory in Quantum Field Theory. The basic steps of the perturbative analysis of Yang-Mills theories have been outlined in Sect.I, where we have seen that the perturbative dynamics describes the (weak) interactions among the quantum fluctuations (free fields) of the gauge-fields around the classical state of minimum energy, the classical ground state. Due to the quadratic nature of the Hamiltonian density of SU(N)

$$H(x) = \frac{1}{2} \sum_{a=1}^{N^2-1} (\vec{E}_a^2(x) + \vec{B}_a^2(x)), \qquad (II.2.1)$$

the classical ground state is clearly the state of zero gauge-fields $F_{\mu\nu}^a(x)=0$; and the perturbative ground state is just the quantum state that describes the "zero-point" fluctuations of the gauge-fields around their classical value $F_{\mu\nu}^a=0$. When the gauge coupling constant g goes to zero, these fluctuations -described by N^2-1 massless, coloured, transverse vector particles, the gluons- completely decouple and the theory becomes trivial and solvable, while for g small perturbation theory produces a systematic (in powers of g) solution of the quantum field equations. But in order for PT to make sense it is necessary that the perturbative ground state be "essentially" stable, i.e. there exists a finite distance d such that for space-time volumes of dimension smaller than d, the perturbative ground state is a good approximation of the real ground state. Note that the requirement that a finite length d exists is rather weak, for it can well coexist with the expectation that the theory is confined at distances much larger than d. It is clear that, if d exists, for distances smaller than d one can make full use of perturbation theory for computing the dynamics of QCD. This is just the strategy that is commonly followed to calculate short distance processes.

Reasonable as it may seem, the idea that such a d exists has never been demonstrated to hold. It was the purpose of the investigations [30,31] to ascertain, in a non perturbative fashion, whether d existed and one could apply, as almost anybody does, perturbation theory to the analysis of the physics at short distances.

Following an idea of G.K.Savvidy (1976) [31], later developed by N.K.Nielsen and P.Olesen (1977) [32], we asked ourselves what happens if we analyse the quantum fluctuations of the gauge-fields not around the classical ground state $F_{\mu\nu}^a = 0$, but around a non-trivial solution of the Yang-Mills field equations [from now on for simplicity we work in SU(2)]

$$A_\mu^a(x) = f_\mu^a(x) = \delta^{a3} g_{\mu 2} \times B, \qquad (II.2.2)$$

describing a homogeneous magnetic field, of intensity B, pointing in the z-direction in space and in the 3-direction in isospace. Classically the field (II.2.2), carries a non-zero energy density $E_{c\ell}$, given by

$$E_{c\ell} = B^2/2, \qquad (II.2.3)$$

and therefore it lies <u>above</u> the classical ground state.

It would then seem natural that also the quantum state, representing quantum fluctuations around (II.2.2), which we shall call the "Savvidy state" $S_{23}(B)$ lies <u>above</u> the perturbative ground state. It is the surprising discovery of G.K. Savvidy [31] that this is not so. Indeed by performing a perturbative analysis of the quantum corrections to the effective potential of a SU(2) Yang-Mills theory Savvidy found that the energy density of the state $S_{\hat{z}3}(B)$ is :

$$E(B) = E(0) + \frac{B^2}{2} - \frac{11}{48\pi^2} g^2 B^2 \ln(\frac{\Lambda^2}{gB}) + O(g^2 B^2), \qquad (II.2.4)$$

E(0) being the energy density of the perturbative ground state (B=0). Thus, according to Savvidy, the classical effective potential $B^2/2$ is modified by the quantum fluctuations in such a way as to generate a minimum away from B=0 [See FIG.22].

Indeed, from (II.2.4)

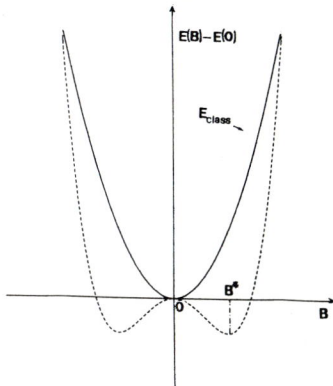

FIGURE 22
Quantum fluctuations modify the classical potential E_{class}.

$$gB^* = \frac{\Lambda^2}{e^{\frac{1}{2}}} \exp - \frac{48\pi^2}{11g(\Lambda^2)} \quad , \tag{II.2.5}$$

and the energy density at the minimum is

$$E(B^*) - E(B) = - \frac{(gB^*)^2}{96\pi^2} . \tag{II.2.6}$$

Thus in a non abelian gauge theory the perturbative ground state, though unstable (B=0 is a maximum, of the effective potential) is "essentially" stable, for one can fix the coupling constant $g(\Lambda)$ in such a way as to render $gB^* = \mu^2$, μ^2 being a finite scale (μ is usually called Λ_{QCD}). If this is done, it is clear that for distances d much smaller than $1/\mu$ the new ground state is practically undistinguishable from the perturbative ground state; thus d exists and it is just equal to $1/\mu$. It is also interesting to note that (II.2.5) can be rewritten as:

$$g^2(\Lambda) \stackrel{\sim}{=} \frac{48\pi^2}{11 \ln(\frac{\Lambda^2}{\mu^2})} \tag{II.2.7}$$

which is precisely the law of evolution, with the ultraviolet cut off Λ, of the "running" coupling constant, predicted by perturbation theory. The calculation of Savvidy is interesting and important for two reasons: (i) it shows that some magnetic condensation occurs at long distances, giving the ground state a magnetic structure very different from the perturbative ground state (confinement?); (ii) it implies that at small distances the vacuum magnetism is negligible and one gets back to the perturbative phase. Both points are perfectly consistent with the generally accepted view of QCD.

Unfortunately, as it was first pointed out by Nielsen and Olesen, Savvidy's calculations contained a serious flaw, in that it neglected the contribution to the energy density of a number of modes of the gauge field fluctuations that in the approximation he adopted had the peculiarity of being unstable. Savvidy's pitfall can be understood in the following way. Let's take the action [i.e. the spacetime integral of the lagrangian (I.2.17)] S[A] of the SU(2) Yang-Mills theory, and set

$$A_\mu^a(x) = f_\mu^a(x) + \eta_\mu^a(x), \qquad (II.2.8)$$

where $f_\mu^a(x)$ is the "background" field (II.2.2) and $\eta_\mu^a(x)$ is a fluctuation. For small η_μ^a one can write

$$S[A] = S[f] + \frac{1}{2} \int \left. \frac{\delta^2 S[A]}{\delta A_\mu^a(x) \delta A_\nu^b(y)} \right|_{f=A} \eta_\mu^a(x) \eta_\nu^b(y) + \ldots \qquad (II.2.9)$$

where the absence of the linear term $\left.\frac{\delta S[A]}{\delta A_\mu^a}\right|_f$ is be noticed, due to the fact that $f_\mu(x)$ is a solution of classical field equations. The stability of the Savvidy state for small fluctuations η is thus determined by the positivity of the operator $1/2 \left.\frac{\delta^2 S[A]}{\delta A_\mu^a \delta A_\nu^b}\right|_{A=f} = O_{\mu\nu}^{ab}(x,y)$. If and only if $O_{\mu\nu}^{ab}(x,y)$ has a positive spectrum, we are guaranteed that the fluctuactions remain small and that the corrections to the energy density due to such fluctuations are $O(g^2 B^2 \ell n \, \Lambda^2/gB)$, just as found by Savvidy. It turns out that the operator $O_{\mu\nu}^{ab}(x,y)$ was diagonalized more than 50 years ago by Landau, who showed that its spectrum is just the spectrum of a relativistic charged spin-one particle in an external magnetic field, which is given by:

$$E_n(p_3)^2 = p_3^2 + gB(2n+1) - 2gBS_3, \qquad (II.2.10)$$

where p_3 is the momentum in the field direction, n is a discrete quantum number describing the confined transverse motion, and S_3 is the spin projection along the field direction. Now for a gauge field $S_3 = \pm 1$ (vector "gluons"), so that for n=0 $S_3 = +1$ and $p_3 \leq (gB)^{\frac{1}{2}}$ (II.2.10) is negative. The field modes for which this happens will be called U-modes, while for the other modes we shall use the term S-modes.

In the calculation of Savvidy only the S-modes were included, for in the effective potential formalism the U-modes, as pointed out in Ref. 32, would contribute an imaginary part to the energy density, thus signalling an instability. The moral of all this is that while the Savvidy state $S_{23}(B^*)$ has an energy density lower than the perturbative ground state, it is itself unstable, so that the physics will migrate somewhere else, to a configuration of still lower energy density. But where will it go? This question has remained without answer for quite some time, until a new way to analyze the problem was finally envisaged. The idea was to make use of a result of Cornwall, Jackiw and Tomboulis [33] that showed that the problem of calculating (non-perturbatively) the effective potential is equivalent to a variational evaluation of the hamiltonian density (II.2.1) on a class of gaussian gauge-field functionals which are approximately gauge-invariant. The construction of such class as well as the evaluation of the hamiltonian density had to go through the solution of a few tricky problems, in particular of the correct handling of ultraviolet divergences; the final result is however simple, physically transparent and rather striking.

Let's see what happens. As I have already remarked, if the U-modes did not exist (like in the perturbative case, B=0) the stability of the quantum fluctuations would imply a correction to the classical energy density $B^2/2$ of order $(gB)^2 \ln \Lambda^2/gB$, for the fluctuation $|\eta|$ would be of O(1) and when $g \to 0$ the correction would obviously go to zero. In presence of an instabiliy, however, $|\eta|$ is no more bounded when $g \to 0$, and an explicit evaluation [29,30] shows that in the U-sector $|\eta|_U \sim 1/g$, thus from the U-modes a contribution arises to the energy density of $O(B^2)$, just comparable to the classical term! Note that no $\ln\Lambda^2$ term can arise from the U-modes, for these modes are spread out over large distances and are thus totally insensitive to the space-time granularity implied by Λ. Analysing the U-modes' contribution to the energy density as well as its renormalization induced by their coupling to the S-modes, leads to the striking result that <u>the classical term $B^2/2$ is completely compensated by the U-modes, contribution</u>, so that Savvidy's result (II.2.4) is drastically modified to:

$$E(B) = E(0) - \frac{11}{48\pi^2} g^2 B^2 \ln(\frac{\Lambda^2}{gB}) + O(g^2 B^2) \ . \tag{II.2.11}$$

The effective potential of FIG.22 becomes now the one reported in FIG.23, where the minimum $gB^* \sim \Lambda^2$ and the value $E(B^*)-E(0) \sim -\Lambda^4$. The fact that both gB^* and $E(B^*)-E(0)$ diverge when $\Lambda \to \infty$, means that the distance $d \sim 1/\Lambda \to 0$, and that the perturbative phase of QCD is <u>dynamically irrelevant</u>. In other words, under no circunstance, not even at very short distances, the perturbative ground state and perturbation theory can give a correct representation of the dynamics of QCD.

I should add that the result (II.2.11) has now been checked also in SU(2) lattice gauge theory [34], thus providing a further demonstration that the

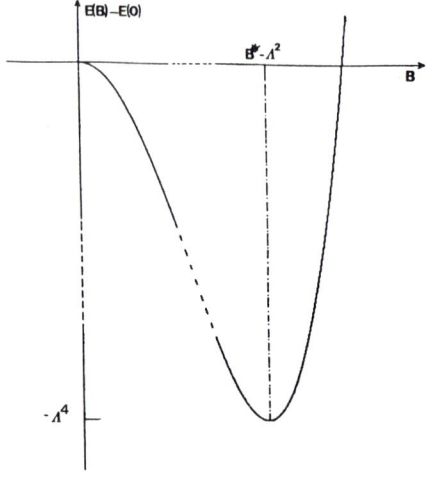

FIGURE 23
The effective potential in the non-perturbative variational approximation of Ref.31.

perturbative vacuum is "essentially unstable". This lattice result also shows that the strategy followed so far to derive results for the continuum QCD from lattice Montecarlo simulations is doomed to fail. Indeed, by using lattices of sizes that can be reasonably managed today or in the foreseeable future ($L \lesssim 100$), the physics that can be simulated is drastically different from the physics that emerges out of the infinite (when $\Lambda \to \infty$) "gluon" condensation implied by (II.2.11), and that will be discussed next.

II.2.b THE ORIGIN OF COLOUR CONFINEMENT

It is clear that a given Savvidy state of minimum energy $S_{u\alpha}(B^*)$ cannot represent the real QCD ground state, for it obviously violates both colour and rotational invariance, neither of which appears to be broken in the real world. However, as I shall demonstrate in a while that these states are an essential ingredient of the QCD ground state, it is worthwhile to investigate whether colour confinement holds in $S_{u\alpha}(B^*)$. To see this, let us put a colour charge in the Savvidy state $S_{\tilde{2}3}(B^*)$, the gauge field $A^i_\mu(x)$ can then be written:

$$A^i_\mu(x) = f^i_\mu(x) + a^i_\mu(x) \quad , \qquad (II.2.12)$$

with

$$f^i_\mu(x) = \delta^{i3} g_{\mu 2} x_1 \bar{B}, \qquad (II.2.13)$$

where $\bar{B}$ is a divergent (when $\Lambda \to \infty$) chromomagnetic field, and $a^i_\mu(x)$ is the part of the gauge-field that is generated by the external current $J^i_\mu(x)^{ext}$, associated with our external charge. The finiteness of the difference $F^i_{\mu\nu}(A) - F^i_{\mu\nu}(f)$ [$F^i_{\mu\nu}(h) = \partial_\mu h^i_\nu - \partial_\nu h^i_\mu + g\epsilon^{ijk} h^j_\mu h^k_\nu$] clearly requires $a^i_\mu(x)$ to point in isospace in the direction of the "background" field (II.2.13), i.e. one must have

$$a^i_\mu(x) = \delta^{i3} a_\mu(x), \qquad (II.2.14)$$

thus rendering $a^i_\mu(x)$ an effectively abelian gauge-field. The colour dynamics in our "magnetic" medium $S_{\hat{z}3}(B^*)$ is thus described by the classical effective lagrangian

$$L_{eff}(x) = -\frac{1}{4} f_{\mu\nu}(x) F^{\mu\nu}(x) + g a^\mu(x) J^3_\mu(x)^{ext}, \qquad (II.2.15)$$

where $f_{\mu\nu}(x) \equiv (\vec{D}, \vec{H})$ is the "magnitude" tensor and $F_{\mu\nu}(x) \equiv (\vec{E}, \vec{B})$ is the "intensity" tensor, whose meaning is well known from the electrodynamics of continuous media. Note that we have:

$$\vec{B} = \vec{B}_{\hat{z}} + \vec{B}_\perp, \qquad (II.2.16)$$

i.e. the full magnetic field $\vec{B}$ has a divergent z-component. In a homogeneous medium, such as $S_{\hat{z}3}(B^*)$, we may set:

$$f_{\mu\nu}(x) = \epsilon^{\alpha\beta}_{\mu\nu}(\hat{z}) F_{\alpha\beta}(x), \qquad (II.2.17)$$

where

$$F_{\alpha\beta}(x) = \partial_\alpha A_\beta(x) - \partial_\beta A_\alpha(x), \qquad (II.2.18)$$

$f_{\mu\nu}(x)$ obeys the second set of Maxwell equations

$$\partial^\mu f_{\mu\nu}(x) = g J^{ext}_\nu(x), \qquad (II.2.19)$$

and $\epsilon^{\alpha\beta}_{\mu\nu}(\hat{z})$ is the "anisotropy" tensor(*), depending on the direction $\hat{z}$, describing the electric and magnetic polarizability of the medium.

(*)The "anisotropy" tensor plays a crucial role in ACD [13].

We can now completely determine, within an inessential scale factor, the structure of the "anisotropy" tensor by making use of the energy-momentum density associated with the effective lagrangian (II.2.15) in the following way.

We have [35]

$$H(\vec{x}) = \frac{1}{2}[\vec{E}\cdot\vec{D} + \vec{B}\cdot\vec{H}] \qquad (II.2.20)$$

$$\vec{P}(x) = \frac{1}{2}[\vec{D}\times\vec{B} + \vec{E}\times\vec{H}], \qquad (II.2.21)$$

which in view of (II.2.16) can remain finite if and only if:

$$\vec{D}(x) || \hat{z} \quad \text{and} \quad \vec{H}(x) \perp \hat{z}. \qquad (II.2.22)$$

These conditions, setting the background field in the generic direction u, determine the "anisotropy" tensor as:

$$\epsilon_{\mu\nu}^{\alpha\beta}(\vec{u}) = -\frac{1}{2}(\delta_\mu^\alpha n_\nu n^\beta + \delta_\nu^\beta n_\mu n^\alpha - \delta_\mu^\beta n_\nu n^\alpha - \delta_\nu^\alpha n_\mu n^\beta), \qquad (II.2.23)$$

with $n_\mu \equiv (0,\vec{u})$. It is remarkable that the effective lagrangian (II.2.15) with the form (II.2.23) of the anisotropy tensor just coincides for $\vec{u}$ fixed with the lagrangian of Anisotropic Chromo Dynamics. According to (II.2.22) and the Gauss theorem a static charge g in $S_{\hat{z}3}$ (B*) gives rise to the field structure depicted in FIG.24(a), while a charge-anticharge pair produces the configuration in FIG.24(b).

This is just a typical manifestation of colour confinement. We may thus conclude that: <u>in a Savvidy state of minimum energy S (B*) colour is confined</u>.

We have already noted that a given Savvidy state cannot describe the QCD ground state which, as far as we know, is both colour and Lorentz invariant. We have however the strong suspicion that mechanisms as effective in lowering the energy of the perturbative ground state as those operative in a Savvidy state presumably do not exist(*), thus it appears extremely reasonable that the family $\{\vec{u},\alpha\}$ of the Savvidy states of lowest energy is involved in the construction of the real QCD ground state. But, how?

(*) In the last few years several theoretical proposals have been advanced of non-perturbative mechanisms to lower the energy density of the perturbative ground state, like instantons, merons etc., but they all lead to an energy gap satisfying, like Savvidy's, the renormalization group equations (II.2.5), (II.2.6).

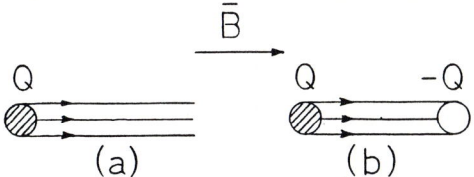

FIGURE 24
The lines of force of the electric diplacement vector $\vec{D}$, for an isolated charge g(a), and for a dipole (b).

The problem we have is to find out in which way the different Savvidy states "mix" to produce a completely honest quantum state that preserves both colour and Lorentz invariant. Let us first make a relevant observation. If

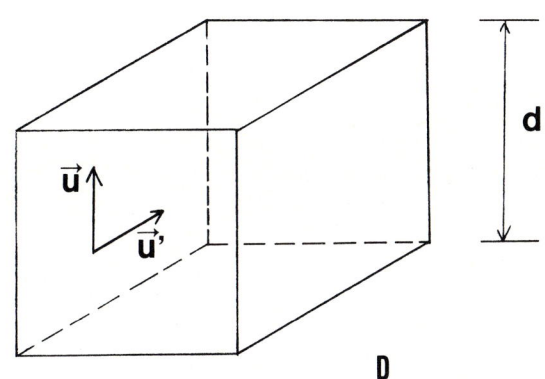

FIGURE 25
The scalar product between the two Savvidy states $S_{\vec{u}\alpha}(B^*)$ and $S_{\vec{u}'\alpha'}(B^*)$ restricted to the space-domain D.

one takes two different Savvidy states $(\vec{u},\alpha)$ $(\vec{u}',\alpha')$ and restricts them to a finite space domain D (See FIG.25), over D these two states will "mix" in order to lower their (degenerate) energy density if and only if their scalar product is finite. This is what happens for well known systems like the

Heisenberg ferromagnet, where one observes the formation of "zones". It is now easy to convince ourselves that, due to the divergence of the "background" fields pointing in two different directions ($\vec{u},\alpha$) and ($\vec{u}',\alpha'$), the scalar product of the two states over D vanishes no matter how small (but finite) is D. To this phenomenon, which as far as I know is totally new, I have given the name of "essential orthogonality". Thus the family $\{\vec{u},\alpha\}$ of Savvidy states of minimum energy turns out to be "essentially orthogonal".

What is the physical meaning of the essential orthogonality? The situation is analogous to what happens in ordinary quantum mechanics with a potential of the type in FIG.26. For small values of the barrier height Δ

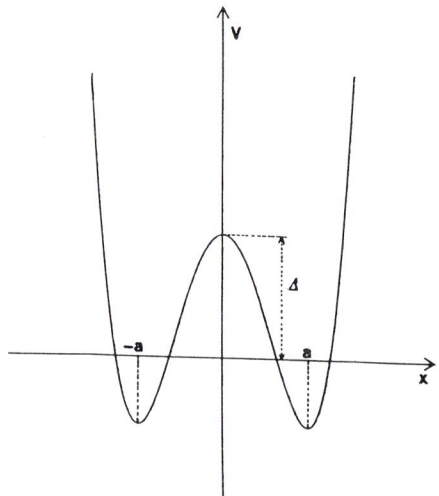

FIGURE 26
The two-minima potential of ordinary quantum mechanics.

the ground state is delocalized, i.e. there is a non zero probability for the particle of being around both a and -a. When Δ increases beyond any bound the physics around a gets decoupled from that around -a, and the corresponding quantum mechanical systems become <u>independent</u>. One can easily convince oneself that the "essential orthogonality" just found for any two states ($\vec{u},\alpha$) and ($\vec{u}',\alpha'$) is nothing but the manisfestation of the existence of our infinite barrier between the two field configurations when restricted to any finite space-domain D. We may thus say that: <u>each state is the ground state of an independent quantum mechanical system</u>, so that the real ground state $|\Omega\rangle$ of QCD can be written:

$$|\Omega\rangle = \prod_{\alpha\vec{\mu}} S_{\alpha\vec{u}}(B^*) \qquad (II.2.24)$$

the product of all the quantum-mechanically independent Savvidy states, which is now obviously colour and Lorentz invariant.

The Lagrangian governing the dynamics of colour confinement is then given by the sum over $\vec{u}$ and α of lagrangians of the type (II.2.15) and (II.2.23). But this is just the ACD lagrangian [13] !

The quantum mechanics of QCD around the ground state $|\Omega>$ goes however somewhat beyond ACD. Indeed, it can be demonstrated [36] that, while the S-modes of the quantum fluctuations of the Savvidy state $S_{\alpha u}^{\rightarrow}(B^*)$ acquire a mass $\mu S \sim gB^* \to \infty$, and get therefore trapped in the process of magnetic condensation, the U-modes, on the other hand, exhibit a non-trivial dynamical behaviour. In the Savvidy state $S_{3\hat{z}}(B^*)$, for instance, they correspond to free charged vector particles moving along the z-direction (any transverse motion is obiously squeezed to nothing by the divergent magnetic field B^*), obeying the dispersion relation:

$$\omega(p_3) = [p_3^2 + \mu_U^2]^{\frac{1}{2}} \qquad (II.2.25)$$

with

$$\mu_U^2 = 4gB^* \exp - \frac{8\pi^2}{g^2(1)} \qquad (II.2.25')$$

which can be rendered finite by adjusting $g^2(\Lambda)$.

These modes will thus form the basis of quantum fluctuations around the ground state $|\Omega>$ that are :

a. massive, as implied by the finiteness of μ_U^2 [Eq.(II.2.25')];

b. abelian, due to the non-existence on each Savvidy state of the 3-gluon coupling;

c. confined, like all coloured quanta in ACD;

d. vector-like, due to their gauge-boson nature.

Thus we have found that the U-modes of the Savvidy states give rise on the "real" QCD ground state to "gluons" of peculiar characteristics. In particular points a) and b) are to be noted, for much of the gluon-phenomenology seems to imply their being both massive and coupled to quarks only(*).

To conclude this Section, I would like to briefly summarize the main points of my discussion.

As for the electroweak part $SU(2)_L \times U(1)_Y$ of the Standard Model, in spite of the remarkable successes obtained, crowned by the recent discovery of the

---
(*) These features are corroborated by the phenomenological analysis of e^+e^--annihilation carried out in Ref.16.

$W^{\pm}$ and Z^0, we are still in need of a full theory of these interactions. Indeed, the crucial problem of the generation of the large masses of the intermediate vector bosons still lacks a consistent solution, the Higgs-mechanism being unacceptable within the prevailing framework of local field theories. It is unlikely that these serious problems will be solved without some informations that only the experimentation at the highest energies will be able to afford.

The situation of the hadronic sector, $SU(3)_c$, looks to my mind much brighter. Colour confinement, the crucial feature of quark dynamics, appears a natural consequence of a strange phenomenon of "infinite" magnetic condensation that non-abelian gauge theories seem to go through, due to the peculiar instability that I have discussed at length. It is a pleasing result of the analysis of the quantum physics of such condensed states (Savvidy states) that they confine and the confinement dynamics is governed by the ACD lagrangian, that I proposed a few years ago to describe in field theoretical terms the phenomenology of hadronic physics. This, together with the emergence of effectively abelian gluons, whose phenomenological relevance is known since the discovery of the "third jet" in ee-annihilation(*), justifies the cautiously optimistic feeling that the hard problem of hadronic behaviour is on the good way of being solved.

CONCLUSIONS

In these lectures I have tried to convey the main ideas of the Standard Model, as well as the problems, still unsolved, that those ideas meet in their impact with the physical reality. I am fully aware that many of the passages of my discourse are rather hard and not sufficiently illustrated and motivated. For this the limited space and time at my disposal carry some if not all the responsibility. Be as it may, I hope that one thing will impress the minds of my readers, that the Standard Model is a great achievement of the human mind in quest for unity (and beauty) at the frontiers of the incredibly small and fast. Even though there remain many unsolved problems, it is very likely that these problems will greatly contribute to a new perception of the field theories (gauge-theories) which lie at the basis of the understanding so far achieved. And in one instance, in the $SU(3)_c$ sector, some rather new physics already appears to be emerging.

(*) Such discovery was reported in 1979 from the different groups working at PETRA, and was construed as a strong corroboration of perturbative QCD.

REFERENCES

1. For a review of the different stages of this important research programme. see M.Gell Mann, The eightfold way, Benjamin N.Y. (1984).
2. G.Zweig, CERN Preprints TH401, 412 (1964) (unpublished).
3. O.W.Greenberg, Phys.Rev. Letters 13,598 (1964)
4. E.Fermi, Riv.Sci.4, 491 (1933);
5. A.Salam, Elementary Particle Theory: Relativistic groups and Analyticity (Nobel Symposium n.8) edit by N.Swartholm, Almqvist and Wiksell, Stockholm 1968; S.Weinberg, Phys. Rev. Letters 19, 1264 (1967).
6. S.L. Glashow, Nucl. Phys. 22, 579 (1961)
7. H. Weyl, Berl. Ber. (1918) 465; Raum-Zeit-Materie, pp.242 (1920).
8. H. Weyl, The theory of groups and Quantum Mechanics, Dover (N.Y.) (1948)
9. C.N.Yang and R.L. Mills, Phys. Rev. 96, 191 (1954).
10. See for instance G.Preparata Proceedings of the EPS Conference on High Energy Physics, Geneva (1979), p.442.
11. G.Preparata, Phys. Rev. D7, 2973 (1973).
12. G.Preparata and N. Craigie, Nucl. Phys. B102, 478 (1976); for a review see G. Preparata in: The why's of subnuclear Physics, A. Zichichi ed. p.727 , Plenum Press (1979).
13. G. Preparata, Phys. Letters 102B, 327 (1981); for a review see G. Preparata, in Fundamental interactions: Cargese 1981, Levy, Basdevant, Speiser, Jacob, and Gastmans, ed. p. 421 Plenum N.Y. 1982.
14. G.Preparata, in: Lepton hadron structure, ed. A. Zichichi (Plenum, New York, 1975) p.54.
15. L.Angelini, L.Nitti, M.Pellicoro, G.Preparata and G.Valenti, Fire string theory of e^+e^--annihilation, Rivista del Nuovo Cimento 6,1 (1983).
16. L.Angelini, L.Nitti, M.Pellicoro, G.Preparata and G.Valenti, Phys. Letters 123B, 246 (1983).
17. G.Preparata, Partons vs. Hadrons: A physical theory of fragmentation, Proceedings of the 1984 SLAC Summer Institute. Stanford (1985)
18. P.Cea, G.Nardulli and G.Preparata, Zeit. fur Phys. C16, 135 (1983).

19. A.Giannelli, L.Nitti, G.Preparata and P.Sforza, Phys.Letters 150B, 214 (1985).

20. Additional references to ACD are: G.Preparata, Phys. Letters 108B, 187 (1982); Nuovo Cim. 66A, 205 (1981); J.L.Basdevant, G.Preparata, Nuovo Cim. 67A, 19 (1982); J.L. Basdevant, P.Colangelo and G.Preparata, Nuovo Cim. 71A, 445 (1982).

21. P.Castorina, P.Cea, P.Colangelo, G.Nardulli and G.Preparata, Phys. Letters 115B, 487 (1982).

22. P.Cea, P.Colangelo, G.Nardulli and G.Preparata, Phys. Letters 115B, 310 (1982).

23. L.Cosmai, M.Pellicoro and G.Preparata, Nucl. Phys. B228, 31 (1983).

24. P.W.Higgs, Phys. Rev. Letters 13, 508 (1964); Phys Rev. 145, 1156 (1966).

25. K.G. Wilson, Phys. Rev. B4, 3184 (1971), was the first to point this problem out, which has now been corroborated by numerical computations on the lattice version of this theory.

26. D.J. Gross and F.Wilczek, Phys. Rev. Letters 30, 1342 (1973); H.D. Politzer, Phys. Rev. Lett. 30, 1376 (1973).

27. For a review see M.Creutz, L. Jacobs and C. Rebbi, Phys. Report 95, 201 (1983);

28. Such as APE: P.Bacilieri et al., preprint IFUP-TH-84/40 (1984).

29. M.Consoli and G.Preparata, Phys. Letters 154B, 411 (1985).

30. G.Preparata, Essential Quantum Instability of the perturbative Yang-Mills vacuum, preprint BA-GT?85-18 (1985).

31. G.K. Savvidy, Phys. Letters 71B, 133 (1977).

32. N.K.Nielsen and P.Olesen, Nucl. Phys. B144, 376 (1978);

33. J.M. Cornwall, R.Jackiw and E.T. Tomboulis, Phys. Rev. D10, 2421 (1974)

34. L.Cosmai and G.Preparata, Evidence against Asymptotic Freedom from SU(2) lattice gauge theory, preprint BA/GT-86/06 (1986).

35. W.Pauli, Relativitäts theorie, Enc. math. Wiss. 5, 19, Lipsia (1921).

36. G.Preparata, Understanding the Yang-Mills ground state; the origin of colour confinement; preprint LNF-86(2)P(1986).

HADRON SPECTROSCOPY – A RELATIVISTIC QUARK-DIQUARK MODEL

Enrico PREDAZZI

Dipartimento di Fisica Teorica – Università di Torino
Torino – Italy
and
INFN – Sezione di Torino

I. INTRODUCTION

The quark language[1] has been used for now twenty years in hadronic spectroscopy[2] but it has become a widely accepted tool only in recent years after its success in studying charmonium[3] ($c\bar{c}$ bound states) by means of the non-relativistic Schrödinger equation with a potential inspired by QCD. That the extension of such ideas from the case of heavy quarks (charm) to the light ones (u,d,s) necessitates the need of a relativistic treatment was realized rather soon[4].

In the impossibility of giving a detailed review of all the papers that have appeared on this subject, in what follows we will concentrate on one particular approach not only because it is the one in which the author is involved[5] but also and principally because it also covers the baryonic sector with the same parameters obtained from fitting the mesons. The algorithm used is conceptually very simple and flexible so that it can, in principle, be extended also to the case of multiquark bound states. This is obtained with the use of relativistic kinematics and with the introduction of "diquarks" (colored "bound" states made of two quarks). The notion of diquark, we recall, has originally been introduced[6] just to deal with the case of baryons.

The basic philosophy is the following: once the scheme is set, the parameters of the model are fixed using few meson masses as input. After the mesons have been accounted for, all the baryons are then constructed as iteration of the previous scheme with no additional free parameters.

Given the economy of the scheme, the success is remarkable, as we shall see. Ambiguity, complications and open problems will be discussed.

II. MOTIVATION AND GENERAL CONSIDERATIONS ON THE POTENTIAL.

As mentioned above, the renewal of interest in the use of the non-relativistic Schrödinger equation originates from its application to charmonium[3], i.e. to the use of QCD motivated potentials to study the bound states of $c\bar{c}$ quarks. Given the large mass of charmed quarks (~ 1500 MeV), the use of a non relativistic scheme is quite adequate and this explains the success in explaining $c\bar{c}$ and $b\bar{b}$ states (b for beauty) with basically any potential obeying the following general condition: i) of having a Coulomb-like behavior ($1/r$) at short distances (corresponding to the exchange of a particle, the gluon, which, aside for colour, is the analog of a photon), ii) of growing with r so as to be confining at large distances (implying that quarks cannot be liberated).

The key to extend the application of this scheme to light quark bound states lies in the further property suggested by QCD namely, iii) the potential should be flavor independent (i.e., the same for all interacting quarks irrespective of their flavors and masses).

This extension, however, raises the preliminary question of devising a relativistic formulation to deal with light quarks. This problem will be considered in Sec. III.

In analyzing the general properties of the potential, one crucial point is the energy level ordering found experimentally in charmonium

(II.1) $$E(1S) < E(1P) < E(2S) < E(1D) < E(2P) < \ldots$$

which is neither that of the hydrogen atom ($V \sim \alpha/r$)

(II.2) $$E(1S) < E(1P) = E(2S) < E(3S) = E(2P) = E(1D) < \ldots$$

nor that of a harmonic potential ($V \sim \beta r^2$)

(II.3) $$E(1S) < E(1P) < E(2S) = E(1D) < \ldots$$

The answer to the theoretical question of which classes of potentials can give rise to the ordering (I.1) has been given by Martin et al.[8] who have shown that

(II.4) $$\Delta U(r) > 0 \qquad (\forall r)$$

is a sufficient condition for

(II.5) $$E_{n,l} > E_{n-1,l+1} \qquad (\forall n, l)$$

and

(II.6) $$D\,U(r) < 0 \qquad (\forall r)$$

for

(II.7) $$E_{n,l} < E_{n-1,l+2} \qquad (\forall n,l)$$

In (II.4) Δ is the Laplacian operator whereas D in (II.6) is

(II.8) $$D = \frac{d^2}{dr^2} - \frac{1}{r}\frac{d}{dr}$$

For what said above, the theoretical prejudice is that at short distances the potential is given by one gluon exchange and is therefore Coulomb behaved (up to logarithms)

(II.9) $$U(r) \underset{r \to 0}{\sim} -\frac{\alpha_s}{r} \qquad (\alpha_s > 0)$$

where α_s is the "running coupling constant".

At large distances, on the other hand, we expect quarks to be confined or the potential to grow like some power of r (up to log's)

(II.10) $$U(r) \underset{r \to \infty}{\sim} \beta\, r^\delta \qquad (\beta, \delta > 0)$$

It is easy to check that both (II.4,6) are satisfied by an arbitrary super-position of (II.9,10)

(II.11) $$U = -\frac{\alpha_s}{r} + \beta\, r^\delta + \gamma$$

so long as $0 < \delta \leq 2$ (and $\alpha, \beta > 0$).

Actually, the value of δ in (II.11) can be further constrained by observing that

(II.12) $$U''(r) \begin{matrix} >0 \\ <0 \end{matrix} (\forall r) \Rightarrow \left|\frac{R_{2s}(0)}{R_{1s}(0)}\right| \begin{matrix} >1 \\ <1 \end{matrix}$$

The data on the leptonic width of $J/\psi(3097)$ and $\psi'(3684)$ show that the lower inequality in (II.12) must be satisfied. This require in (II.11)

(II.13) $$0 < \delta \leq 1$$

Naturally, any interpolation between (II.9) and (II.10) obeying (II.4), (II.6) and (II.12b) will be equally good.

We now have to consider the various problems to be solved to get a consistent scheme in which to attack the problem of hadronic spectroscopy in a global way i.e. both for mesons ($q\bar{q}$ bound states) and baryons (qqq bound states) for light and heavy hadrons at the same time.

The steps toward defining such a scheme are the following:

A) definition of a suitable relativistic formulation of the problem (Sec. III);

B) definition of a suitable scheme for dealing with three body bound states (Sec. IV);

C) introduction of the spin (Sec. V).

When the above steps will have been taken, we will be able not only to tackle the spectroscopic problem of ordinary hadrons (Sec. VI), but we could actually venture to discuss also the formation of unusual hadrons (such as baryonium, dibaryons and exotic multiquark states). We shall comment in the conclusion (Sec. VII) about some of the difficulties connected with the latter problem.

III. A RELATIVISTIC SCHEME FOR INTERACTING QUARKS.

Fully relativistic schemes, such as the Bethe Salpeter equation have been considered[9] but their treatment is complicated by a series of conceptual and practical difficulties. It is therefore advisable to take a less ambitious starting point on which, however, we may have a better control. To this aim, we begin by writing the relativistic energy of two free particles in the c.m. system of masses m_1 and m_2

(III.1)
$$W = (\vec{p}^2 + m_1^2)^{1/2} + (\vec{p}^2 + m_2^2)^{1/2}$$

We shall not worry here to prove (as it can be done) that the above operator is mathematically well defined.

Squaring (twice) to eliminate square roots and solving for $\vec{p}^2$ we get

(III.2)
$$\vec{p}^2 = \frac{W^2}{4} - \frac{m_1^2 + m_1^2}{2} + \frac{(m_1^2 - m_2^2)^2}{4W^2}$$

At this point, there are (at least) two ways of going further that are found in the literature to derive dynamical equations with relativistic kinematics.

a) <u>Krolikowski equation</u>[10].

If we have a four-vector and a scalar potential, $A_\mu = (\vec{A}, V(r))$ and $S(r)$

respectively, we replace p,W and m_i in (III.2) by

(III.3)
$$\vec{p} \to \vec{p} - \vec{A}$$
$$W \to W - V(r)$$
$$m_i \to M_i(r) = m_i + \tfrac{1}{2}S(r)$$

so that from (III.2) we get

(III.4)
$$(\vec{p}-\vec{A})^2 = \tfrac{1}{4}(W-V(r))^2 - \tfrac{1}{2}[(m_1+\tfrac{S}{2})^2 + (m_2+\tfrac{S}{2})^2]$$
$$+ \frac{1}{4(W-V(r))^2}[(m_1+\tfrac{S}{2})^2 - (m_2+\tfrac{S}{2})^2]^2$$

where we shall, as usual, take

(III.5)
$$\vec{p} = -i\hbar\vec{\nabla}$$

b) <u>Todorov equation</u>[11].

The prescription is different since we first set

(III.6)
$$E_W = \frac{W^2 - m_1^2 - m_2^2}{2W} \quad ; \quad m_W = \frac{m_1 m_2}{W}$$

so that eq. (III.2) becomes

(III.7)
$$\vec{p}^{\,2} = E_W^2 - m_W^2$$

i.e., formally, the energy-momentum relation for a relativistic particle of mass m_W. At this point we substitute

(III.8)
$$\vec{p} \to \vec{p} - \vec{A}$$
$$E_W \to E_W - V(r)$$
$$m_W \to m_W + S(r)$$

and the dynamical equation[11] becomes

(III.9)
$$(\vec{p}-\vec{A})^2 = (E_W - V(r))^2 - (m_W + S(r))^2$$

where, again, (III.5) will have to be used.

In both cases (III.4) and (IV.9), the energy eigenvalues will have to be determined numerically. It is, however, remarkable that, in practice, the eigenvalues obtained from the previous two equations are not very different, as we have checked numerically.

In the following, we will report the values obtained with (III.4) only.

One of the reasons to show the derivation of both (III.4) and (III.9) is to exhibit clearly one of the (many) ambiguities inherent in the approach. Presumably more, equally plausible, derivations of other equations could have been given.

A further ambiguity arises at the level of defining the potential to be used in the above equations. While, in fact, there is a general consensus that one should take $\vec{A}=0$, it is not obvious how the full potential U(r) should be split between the fourth component V(r) of $A\mu$ and the scalar part S(r).

In the following, for economy of parameters we shall choose to interpolate between (II.9) and (II.10) using the two-parameter potential[12]

(III.10) $$U(r) = -\vec{F}_1 \cdot \vec{F}_2 \left[\frac{6\pi}{27} \frac{(1-\lambda r)^2}{r \ln \lambda r} + U_o \right]$$

where $\vec{F}_1 \cdot \vec{F}_2$ is the color factor given by

(III.11) $$\vec{F}_1 \cdot \vec{F}_2 = -\frac{4}{3} \quad \text{for} \quad 3 \otimes \bar{3} = 1 \oplus \ldots$$

(i.e. proper for a $q\bar{q}$ bound state) and by

(III.12) $$\vec{F}_1 \cdot \vec{F}_2 = -\frac{2}{3} \quad \text{for} \quad 3 \otimes 3 = \bar{3} \oplus \ldots$$

(i.e. proper for a qq bound state).

λ is a (more or less) free parameter related to the QCD cut off parameter Λ by

(III.13) $$\lambda = e^{\gamma} \Lambda$$

where γ is the Euler-Mascheroni constant.

Obviously, the form (III.10) satisfies the various conditions (II.4,6,12b) required to give rise to the charmonium spectrum (II.5,7).

We shall now decompose (II.10) as

(III.14) $$U(r) = V(r) + S(r)$$

where, based on the analogy with QED we shall identify V(r) with the one gluon exchange responsible of the interaction at short distances leaving to S(r) the task of confining, i.e., we shall write

(III.15) $$V(r) = -\vec{F}_1 \cdot \vec{F}_2 \left[\frac{6\pi}{27} \frac{1-\lambda r}{r \ln \lambda r} + U_o \right]$$

(III.16) $$S(r) = -\vec{F}_1 \cdot \vec{F}_2 \left[\frac{6\pi}{27} \frac{\lambda(\lambda r-1)}{\ln \lambda r} \right]$$

which we shall use in eq. (III.4) to study the two-body bound state problem.

Without insisting further on all the questions connected with the derivation of the dynamical equation (III.4) (or (III.9)) or of the potentials $V(r)$ and $S(r)$ to be used in it, we shall just notice that the free parameters in our scheme are a) the quark masses and b) the two parameters in the potential λ and U_o. On all these parameters (U_o excepted), we have, actually, some control in the sense that we "know" in which interval they can, reasonably, be expected to vary. Typically, we expect these parameters (all expressed in MeV) to be, roughly, in the following ranges:

$m_n \simeq 5$, $m_d \simeq 9$, $150 \lesssim m_s \lesssim 200$, $1350 \lesssim m_c \lesssim 1600$, $4600 \lesssim m_b \lesssim 4900$ and $200 \lesssim \Lambda \lesssim 400$.

IV. HOW CAN WE TREAT A THREE-QUARKS BOUND STATE?

One of the most difficult points in dealing with hadron spectroscopy is how to handle the case of three quarks bound to give a baryon.

We propose to show that it is reasonable to treat a baryon as the bound state of a quark and a diquark where by the latter we mean the "bound state" of two quarks. Given that two objects in the fundamental representation 3 (two quarks) combine according to

(IV.1) $\qquad 3 \otimes 3 = \bar{3} \oplus 6$

and that (III.12) in the $\bar{3}$ the colour coupling constant $\vec{F}_1 \cdot \vec{F}_2$ is attractive (whereas it is repulsive in the 6 representation), we can, from this point of view, assimilate a diquark to an antiquark in the fundamental representation $\bar{3}$. In this case, a baryon has, formally, the same structure of a meson $q\bar{q}$ where the $\bar{q}$ is replaced by a diquark.

This makes sense, clearly, only if we can treat the diquark as an elementary constituent while, on the other hand, we have just agreed to consider it as the (colored) bound state of two quarks. This inherent contradiction is partially removable if the diquark has a sufficiently low mass as to allow us to consider it as an elementary object. On the other hand there will be no contradiction, (by definition!) if the scheme will be successful.

Thus, the scheme we suggest is an iterative one whereby, using the same equation, with the same potential and the same parameters used for the meson sector, we will first contruct the "diquark spectrum" and then the baryonic spectrum combining quarks with diquarks. Schematically

(IV.2) $\qquad B \equiv (qq)q = Dq$

The only formal difference will be that for the mesons

(IV.3) $$M = q\bar{q}$$

we use the color factor (III.11) whereas for the baryons the first step requires the color factor (III.12) while for the second we have to go back to (III.11).

Obviously, we will have to take into account that a diquark is the same as an antiquark only from the point of view of colour. From the point of view of spin, for instance, a diquark can either be a scalar (spin 0) or a vector (spin 1) while an antiquark has spin 1/2. On the other hand a diquark is the equivalent of a meson from the spin point of view but is colored and can not, therefore, be found free in nature (just as a quark or an antiquark).

It is clear that only the success (or the insuccess) of a scheme of this kind can legitimate it (or invalidate it). If, however, the scheme is successful, as we shall prove it to be, it is potentially an extremely promising one since it unifies the treatment of mesons and baryons reducing the latter to an iteration of the former with a great economy of parameters. This raises the question of whether one could not iterate further the scheme: for instance, a diquark-antidiquark combination would be what is known as a baryonium state

(IV.4) $$D\bar{D} = (qq)(\overline{qq}) = qq\bar{q}\bar{q}$$

whereas a 3 diquark system (6 quarks) would originate what is known as a dibaryon state

(IV.5) $$DDD = (qq)(qq)(qq) = qqqqqq$$

However, even if we suceed in constructing the baryons as quark-diquark bound states, it is not obvious that we will also be successful when combining together several bound states (diquarks) considered as elementary or, which is the same, when subsequently iterating the procedure which was giving good results after one interation only.

A last consideration in favor of the diquark scheme comes from Martin who has shown[13] both classically as well as relativistically that a bound state of three components held together by a confining force is in its state of minimal energy when two of these constituents are close together, that is when, in turn, they form a bound state. This fact is probably the physical basis of the success of the model. A similar success is encountered in related models[14].

V. SPIN PROBLEMS.

This is probably the point in which there are more practical ambiguities. Implicitely we have assumed the potential to be spin independent but spin plays actually a most important role in telling us how a given bound state should be identified with a given particle.

The interaction terms due to the spin are usually derived[15] perturbatively in the non-relativistic limit from the Bethe-Salpeter equation. This procedure gives rise to various terms: spin-orbit, tensor and spin-spin coupling. In the following, we take into account the last such term only which in the nonrelativistic limit is given by

(V.1) $$V_s^{n.r.} = g_1 g_2 \frac{\nabla^2 V}{m_1 m_2} \vec{S}_1 \cdot \vec{S}_2$$

where g_i are the giromagnetic color factors (g=2 for a quark), V(r) is the potential introduced previously and $\vec{S}_i$ are the spin operators of the particles of masses m_i.

It is immediate to see that the limit (V.1) is too naive. Not only it is singular in the limit $m_i \to 0$ (not surprising given its non-relativistic nature) but it is too singular at r=0 if V(r) is Coulomb bahaved (not surprising since it is a perturbative approach). An alternative recipe to (V.1) is provided by an application[16] of Dirac's constraint method[17] that gives

(V.2) $$V_s = - \frac{g_1 g_2}{\mu} \vec{S}_1 \cdot \vec{S}_2 \nabla^2 \ln \left[1 - \frac{V(r)}{3(m_1 + m_2 + S(r))} \right]$$

where μ is the reduced mass. Not only (V.2) is much less singular than (V.1) but it reduces to the latter in the nonrelativistic limit.

We shall use (V.2) to replace

(V.3) $$\vec{p}^2 = (\vec{p} - \vec{A})^2 \Longrightarrow \vec{p}^2 + 2\mu V_s$$

in eqs. (III.4) (or (III.9)) of which we shall discuss the numerical eingenvalues in the next Section. It is to be noticed, however, that the recipe (V.2) is far from being free of ambiguities or fully satisfactory; for example, the singularity in r=0 is not cured entirely by (V.2) and to perform numerical calculations we will have to introduce an effective cut off. This will be done by replacing $r \to (r^2 + a^2)^{1/2}$ where $\underline{a}$ is, to all effects, same effective inter-quark radius of iteration.

VI. THE HADRONIC SPECTRUM.

a) <u>The mesonic sector</u>.

The numerical solution of (III.4) with the potential (II.15,16) and the spin term (V.3,2) gives us the $q\bar{q}$ bound states (i.e. the meson spectrum) when the color factor (III.11) is used. To this aim we have to fix first the free parameters (see end of Sec. III). To reduce the degrees of freedom, we choose $m_u = m_d = 5$ MeV from the start. Then we use as input the following mesons

(VI.1) $\pi(138)$, $\rho(770)$, $\phi(1020)$; $J/\psi(3097)$; $\psi'(3684)$; $\Upsilon(9460)$

to fix the remaining six parameters (m_s, m_c, m_b, λ, U_o and a).

We also use in (V.2)

(VI.2) $g_1 \cdot g_2 \, \vec{S}_1 \cdot \vec{S}_2 = \begin{cases} 1 & \text{for spin 1} \\ -3 & \text{for spin 0} \end{cases}$

Best fitting the free parameters to the masses of the input mesons (VI.1) we find the values in Table I

TABLE I
Parameters of the best fit. All values in MeV except a in Fermi

m_u (input)	m_d (input)	m_s	m_c	m_b	U_o	λ	a
5	5	183	1416	4794	-120	740	0,041

which are all very reasonable and in the expected range (see end of Sect. III). $\lambda=740$ MeV corresponds to $\Lambda=411$ MeV (see (III.13)).

The effective constituent masses one gets from the definition

(VI.3) $M_i = \langle M_i(r) \rangle = m_i + \frac{1}{2} \langle S(r) \rangle$

are also very plausible (see Table II).

TABLE II
Effective quark masses (in MeV)

$M_u = M_d$	M_s	M_c	M_b
360	530	1650	5000

Comparing now the masses of the ground state mesons calculated in our scheme for the vector mesons $J^P=1^-$ (Table III) and for the pseudoscalars 0^- (Table IV) with the experimental data [18], the overall picture is very good. If some problem arises is in the lowest masses (K, for instance, which we could have taken as an input but we have not).

The scheme is presumably not going to the so accurate for the higher recurrences (N<1) which we have considered only for the high mass mesons. For more details, see ref. 5.

Table III

Vector Mesons $J^P=1^-$ (Masses in MeV)

Particle	N	Theory	Experiment
ρ	1	(input)	770
ϕ	1	(input)	1020
K^*	1	896	892
D^*	1	1959	2010
F	1	2116	2140
J/Ψ	1	(input)	3097
	2	(input)	3686
	3	4072	4030
	4	4380	4415
B^*	1	5235	5270
Υ	1	(input)	9460
	2	10043	10025
	3	10394	10355
	4	10664	10575
	5	10891	

Table IV

Pseudoscalar Mesons $J=0^-$ (Masses in MeV)

Particle	N	Theory	Experiment
π	1	(input)	138
K	1	564	496
D	1	1880	1870
F	1	2005	1971
η_c	1	3057	2981
	2	3660	3590

b) The baryonic sector.

Having found so good an agreement with the mesons we now go to the baryons. For this, we start building the diquarks using the same equation, the same potential, the same spin factor, the same parameters determined previously but the color factor (III.12). The diquark spectrum is given in Table V

Table V
Diquark spectrum (masses in MeV)

Quark content	Spin	Diquark mass
uu=dd=ud	0	321
	1	642
ud=ds	0	671
	1	771
ss	1	908
uc=dc	0	1889
	1	1909

Before proceeding to the baryons according to the scheme outlined in Sec. IV we have, preliminarily, to agree on what combination of spin 0 and spin 1 diquarks to use when studying the spin ½ baryons. Lacking a better prescription we will use SU(6) combinations but this will have to be reconsidered at the light of the results we will find. Notice that this complication does not exist for spin 3/2 baryons to which only spin 1 diquarks contribute. The latter represent, therefore, a much more crucial test of our dynamical model.

Stressing once again that at this stage we have no further adjustable degrees of freedom, we compare the result of the model with the experimental masses for the 3/2 spin baryons (Table VI) and for the sin 1/2 ones (Table VII)

Table VI
Spin 3/2 Baryons (all masses in MeV)

Particle	Theory	Experiment
Δ	1286	1232
Σ^*	1403	1384
Ξ^*	1519	1533
Ω	1652	1672
Σ_c^*	2449	2430

Table VII

Spin ½ baryons (all masses in MeV)

Particle	Theory	Experiment
N	1015	939
Λ	1176	1116
Σ	1239	1193
Ξ	1369	1318
Λ_c	2254	2272
Σ_c	2360	--

First of all, we remark the excellent predictions for the spin 3/2 baryons (errors less than 5% even at the low end of the spectrum). This as pointed out before, is the crucial test for the model since no other ingredients are needed to get these results. The situation is much more delicate for the spin ½ baryons which required the extra input of taking SU(6) combinations of spin 0 and spin 1 diquarks. Given the considerable mass difference of the lowest diquarks, it is quite clear from Table VII than the situation could have been drastically improved (especially for the low lying baryons) should we have taken a somewhat larger contribution of spin 0 than of spin 1 diquarks. This point is presently under consideration. Even so, Tables VI and VII show that given the economy of parameters in the scheme, this is highly successful.

VII. CONCLUSIONS.

As we have seen, a relativistic scheme with a potential suggested by QCD is capable of reproducing the entire hadronic spectrum of mesons and baryons in which the latter are seen as quark-diquark bound states. Given that a total count of six parameters (all more or less constrained to be where they are found) takes care of over 30 particles with quite acceptable errors (and that when the errors are larger than acceptable it is quite clear why this happens), we conclude that the scheme is quite good. Add to this that also the magnetic moments are quite well reproduced[19].

Several points remain to be investigated.

First of all the problem of the combination of spin 0 and spin 1 diquarks in the baryons deserves being understood better. When this point will be cleared out, we plan to investigate the formation of exotic states like baryonia and dibaryons about which preliminary results show a promising trend. Presumably, however, new ingredients will be needed such as a coupled channel formalism and a unitarization procedure. Work in under way in this direction.

REFERENCES

1) M Gell-Mann: Phys. Lett. $\underline{8}$ 214 (1964) e G. Zweig: CERN preprint 8409/Th. 412 (1964).

2) See, for instance, R.H. Dalitz in "Meson Spectroscopy" (Ed. C. Baltay e A. Rosenfeld), Benjamin (New York 1968) p. 497 and references therein.

3) T. Appelquist and H.D. Politzer: Phys. Rev. Lett. $\underline{34}$ 43 (1975).

4) J.L. Basdevant and G. Preparata: Proc. Int. Conf. on High Energy Physics Lisbon 1981 (ed. J. Dias de Deus and J. Soffer) p. 965.

5) D.B. Lichtenberg, W. Namgung, E. Predazzi and J.G. Wills: Phys. Rev. Lett. $\underline{48}$ 1653 (1982) and Zeit. f. Physik $\underline{C19}$ 19 (1983).

6) D.B. Lichtenberg and L.J. Tassie: Phys. Rev. $\underline{155}$ 1601 (1967).

7) As a general reference, see E. Leader ed. E. Predazzi: "An Introduction to Gauge Theories and The New Physics"; Cambridge Univ. Press, London (1982).

8) B. Baumgartner, H. Gosse and A. Martin: Nucl. Phys. $\underline{B254}$ 528 (1985).

9) A.N. Mitra: Zeit. f. Physik $\underline{C8}$ 25 (1981) and Zeit. f. Physik $\underline{C8}$ 33 (1981).

10) W. Krolikowski: Acta Phys. Pol. $\underline{B11}$ 387 (1980), $\underline{B12}$ 793 (1981).

11) I.T. Todorov: Phys. Rev. $\underline{D3}$ 235 (1971).

12) D.B. Lichtenberg and J.G. Wills: Lett. Nuovo Cimento $\underline{32}$ 86 (1981).

13) A. Martin: Cern preprint Th. 4259/85.

14) J.L. Basdevant and S. Boukraa: Zeit. f. Physik $\underline{C28}$ 413 (1985).

15) D. Gromes: Nucl. Phys. $\underline{B131}$ 80 (1977); H.J. Schnitzer: Phys. Rev. $\underline{D13}$ 74 (1966).

16) H. Crater and P. Van Alstine: Phys. Lett. $\underline{100B}$ 166 (1981); Jour. Math. Phys. $\underline{23}$ 1697 (1982) and private communication.

17) P.A.M. Dirac: Lectures in Quantum Mechanics: Belfer Graduate School of Science: Yeshiva Univ. (New York, 1964).

18) Review of Particle Properties, April 1984 Edition; Rev. Mod. Phys. <u>56</u> n. 2 (1984).

19) D.B. Lichtenberg, E. Predazzi and J.G. Wills: Zeit. f. Physik <u>C17</u> 57 (1983).

QUARK AND LEPTON FLAVOR MIXING: ROLE OF INTERMEDIATE ENERGIES

Nello PAVER
Dipartimento di Fisica Teorica, University of Trieste, Italy and
INFN - Sezione di Trieste, Italy

We briefly review, motivated by recent projects of high intensity, intermediate energy facilities, the accelerator searches of neutrino masses and the precision measurements of CP-violation in the $K^°$ - $\bar{K}^°$ system.

1. INTRODUCTION

There has been an impressive accumulation of experimental data, culminating in the observation of the W and Z vector bosons at the predicted masses, which confirm the various aspects of the Standard Model[1] description of electroweak phenomena. There are however a number of compelling theoretical reasons why we should not consider the Standard Model as the ultimate theory of particle physics. To enumerate them very briefly let me just recall the elementary "building blocks" of the "minimal" Standard Model:

- Spin ½ fermions (quarks and leptons), organized in three "generations" with identical quantum numbers, except for the mass:

$$\begin{pmatrix} \nu_e \\ e \end{pmatrix}_L , \begin{pmatrix} u_i \\ d_i \end{pmatrix}_L ; (e)_R , (u_i)_R , (d_i)_R \qquad m \sim \frac{MeV}{c^2}$$

$$\begin{pmatrix} \nu_\mu \\ \mu \end{pmatrix}_L , \begin{pmatrix} c_i \\ s_i \end{pmatrix}_L ; (\mu)_R , (c_i)_R , (s_i)_R \qquad m \sim 0.1 \div 1 \frac{GeV}{c^2}$$

$$\begin{pmatrix} \nu_\tau \\ \tau \end{pmatrix}_L , \begin{pmatrix} t_i \\ b_i \end{pmatrix}_L ; (\tau)_R , (t_i)_R , (b_i)_R \qquad m \sim 2 \div 50 \frac{GeV}{c^2}$$

where i is the color index, L,R denote left- and right-handed components respectively, and some indicative values of the masses are reported for each family. Neutrinos are strictly massless, $m_\nu = 0$, as there is no ν_R. Indeed the present limits on neutrino masses are [2]

$$m_{\nu_e} < 46 \text{ eV} \quad [20 < m_{\bar{\nu}_e} < 45 \text{ eV}]$$

$$m_{\nu_\mu} < 250 \text{ KeV} \tag{1}$$

$$m_{\nu_\tau} < 70 \text{ MeV} .$$

where in brackets there is the famous ITEP result [3], still waiting for a confirmation.

- Bosons. These are:

i) The $W^\pm$, Z^0 and γ, with spin 1, $m_\gamma = 0$ and $M_{W,Z} \simeq 80\text{-}90$ GeV/c^2 (theoretically predictable!), which are the carriers of the electroweak force.

ii) The Higgs, with spin 0, which plays an essential role in the dynamical mechanism generating $M_{W,Z}$ and $m_{fermions}$, but whose mass is not predictable theoretically (although M_H should not exceed a value of the order of 1 TeV/c^2).

iii) The gluons, with spin 1 and zero mass, only coupled to quarks, which carry the QCD confining color force.

- The electroweak mixing angle θ_W (related to $M_{W,Z}$ by the well-known relation $\cos \theta_W = M_W/M_Z$) and the Kobayashi-Maskawa quark mixing matrix

$$U_{KM} = \begin{pmatrix} U_{ud} & U_{us} & U_{ub} \\ U_{cd} & U_{cs} & U_{cb} \\ U_{td} & U_{ts} & U_{tb} \end{pmatrix} = \begin{pmatrix} c_1 & -s_1 c_3 & -s_1 s_3 \\ s_1 c_2 & c_1 c_2 c_3 - s_2 s_3 e^{i\delta} & c_1 c_2 s_3 + s_2 c_3 e^{i\delta} \\ s_1 s_2 & c_1 s_2 c_3 + c_2 s_3 e^{i\delta} & c_1 s_2 s_3 - c_2 c_3 e^{i\delta} \end{pmatrix} \tag{2}$$

where $c_i = \cos \theta_i$, $s_i = \sin \theta_i$ ($i = 1,2,3$) are the mixing angles and δ is a (CP-violating) phase (in the following we will similarly use c_δ for $\cos \delta$ and s_δ for $\sin \delta$). The complete experimental determination of U_{KM} has not been fully achieved yet. The following constraints are found [4]:

$$\begin{aligned} \sin \theta_1 &= 0.231 \pm 0.003 \\ 0.025 &< \sin \theta_2 < 0.06 \\ \sin \theta_3 &< 0.02 , \end{aligned} \tag{3}$$

while δ is still unconstrained. There is no lepton mixing as long as $m_\nu = 0$.

- There finally are the electroweak coupling constant g and the QCD coupling constant g_s.

The problem with the Standard Model is then that, although being so successful, it contains too large a number of parameters (masses, mixings, coupling constants) which we must fit from experiment, whereas we would like to be able to determine their values theoretically from a much reduced number of fundamental constants. Particularly puzzling is the repetition of fermions in three generations, their observed masses and mixing angles and the origin of the correlation between quarks and leptons. Also, the Higgs is disappointing, on the one hand because it is elusive experimentally and on the other because its mass is unstable in perturbation theory.

Clearly any answer to these questions can only come from new physics, outside the Standard Model. In fact, many directions in extending the Standard Model have been pursued. To give an impression, I just try a list of them, with for each, the related problems of the Standard Model in parentheses and some review papers in the references: Left-right symmetric models [5] (parity violation); Technicolor models [6] (Higgs); Boson-fermion supersymmetry [7] (Higgs, mass hyerarchies); Grand-unified theories [8] (number of parameters, quark-lepton correlation, unification of fundamental forces); Horizontal symmetries [9] (generations); Composite models of quarks and leptons and/or W,Z, Higgs [10] (generations); and combinations thereof. In the absence of experimental indications, none of them seems completely satisfactory at present. The important point is however that these models suggest the existence of new mass scales $\Lambda \gg M_W$ and of scenarios where, in addition to the familiar Standard Model phenomenology, new "building blocks" and new interactions, governed by those mass scales Λ, become effective. In particular, there can be massive neutrinos, with $m_{\nu_L} \sim \frac{m_{lepton}^2}{\Lambda} \ll m_{lepton}$, in agreement with the pattern in (1) and $m_{\nu_R} \sim \Lambda$, and new kinds of mixings inducing transitions which would be rigorously forbidden by the Standard Model selection rules. Consequently those mass scales Λ should be regarded as new constants of nature which we should try to measure experimentally. Since some of them might well be outside the reach of the biggest machines, precision measurements at the planned intermediate energy, high intensity accelerators [11] could produce glimpses on the prospected new physics, capable as they should be to allow decisive tests of the Standard Model and improved determinations of fermion masses and mixing angles. In this spirit I will devote my lecture to accelerator searches of neutrino masses and to the determination of the CP-violation parameters in the $K^\circ - \bar{K}^\circ$ system, two longstanding problems in weak interactions. Since I prefer to emphasize the progress obtainable in principle from intense, dedicated beams rather than to try an exhaustive theoretical discussion, the material will be presented here at a simple, introductory level. More details can be found by the interested reader in the review articles listed in the references.

2. NEUTRINO MASSES

The possibility of nonvanishing neutrino masses is one of the most fascinating issues in particle physics, which is receiving ever increasing attention. In summary, the main motivations to look for massive neutrinos are the following:
- As said, new physics wants $m_\nu \neq 0$ neutrinos [12], either much lighter than the corresponding lepton or maybe very heavy.
- Some cosmological problems, in particular the so-called hidden mass of the universe, might be understood if neutrinos had small masses, of the order of some eV.
- There is the challenging ITEP result [3].

Of particular importance is the question whether the neutrino should be a Dirac or a Majorana neutral fermion, namely whether neutrino and antineutrino are different or are identical (which would lead to lepton number violation). Also very important is the determination of the neutrino mass spectrum and mixing angles.

If $m_\nu \neq 0$ neutrino mixing can occur à la Kobayashi-Maskawa:

$$|\nu_\ell\rangle = \sum_i U_{\ell i} |\nu_i\rangle \quad , \quad U^+ U = 1 \tag{4}$$

where $|\nu_i\rangle$ are the mass eigenstates with $m_i = m_\nu$, $|\nu_\ell\rangle$ are the "weak" eigenstates ($\ell = e, \mu, \tau$ is the "flavour") and U^i is a unitary matrix. Depending on the value of m_ν, different kinds of experiments are most sensitive.

2.1. β- and double β-decay experiments

These allow the determination of m_{ν_e} directly from kinematics. In particular the double β-decay [14] is a decisive test of Majorana neutrinos.

2.2. Search for neutrino oscillations

In the situation where neutrinos are massive and mix with each other, they can change their flavour by undergoing transitions of the form of oscillations between the different species [15]. Limiting for simplicity to the case of only two generations, so that the flavour $\ell = e, \mu$, the mixing in (4) simplifies to

$$|\nu_e\rangle = \cos\alpha |\nu_1\rangle - \sin\alpha |\nu_2\rangle$$
$$|\nu_\mu\rangle = \sin\alpha |\nu_1\rangle + \cos\alpha |\nu_2\rangle \quad , \tag{5}$$

where α is a mixing angle. A pure $|\nu_\mu\rangle$ state emitted by a neutrino source at $t = 0$ will then evolve in time to the combination of mass eigenstates

$$|\nu_\mu\rangle_t \to \sin\alpha\, e^{-iE_1 t}|\nu_1\rangle + \cos\alpha\, e^{-iE_2 t}|\nu_2\rangle \quad , \tag{6}$$

which contains a nonvanishing component of $|\nu_e\rangle$ if $m_1 \neq m_2$ (so that $E_1 \neq E_2$) and oscillations $\nu_\mu \to \nu_e \to \nu_\mu$ and so on will occur. Of course the same kind of phenomenon will happen starting with a pure $|\nu_e\rangle$ beam. If we place a detector at a distance L from the source of ν_μ, the probability to observe ν_e's (appearance experiments) is given by

$$P(\nu_\mu \to \nu_e) = \sin^2 2\alpha \cdot \sin^2\left[\frac{1.27\,\Delta m^2\, L}{E_\nu}\right] \tag{7}$$

while the probability to observe ν_μ's (disappearance experiment) is

$$P(\nu_\mu \to \nu_\mu) = 1 - P(\nu_\mu \to \nu_e) \tag{8}$$

In (7), (8) $\Delta m^2 = m_1^2 - m_2^2$ in eV^2, E_ν is the energy of the neutrino beam in MeV, and the distance L is in metres. In practice eqs.(7) and (8) should be suitably averaged over the dimensions of the source (and of the detector), the time and the neutrino energy spectrum. Eqs.(7) and (8) show that the necessary conditions for neutrino oscillations to occur are i) that neutrino mass eigenvalues are different from each other ($\Delta m^2 \neq 0$) and ii) that in addition lepton mixing exists ($\alpha \neq 0$). Notice that, in fact, this kind of phenomenon involves lepton number nonconservation, and consequently the existence of a dynamical mechanism strictly beyond the Standard Model. As one can see from (7) and (8) the sensitivity of a neutrino oscillation experiment is determined by the value of the ratio L/E_ν, which for small neutrino masses must be as large as possible, requiring thus large L and small E_ν. In the table below I report some figures of merit, which show the sensitivities to Δm^2 obtainable in experiments with different neutrino sources

	sun	cosmic rays	reactors	accelerators
$\Delta m^2 (eV^2)$	10^{-11}	10^{-5}	10^{-2}	$10^{-1} - 10^3$

In Fig. 1 some typical results are shown, including accelerator and reactor experiments. As there is no well established indication of neutrino oscillations, the experimental results are usually presented as bounds on Δm^2 versus $\sin^2 2\alpha$ or viceversa. Notice that while reactor experiments are sensitive (in principle) to smaller masses, accelerator experiments have a better sensitivity to small mixing angles. Of particular importance, in the latter case, would be the realization of high flux neutrino beams with "moderate" E_ν. This would allow, trading the intensity with the distance, much larger L/E_ν hence one or two orders of magnitude improvements in the sensitivities to Δm^2 (and $\sin^2 2\alpha$). Also, ν_e (and $\bar{\nu}_e$) initiated oscillations (for the moment limited to reactors) could presumably be studied, the interest for this being that the only reported positive result for neutrino mass refers to $\bar{\nu}_e$.

2.3 Search for massive neutrinos in π and K decays

For m_ν of the order of, say, one to a hundred MeV, precision measurements of the decays $\pi \to \ell\nu$ and $K \to \ell\nu$ should represent a sensitive test.[16] In fact the contribution of a mixed massive neutrino ν_i to e.g. $K \to e\nu$ can be expressed as

$$\frac{\Gamma(K \to e \, \nu_j)}{\Gamma(K \to e \, \nu_e)} = |U_{ei}|^2 \, \rho(\delta_i) \,, \tag{9}$$

where ν_e refers to the "conventional" massless neutrino and $\rho(\delta_i)$ is a phase-space factor, with $\delta_i = (m_{\nu_i}/m_K)^2$. Notice that for massive neutrinos the familiar helicity suppression for $K \to e \, \nu_e$ (and $\pi \to e \, \nu_e$) no longer works.

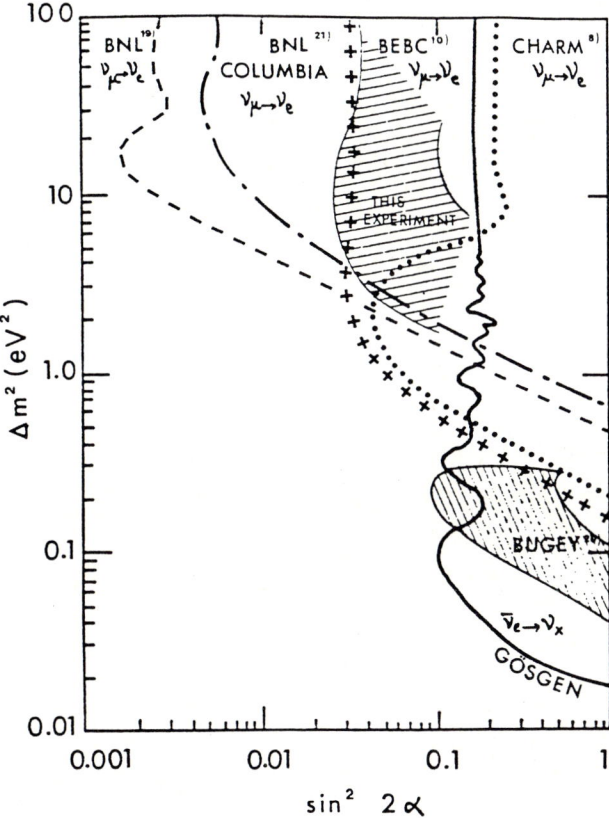

Figure 1

This "Mℓ_2" test should then consist in the search for additionals peaks or kinks in the spectrum of the recoil charged lepton, like in Fig. 2. Some experimental results, in the form of correlated limits on neutrino masses and mixings, are

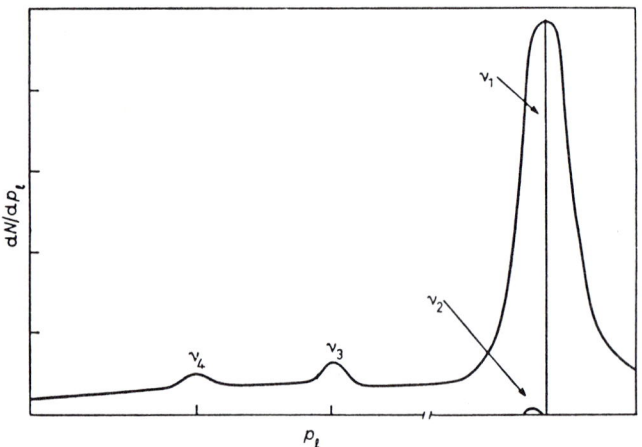

Figure 2

shown in Fig. 3.

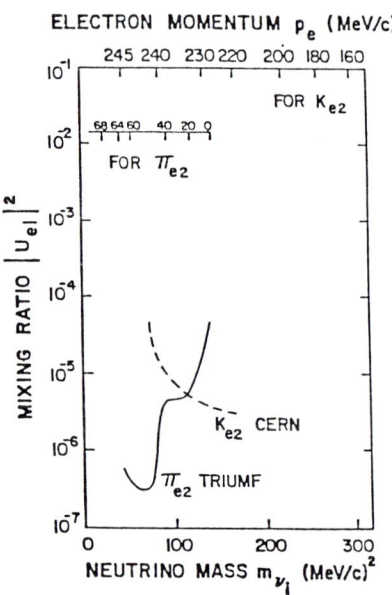

Figure 3

As one can see, the sensitivity reached so far for the mixing matrix elements is rather impressive (can be as low as 10^{-3}), and certainly further improvements will be achieved in the future.

2.4. Search for massive unstable neutrinos

The idea is that neutrinos, if heavy enough, may decay in the neutrino detector with a reasonable probability [17]. For masses much larger than the electron mass the expected decay will be $\nu_H \to e^+ e^- \nu_e$, where ν_H denotes the heavy neutrino with mass m_{ν_H}. The probability for such a neutrino to be present in a neutrino beam from, e.g., $K \to \mu\nu$ or $\pi \to \mu\nu$ is given by $\rho |U_{\mu_H}|^2$ as in eq.(9). The decay lifetime for $\nu_H \to e^+ e^- \nu_e$ is given by

$$\tau_{\nu_H} \sim |U_{eH}|^2 \left(\frac{m_\mu}{m_{\nu_H}}\right)^5 \tau_\mu \quad , \tag{10}$$

where τ_μ is the $\mu \to e\nu\bar{\nu}$ decay lifetime. One thus obtains the ratio:

$$\frac{\text{No. of } \nu_H \to e^+ e^- \nu_e \text{ decays}}{\text{No. of } \nu_\mu \text{ interactions}} \sim |U_{\mu H} U_{eH}|^2 \frac{m_{\nu H}}{E_\nu} \frac{(m_{\nu H}/m_\mu)^5}{\sigma \mathcal{N} c \tau_\mu} \tag{11}$$

where σ is the total neutrino-nucleon cross-section, $\mathcal{N}$ is the target density and E_ν is the energy of the neutrino beam. Typically (11) gives

$$\frac{\text{No. of decays}}{\text{No. of } \nu_\mu \text{ interactions}} \sim 10^8 |U_{\mu H} U_{eH}|^2 \frac{(m_{\nu H}/m_\mu)^5}{E_\nu^2 \, (\text{GeV}^2)} \quad . \tag{12}$$

Similar considerations hold for other possible neutrino decay channels, like $\nu_H \to \mu e \nu_e$, $\nu_H \to \pi e$, etc. Notice in (12) the factor 10^8, and the $1/E_\nu^2$ behaviour, which indicates the advantage of using moderate to small energy neutrino beams. In Fig. 4 an example of experimental data is represented, as a limit on the coupling constant versus the neutrino mass. Again, the sensitivity of such experiments looks really good.

In conclusion, the question of neutrino masses is still an open one. Given the high theoretical interest, further experiments are clearly necessary, to improve the explored ranges of masses and mixing angles, and indeed some ambitious projects are being planned for the future [18].

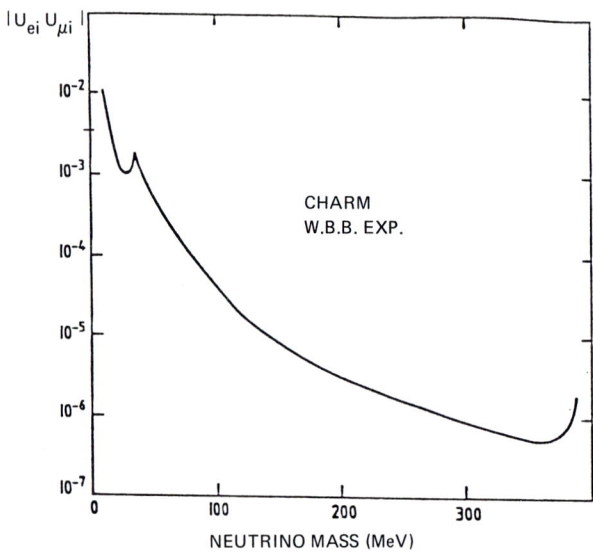

Figure 4

3. CP-VIOLATION IN THE $K^\circ - \bar{K}^\circ$ SYSTEM

This is quite a complicated theoretical problem, because in addition to the perturbative aspects of the Standard Model whose structure we believe to understand, it involves significantly the non-perturbative QCD quark confinement we do not fully control yet, giving rise to large uncertainties in the predictions. Even so, it appears that precision measurements of CP-violation parameters can allow to test the Standard Model in a decisive way.

Let me just recall the definition of the CP-violation parameters for $K^\circ \to 2\pi$ decay [19]. There is ε, the CP impurity in the K_L, K_S wave functions

$$|K_L^\circ> = (1+|\varepsilon|^2)^{-\frac{1}{2}} [(1+\varepsilon)|K^\circ> + (1-\varepsilon)|\bar{K}^\circ>] \frac{1}{\sqrt{2}}$$
$$|K_S^\circ> = (1+|\varepsilon|^2)^{-\frac{1}{2}} [(1+\varepsilon)|K^\circ> - (1-\varepsilon)|\bar{K}^\circ>] \frac{1}{\sqrt{2}} \quad , \tag{13}$$

where $|K^\circ\rangle, |\bar{K}^\circ\rangle$ are the strong interactions eigenstates with strangeness $S = \pm 1$ respectively. Since weak interactions do not conserve S, $\Delta S = 2$ $K^\circ \rightleftarrows \bar{K}^\circ$ oscillations will occur, which have been observed experimentally. If CP were conserved by the weak interactions, then (13) with $\varepsilon \to 0$ would be the unmixed eigenstates of the total hamiltonian (notice $\langle K_L^\circ | K_S^\circ \rangle \sim 2\text{Re } \varepsilon/(1+|\varepsilon|^2)$) with CP = ± 1 respectively, and only $K_L^\circ \to 3\pi$ and $K_S^\circ \to 2\pi$ would be allowed. Instead, $K_L^\circ \to 2\pi$ is seen to occur [20] (at the 10^{-3} level):

$$\eta_{+-} = \frac{A(K_L^\circ \to \pi^+\pi^-)}{A(K_S^\circ \to \pi^+\pi^-)} = (2.274 \pm 0.022) \cdot 10^{-3} \, e^{i(44.6 \pm 1.2)^\circ}$$

$$\eta_{oo} = \frac{A(K_L^\circ \to \pi^\circ\pi^\circ)}{A(K_S^\circ \to \pi^\circ\pi^\circ)} = (2.33 \pm 0.08) \cdot 10^{-3} \, e^{i(54 \pm 5)^\circ} \quad (14)$$

and also a charge asymmetry in $K\ell_3$ decay has been observed:

$$\delta_\ell = \frac{\Gamma(K_L^\circ \to \nu\ell^+\pi^-) - \Gamma(K_L^\circ \to \bar{\nu}\ell^-\pi^+)}{\Gamma(K_L^\circ \to \nu\ell^+\pi^-) + \Gamma(K_L^\circ \to \bar{\nu}\ell^-\pi^+)} = (0.330 \pm 0.012)\% \quad (15)$$

The other CP-violation parameter is ε', characterizing the "direct" breaking of CP invariance in the $\Delta S = 1$ transition amplitude $A(K^\circ \to 2\pi)$. In terms of pure isospin amplitudes

$$A(K^\circ \to (\pi\pi)_I) = i A_I e^{i\delta_I}, \quad I = 0,2 \quad (16)$$

where δ_I are the $\pi\pi$ phaseshifts at $\sqrt{s} = m_k$, and ε' can consequently be expressed as

$$\varepsilon' \simeq \frac{1}{\sqrt{2}} e^{i(\delta_2 - \delta_0 - \frac{\pi}{2})} \frac{A_2}{\text{Re } A_o} \frac{\text{Im } A_o}{\text{Re } A_o} \quad (17)$$

In (17) $\delta_2 - \delta_0$ is experimentally known, $\delta_2 - \delta_0 = 29.2° \pm 3°$, and so is $A_2/R_e A_0 \sim 0.05$ from the $\Delta I = 1/2$ enhancement in e.g. $\Gamma(K^+ \to \pi^+ \pi^0)/\Gamma(K_S^0 \to \pi^+ \pi^-) \sim 2 \cdot 10^{-3}$. In terms of ε, ε' eq.(14) becomes

$$\eta_{+-} \simeq \varepsilon + \varepsilon'$$
$$\eta_{oo} \simeq \varepsilon - 2\varepsilon'$$
(18)

and by comparison we find $\varepsilon_{exp} \simeq 2.28 \cdot 10^{-3} e^{i\pi/4}$ and $\varepsilon'/\varepsilon \ll 1$.

Theories of CP-violation such that $\varepsilon' = 0$ are called "superweak". Those where both ε and ε' are different from zero are called "milliweak". In the Standard Model both ε and ε' are proportional to $\sin \delta$ (eq.(2)). From the preceding considerations it is clear that while ε is connected to the weak non-leptonic effective hamiltonian $H_W^{\Delta S=2}$, ε' is controlled by the weak hamiltonian $H_W^{\Delta S=1}$.

The $\Delta S=2$ weak hamiltonian leading to $K^\circ - \bar{K}^\circ$ mixing is given in the Standard Model by the "box" diagrams in Fig. 5.

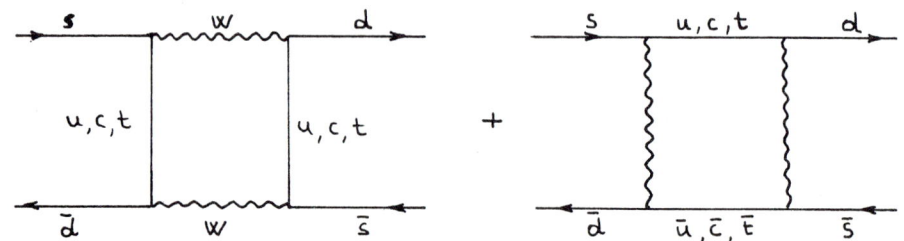

Figure 5

Since M_W should be the largest mass in the game, one can perform a "short-distance" expansion (in $1/M_W^2$) of the diagrams in Fig. 5. The final result, including perturbative QCD corrections to those diagrams, has the form

$$H_W^{\Delta S=2} = \frac{G_F^2}{16\pi^2} \left\{ \eta_1 (U_{cd}^* U_{cs})^2 m_c^2 + \eta_2 (U_{td}^* U_{ts})^2 m_t^2 \right.$$
$$\left. + 2\eta_3 U_{cd}^* U_{cs} U_{td}^* U_{ts} m_c^2 \ln \frac{m_t^2}{m_c^2} \right\} \cdot O^{\Delta S=2} + h.c.$$
(19)

where η_i are numerical coefficients (which would be one neglecting QCD corrections and typically are of the order $\eta_1 \sim 0.8$, $\eta_2 \sim 0.6$, $\eta_3 \sim 0.4$) and where $0^{\Delta S=2}$ is the four-quark operator

$$0^{\Delta S=2} = (\bar{S}_L \gamma_\mu d_L)(\bar{S}_L \gamma_\mu d_L) \quad , \tag{20}$$

to be sandwiched between the hadronic $<\bar{K}°|$ and $|K°>$ states. I have to mention that the procedure to arrive at eq.(19) from the diagrams of Fig. 5, in particular to decouple the "heavy" quarks c,t into the coefficients is rather complicated [21]. ε is finally determined by the imaginary part of the matrix element $(K°|H_W^{\Delta S=2}|K°)$, and in terms of the angles and phase of eq.(2) it can be expressed as:

$$\varepsilon_{Box} \simeq e^{i\pi/4} B \frac{5.96}{(GeV)^2} s_1^2 (c_2 s_2 c_3 s_3) s_\delta \cdot$$

$$\left\{ -\eta_1 m_c^2 + \eta_2 s_2^2 m_t^2 + \eta_3 m_c^2 \ell n \frac{m_t^2}{m_c^2} \right\} \quad , \tag{21}$$

where B is the hadronic matrix element, defined as

$$<K° | 0^{\Delta S=2} | \bar{K}°> = \frac{4}{3} m_K f_K^2 B \quad . \tag{22}$$

One can see that for "reasonable" values of quark masses the top quark gives the dominant contribution to eq.(21).

The calculation of the parameter B is a genuine non-perturbative problem, for which no clear-cut solution exists yet. Indeed, a variety of calculations have been presented [22], the estimated values of B ranging from $B \sim 0.33$ up to $B \sim 1.5$ (depending on the theoretical framework adopted), to be compared with the simplest "vacuum saturation" of (22) giving $B=1$ [23]. There is thus an appreciable numerical uncertainty in the theoretical prediction (21) coming from non-perturbative phsyics. In principle, the real part of (19) would be related to

$\Delta m = m_L - m_S = (3.521 \pm 0.014) \cdot 10^{-12}$ MeV, which might help in such a situation. This identification however would be quite ambiguous, because (19) only represents the "short-distance" part of $H_W^{\Delta S=2}$, to which "long-distance" contributions should be added. Like B, these are obviously of non-perturbative origin, so model dependent. It is found however that long-distance corrections, while substantial in the real part, in general tend to cancel with each other in the imaginary part of $<K°|H_W^{\Delta S=2}|\bar{K}°>$, so that the identification of ε to ε_{box} eq.(21) should be allright anyway.

All in all, using the available experimental information on the Kobayashi-Maskawa angles (3), one obtains for m_t = 40 GeV the bound

$$|\varepsilon_{Box}| \leq 2.4 \cdot 10^{-3} |B \sin \delta|, \qquad (23)$$

consistent with the experimental result in (18) provided $|B \sin \delta| \sim 1$.

Turning now to ε', an analogous expression as (19) can be worked out for $H_W^{\Delta S=1}$, to order G_F and including perturbative QCD corrections, which retains the factorized form of coefficient functions of quark masses, mixing angles, M_W etc., times hadronic four quark operators [24]. The situation is complicated in this case by the fact that there are altogether six independent effective operators. Fortunately it turns out that the CP-violating part of $H_W^{\Delta S=1}$ is dominated by just one of those operators, induced by the ($\Delta I = \frac{1}{2}$) "penguin diagram" in Fig. 6:

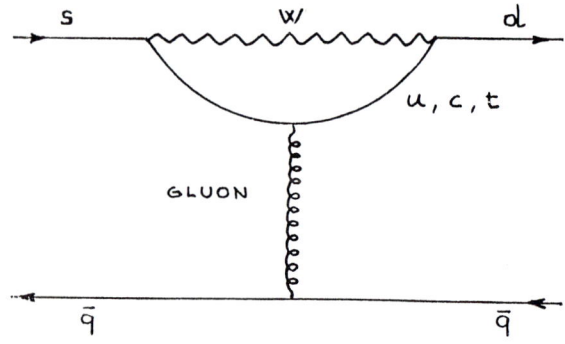

Figure 6

I just mention that this kind of diagrams,[25] combining simultaneously W and gluon exchanges, were in fact advocated as a possibility to understand the enhancement of the $\Delta I = \frac{1}{2}$ selection rule in non-leptonic weak decays. Omitting details, the final result for ε' is typically

$$\varepsilon'|_{th} \simeq -3 \cdot 10^{-2} \, H \, s_2 \, c_2 s_3 s_\delta \, e^{i\pi/5} \quad , \tag{24}$$

where H is a dynamical factor, similar to B, related to the matrix element $<\pi\pi|H_W^{\Delta S=1}|K^\circ>$ (more precisely, that of the four-fermion effective operator mentioned above). Like B, also H is controlled by the non-perturbative confinement physics, and consequently is theoretically uncertain, the estimated values ranging from $H \sim -0.25$ to $H \sim -1$.

Combining (24) with (21) the ratio ε'/ε can be approximately expressed as

$$\frac{\varepsilon'}{\varepsilon}\bigg|_{th} \simeq \left[-\frac{H}{B}\right] \frac{e^{-i\pi/20} \; 0.11 \; (\text{GeV})^2}{\left(-m_c^2 \, n_1 + s_2^2 \, m_t^2 \, n_2 + n_3 \, m_c^2 \, \ell n \, \frac{m_t^2}{m_c^2}\right)} \tag{25}$$

Referring to the literature for an exhaustive phenomenological analysis[22,26], we limit to observe from (25) that i) $\varepsilon'/\varepsilon|_{th}$ is predicted almost real, ii) the smaller H (the larger B) the smaller $\varepsilon'/\varepsilon|_{th}$ and viceversa, and iii) the larger m_t the smaller $\varepsilon'/\varepsilon|_{th}$. As an example we show in Fig. 7 a lower bound on the value of $\varepsilon'/\varepsilon|_{th}$ as a function of m_t.[27] In this Figure the solid line, the dash-dotted line and the dashed line refer to $\tau_b = 0.6 \cdot 10^{-12}$ sec, $0.9 \cdot 10^{-12}$ sec and $1.2 \cdot 10^{-12}$ sec respectively, with τ_b the beauty lifetime. The dependence on this quantity enters through the analysis of the Kobayashi-Maskawa matrix elements.

The results of the complete analysis, taking the various kinds of uncertainties into account, can be somehow summarized by saying that in the Standard Model

$$\varepsilon'/\varepsilon|_{th} \sim (3 \div 12) \cdot 10^{-3} \quad . \tag{26}$$

the two most recent experimental determinations being [28]

$$\left.\frac{\varepsilon'}{\varepsilon}\right|_{th} = \begin{cases} (-4.6\pm5.3\pm2.4)\cdot 10^{-3} \\ (1.7\pm8.2)\cdot 10^{-3} \end{cases} \quad (27)$$

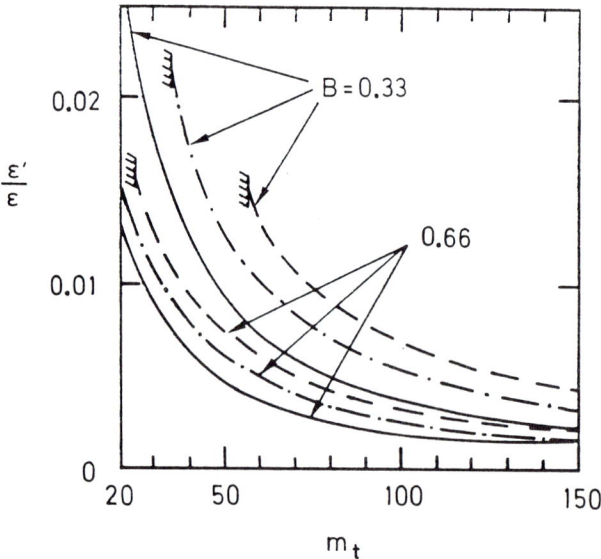

Figure 7

The agreement between (27) and (26) does not appear particularly good. However, keeping in mind the large theoretical as well as experimental uncertainties, any firm conclusion is for the moment premature. What we can say with certainty is that we are very close to the crucial point of testing the Standard Model so stringently, to eventually make a case for new physics, in the case a significant deviation from (26), in particular a sign difference, were confirmed experimentally. In the table below a few examples are given, which show the sensitivity of ε'/ε on the prospected new physics.

To perform such a test it is essential to reduce the uncertainty on the theoretical prediction. This requires on the one hand a more accurate determination of the quark mixing angles, and on the other the formulation of more reliable theoretical devices to quantitatively estimate hadronic matrix elements. In this latter respect encouraging progress seems to be achieved by lattice calculations [29] and QCD sum rules [30]. As is clear from the Table, a sensitivity of at least 10^{-3} on ε'/ε is required to experiments. In this context, very

Model	ε'/ε
Superweak	0
Standard Model	$> 3 \cdot 10^{-3}$
Left-right Models	$O(10^{-3})$ or less
Extra Higgses	$O(10^{-2})$
Supersymmetry	less than Standard Model (possibly < 0)

interesting appears the proposal to use K° and $\bar{K}^\circ$ produced with known momentum, high flux and equal weights in $\bar{p}p$ annihilation nearly at rest [31]. This clearly adds further support to the request of high intensity and good quality hadron beams.

REFERENCES

As the literature is now immense, I find impossible to make an exhaustive list of all the relevant contributions. Regretfully I must thus limit the References to just a few (mainly review) articles, and apologize to the so many authors whose work is not represented here as adequately as it deserves.

1) For a review see for instance: L.B. Okun, Leptons and Quarks (North-Holland, Amsterdam, 1982).

2) F. Vannucci, Proc. of the International EPS Conference on High Energy Physics, eds. L. Nitti and G. Preparata, Bari, 1985.

3) S. Boris et al., Phys. Lett. 159B (1985) 217.

4) See e.g.: R. Rückl, Proc. of the Workshop on the Future of Medium Energy Physics in Europe, eds. H. Koch and F. Scheck, Freiburg, 1985; A.J. Buras, ibidem; E. de Rafael, preprint MPI-PAE/PTh 72/84 (1984); P. Langacker, Aspen Winter Conference Series, preprint UPR-0276T (1985).

5) R.N. Mohapatra and J.C. Pati, Phys. Rev. D11 (1975) 566; R.N. Mohapatra and G. Senjanovic, Phys. Rev. D23 (1981) 165; see for a review H. Harari SLAC Summer Institute on Particle Physics, Stanford 1984, preprint WLS-85/7/Feb-Ph.

6) S. Dimopoulos and J. Ellis, Nucl. Phys. B182 (1981) 505; J. Ellis, M.K. Gaillard, D.V. Nanopoulos and P. Sikivie, Nucl. Phys. B182 (1981) 529; E. Fahri and L. Susskind, Phys. Rep. 74 (1981) 277.

7) P. Fayet and S. Ferrara, Phys. Rep. 32C (1977) 249; H.E. Haber and G.L. Kane, Phys. Rep. 117 (1985) 75.

8) For a review see: P. Langacker, Phys. Rep. 72C (1981) 185; D.V. Nanopoulos, Proc. of the XXII International Conference on High Energy Physics, Leipzig, 1984.

9) See e.g. A. Davidson and K.C. Wali, Phys. Rev. Lett. 26 (1981) 691.

10) For recent reviews on compositeness see e.g. R. D. Peccei, preprint MPI-PAE/PTh 35-84 (1984); H. Harari, Proc. of the 1984 Scottish Universities Summer School, St. Andrews.

11) EHF Project, F. Bradamante, this volume; TRIUMPH Kaon Factory Proposal, 1985; LAMPF II Proposal, Los Alamos, LA-UR-84-3982 (1984);SIN Project, 1985.

12) Theoretical predictions are reviewed e.g. in L. Maiani, CERN-TH-2846 (1980).

13) For a review of experimental results see: D.R.O. Morrison, CERN/EP 86-44 (1986); V.A. Lubimov, Proc. of the XXII International Conference on High Energy Physics, Leipzig, 1984).

14) A recent review can be found in M. Doi, T. Kotani and E. Takasugi, Osaka preprint OS-GE-85-02 (1985).

15) S.M. Bilenky and B. Pontecorvo, Phys. Rep. 41 (1978) 225.

16) R.E. Shrock, Phys. Rev. D24 (1981) 1232.

17) M. Gronau, Phys. Rev. D28 (1983) 2762, and references therein.

18) P. Pistilli, private communication.

19) For details on the formalism see e.g. J. Steinberger, CERN lectures, CERN 70-1 (1970); K. Kleinknecht, Ann. Rev. Nucl. Sci. 26 (1976) 1; L.L. Chan, Phys. Rep. 95C (1983) 1.

20) J.H. Christenson, J.W. Cronin, V.L. Fitch and R. Turlay, Phys. Rev. Lett. 13 (1964) 138; V.L. Fitch, Rev. Mod. Phys. 53 (1981) 367; J.W. Cronin, Re. Mod Phys. 53 (1981) 373.

21) F.J. Gilman and M.B. Wise, Phys. Rev. D27 (1982) 1128.

22) For a survey see for example A.J. Buras, in ref. 4).

23) M.K. Gaillard and B.W. Lee, Phys. Rev. D10 (1974) 897.

24) M.K. Gaillard and B.W. Lee, Phys. Rev. Lett. 33 (1974) 108; G. Altarelli and L. Maiani, Phys. Lett. 52B (1974) 351; the procedure is illustrated with many details in E. de Rafael, ref. 4).

25) M.A. Shifman, A.I. Vainshtein and V.I. Zakharov, Nucl. Phys. B120 (1977) 316; M.B. Wise and E. Witten, Phys. Rev. D20 (1979) 2392; B. Guberina and R.D. Peccei, Nucl. Phys. B163 (1980) 289.

26) G.G. Ross, Proc. of the Conference on Tests of Electroweak Theories, Trieste, 1985.

27) F.J. Gilman and J. Hagelin, Phys. Lett. 133B (1983) 443.

28) See e.g. K. Nishikawa, Proc. of the XXII International Conference on High Energy Physics, Leipzig, 1984.

29) See e.g. L. Maiani, Proc. of the Conference on Tests of Electroweak Theories, Trieste, 1985.

30) B. Guberina, A. Pich and E. de Rafael, preprints CTP-85/P-1787 and MPI-PAE/PTh-36/95 (1985).

31) L. Adiels et al., CERN proposal PSCC/P82 (1985).

SPIN EFFECTS AT SHORT DISTANCES

J. SOFFER

Centre de Physique Théorique - CNRS Luminy - Case 907
13288 MARSEILLE CEDEX 9 - FRANCE

In this lecture we will stress the crucial rôle played by spin effects at short distances, discussing the theoretical implications of striking experimental results recently obtained in exclusive reactions at large angles. We will also review the situation of transverse spin asymmetries in inclusive hyperon production.

1. INTRODUCTION

The rôle of spin in our understanding of hadronic interactions has been considered only of accessory interest over the last twenty years or so. One of the main reason for this situation is the relative scarcity of the data obtained by difficult experiments involving polarized beams and polarized targets. Nevertheless, as we will see, several very unexpected spin effects have been discovered, in various processes and kinematic regions. However there is also a widespread prejudice that spin is an unnecessary complication and most theorists, instead of bearing the constraints it puts on models, prefer to blame striking disagreement, with all kinds of hand-waving arguments, on non-leading contributions too hard to evaluate! On the contrary, we believe that these remarkable experimental results should strongly motivate new theoretical efforts [1] because they contain some unique information which can give clues to underlying hadron spin structure and strong interactions dynamics. Clearly this information is lost in unpolarized measurements and comparing rates in different pure spin states can help us to emphasize terms which are burried in spin average cross sections. Asymmetry measurements which involve difference of number of events clearly require a larger statistics. Major progress has been made to get intense polarized beams[2] and even the possibility to built them at supercollider energies has been seriously considered[3] together with some significant physics issues for a multi-TeV machine[4]. Here we are mainly interested in the energy range of 20-30 GeV and we will argue that a new facility producing an intensity higher by one or two orders of magnitude than the existing ones is very desirable. This will become clear after we will survey the experimental situation for exclusive and inclusive processes and discuss the implications of these results for current theoretical ideas. We will also suggest some experiments which can be done only with a high luminosity machine.

2. EXCLUSIVE REACTIONS AT LARGE ANGLES

Exclusive reactions at large angles are described in terms of hard scattering models and are expected to provide important tests on hadronic wave functions at short distances and on the nature of parton-parton interactions. We will first examine baryon-baryon scattering, and, as a special case, pp elastic scattering, and second meson-baryon scattering which is being carefully studied in the BNL experimental programme.

2.1. Baryon-baryon scattering

Let us recall *three* striking experimental facts for NN elastic scattering in the large angle region. They concern the analyzing power A and the transverse spin-spin correlation parameter A_{NN} defined as

$$\sigma_o A = \text{Im} [(\varphi_1 + \varphi_2 + \varphi_3 - \varphi_4)^* \varphi_5] \quad (1)$$

$$\sigma_o A_{NN} = \text{Re} (\varphi_1 \varphi_2^* - \varphi_3 \varphi_4^* + 2 |\varphi_5|^2) \quad (2)$$

where

$$\sigma_o = 1/2 (|\varphi_1|^2 + |\varphi_2|^2 + |\varphi_3|^2 + |\varphi_4|^2 + 4|\varphi_5|^2) \quad (3)$$

is, up to a normalization, the differential cross section. Here the φ_i's are the five NN helicity amplitudes defined in Ref.1.

i) In np elastic scattering at 6 GeV/c A was measured up to $\theta_{c.m.} = 180°$ and the results [5] are presented in Fig.1. We see that in the large angle region A is negative and becomes as large as 40% or more. Note that for pp elastic scattering A must vanish at 90°, because φ_5 vanishes by symmetry arguments, unlike in the case of np elastic scattering. Unfortunately the energy is too low to draw definite conclusions on the nature of this effect, but it will be remeasured at higher energies with the polarized proton beam on a deuterium target at BNL.

ii) In pp elastic scattering A_{NN} at 90° has a dramatic rise for P_{lab} above 8 GeV/c and reaches the value 60% at the highest energy ever measured as shown in Fig.2.

iii) In pp elastic scattering at 28 GeV/c, A is of the order of 5% or so in the small angle region, but increases to a much higher value for $p_\perp^2 = 6.56 \text{ GeV}^2$ as shown in Fig. 3.

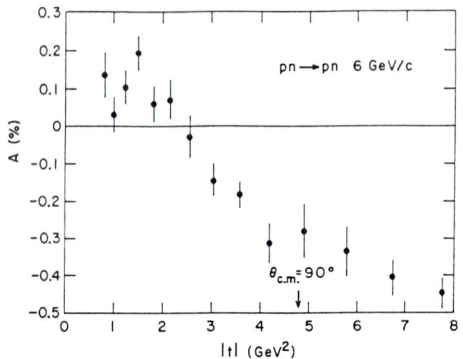

Fig.1 - The analyzing power A for np elastic scattering at 6 GeV/c (taken from Ref. 5)

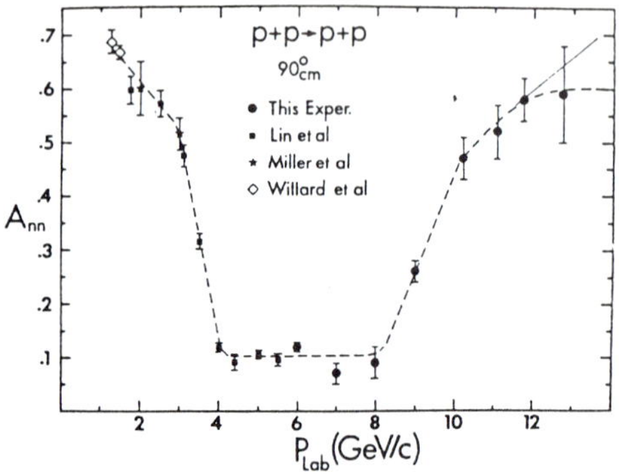

Fig.2 - The A_{NN} parameter for pp elastic scattering at $\theta_{c.m}$ = 90° versus p_{lab} (taken from Ref.6). The curve is hand-drawn to guide the eye.

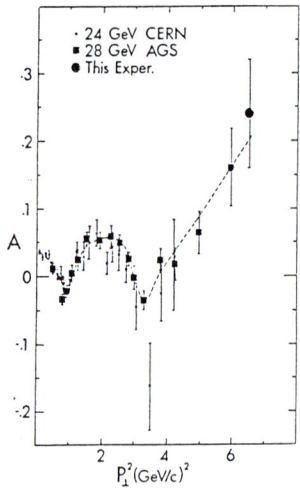
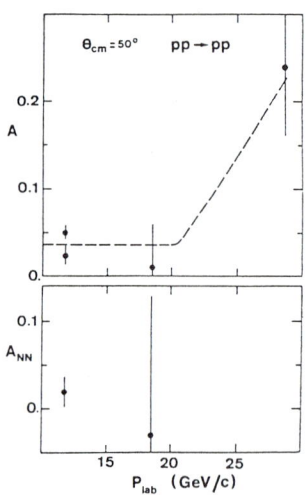

Fig.3 - The analyzing power A for pp elastic scattering versus $p_\perp^2$ (taken from ref.7). The curve is hand-drawn to guide the eye.

Fig.4 - Fixed angle data for A and A_{NN} versus p_{lab} (taken from refs. 6-7). The curve is hand-drawn to guide the eye.

Fixed angle data is reported in Fig.4 showing for A the presence of a new regime above 20 GeV/c or so which is also expected for A_{NN} as we will see later (see Fig. 7).

What theoretical ideas can we advocate to understand these facts ? One possible approach is perturbative QCD requiring quark helicity conservation implied by vector-gluon coupling. This predicts that at short distances the hadronic scattering amplitudes must obey the helicity conservation rule [8]

$$\lambda_a + \lambda_b = \lambda_c + \lambda_d \qquad (4)$$

for any exclusive reaction a+b→c+d. For NN scattering it leads to $\varphi_2=\varphi_5=0$ at large angles, so A should vanish. In addition by symmetry arguments at 90° we have $\varphi_3=-\varphi_4$ and from the quark interchange model $\varphi_3=1/2\varphi_1$. Consequently one gets $A_{NN}(90°)=1/3$ in contradiction with the above experimental result. There are other bad consequences of eq.(4) data that we will examine below and we regard this situation as an evidence for serious need of non-perturbative effects in a kinematic region where perturbative QCD is believed to be relevant. Before we describe another hard scattering model let us go back to a fundamental quantity which is very often neglected, namely, the differential cross section. We recall in fig.5 the best available data[9] at p_{lab}=24 GeV/c from small angles up to $|t|$=6.72 GeV^2. The solid curve represents the large t extrapolation of the impact picture[10(*)] which is clearly below the data for $|t|$ above 4 GeV^2 or so. This is not surprising because diffractive scattering cannot be the "whole story" and we ought to include hard scattering effects at large angles. The

---

(*) We proposed in 1979 this description of the angular distribution of pp and p̄p elastic scattering up to ISR energies and we gave at that time some predictions in the TeV energy range which turned out to be in excellent agreement with the recent CERN p̄p collider data[11]

difficulties of perturbative QCD mentioned above do not appear in the framework of Quark Geometro Dynamics (QGD) and its more refined version Anisotropic Chromodynamics (ACD) whose applications to large angle scattering has been discussed elsewhere[12]. This model displays many features of the naive quark model and in addition it incorporates quark confinement explicitely. The scattering amplitudes are obtained by folding of the hadron vertex functions with the elementary quark-quark amplitudes at large angles. The vertex functions are constructed with the assumption that two baryons, before and after the scattering, have in common two spectators quarks which conserve spin (not helicity), momentum and the internal degrees of freedom. We also assume that two spin -1/2 quarks can exchange in addition to transverse vector states, behaving like a "gluon", a pseudoscalar and a longitudinal vector state ; this set constitutes a richer spin structure than in perturbative QCD. The NN helicity amplitudes given in Table I of ref.12 allow a good description of pp elastic scattering near 90°. By adding the the QGD-ACD contribution normalized at 10 GeV/c and 90° to the impact picture amplitude the resulting cross section is shown in fig.5 by the dashed line which is in better agreement with the data for $|t|>5$ GeV^2 or so. At p_{lab}=28.7 GeV/c at the highest t value , $|t|$=10 GeV^2, we find $d\sigma/dt$=3.7x10^{-8} mb/GeV^2 to be compared with the experimental result (3.5±.7)10^{-8} mb/GeV^2 reported in ref.13. If we now calculate[14] the analyzing power A at 28 GeV/c for $p_\perp^2>5$ GeV^2 we get the solid line in fig.6 with a clear excursion of A up to 30% or so near $p_\perp^2$=7 GeV^2. It is in agreement with the data (see fig. 3) and this large effect is due to a maximum interference between the hard scattering amplitudes which are real and the dominantly imaginary diffractive amplitude. Below $p_\perp^2 = 5$ GeV^2 the cross section is not reliable (see fig.5) and the resulting A (broken line starting with the arrow) does not make sense indeed, compared to the small angle data. More work is needed to understand properly this small angle kinematic region. In fig.6 we also

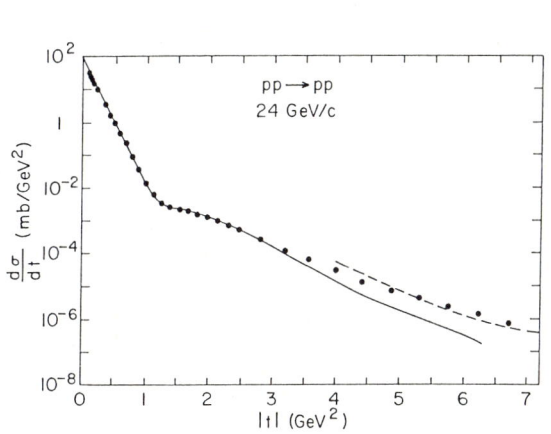

Fig. 5 - pp elastic scattering data from ref. 9. Solid and dash curves are explained in the text.

show our prediction for A at 50 GeV/c and 100 GeV/c. The 50 GeV/c curve is certainly reliable down to $p_\perp^2 = 2$ GeV^2 or so, because the cross section data [15] is well described up to smaller $p_\perp^2$ values. This is a good case for an experimental check. A is expected to be smaller at 100 GeV/c because the interference mechanism becomes less efficient at high energies due to the small-

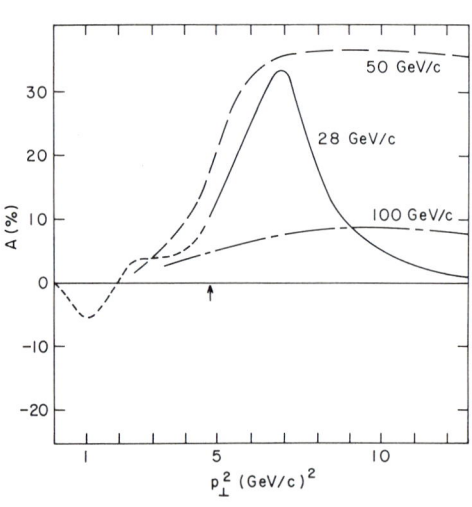

Fig.6 - Our predictions for the analyzing power A at three different energies.

ness of the QGD-ACD amplitudes. Finally we show in fig. 7 our predictions for A_{NN} at 28 and 50 GeV/c which are rising to about 100 % near 90°. We would like to stress that if the predictions for A and A_{NN} turn out to be verified, they would confirm the correctness of the phase of the impact picture and they would indicate the existence of a large hard scattering single flip amplitude φ_5. In ref.14 we give a discussion on the size of φ_2 which is also found large, in contrast with perturbative QCD expectations. The existence of a polarized beam facility would also allow to measure other two spins correlation parameters and with this in mind we have calculated in particular A_{LL}, D_{NN} and K_{NN} defined in ref.1. We find a small A_{LL} of the order of few percents, a large D_{NN} close to 100% and K_{NN} very similar to A_{NN} (see ref.14).

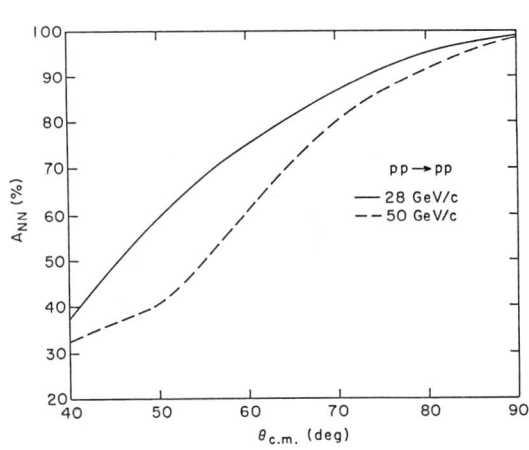

Fig.7 - Our predictions for the A_{NN} parameter at 28 and 50 GeV/c.

It is also worth making predictions for pn elastic scattering [14] and for different $\bar{p}p$ reactions involving annihilation contributions. We hope these speculations will stimulate more theoretical and experimental work in the exciting field of NN elastic polarizations at high energies.

2.2 Meson-baryon scattering

The QGD-ACD approach has been also applied to meson-baryon large angle scattering and it allows to predict the angular distribution of several elastic and inelastic reactions whose relative normalization is being determined by the model[16,17]. For the reactions of the type $0^-1/2^+ \to 0^-1/2^+$ we find in particular at 90° that $d\sigma/dt$ $(\pi^+p \to \pi^+p)$ = $d\sigma/dt$ $(\pi^-p \to \pi^-p)$ while the inelastic channel $\pi^-p \to K^°\Lambda$ is about one order of magnitude smaller and $\pi^-p \to K^+\Sigma^-$ is even much smaller. In fig. 8 we show as an example our predictions for the angular dependence of $\pi^-p \to \pi^-p$ and $\pi^-p \to \rho^-p$.

Another remarkable spin phenomenon that eq. (4) cannot accomodate, but is predicted by QGD-ACD[16], is the ρ^--spin alignment effect that has been recently observed in the reaction $\pi^-p \to \rho^-p$[19]. In the c.m. ρ-helicity frame the angular distribution in terms of the density matrix elements ρ_{ij} of the ρ is

$$(4\pi/3)W(\theta,\varphi) = \rho_{00}\cos^2\theta + \rho_{11}\sin^2\theta - \rho_{1-1}\sin^2\theta \cos 2\varphi$$
$$- \sqrt{2}\,\text{Re}\,\rho_{10}\sin 2\theta \cos\varphi \qquad (5)$$

where θ is polar angle and φ the azimuth of the charged pion produced in the decay. We give below the experimental values[19] and in parenthesis our QGD-ACD expectations

$$\rho_{00} = 0.12 \pm 0.30 \quad (0.096)$$
$$\rho_{11} = 0.44 \pm 0.15 \quad (0.451) \qquad (6)$$
$$\rho_{1-1} = 0.32 \pm 0.10 \quad (0.355)$$
$$\rho_{10} = -0.01 \pm 0.05 \quad (-0.01)$$

Considering the extreme smallness of the theoretical input, the agreement between theory and experiment is really remarkable. Note that as a consequence of eq.(4) one finds in perturbative QCD $\rho_{10} = \rho_{1-1} = 0$, i.e. a flat φ dependence, again at odds with this recent observation. For the reactions of the type $0^-1/2^+ \to 1^-1/2^+$ we find that all vector mesons are produced with a strong spin alignment. We prompt experimentalists to look into the possibility to test our prediction in other reactions in particular ρ^+, ρ^0 or K^{*0} production.

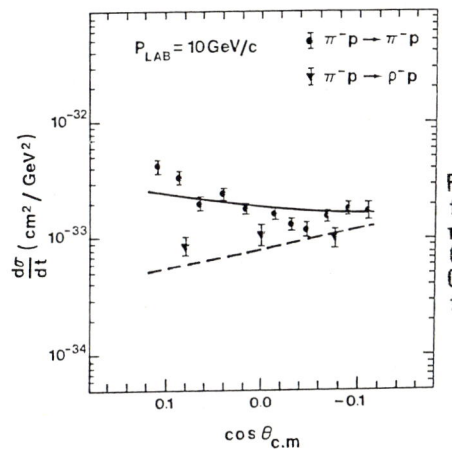

Fig.8 - Comparison of our predictions for the differential cross sections $\pi^-p \to \pi^-p$ (solid curve) and $\pi^-p \to \rho^-p$ (dashed curve) at p_{lab} = 10 GeV/c v.s. CM scattering angle (data points are from Ref. 18).

3. SINGLE TRANSVERSE ASYMMETRIES IN INCLUSIVE REACTIONS

So far we have only considered two-body scattering and we now turn to multiparticle production and more specifically to spin effects in inclusive reactions. First we will recall some hyperon polarization data in hadron fragmentation and we then will discuss theoretical attempts to understand these large single transverse asymmetries whose ultimate origin remains obscure. Experiments using polarized beams might help to clarify this situation.

An inclusive cross section is built from many inelastic channels and since any single transverse asymmetry requires some phase difference between two amplitudes (as an example see eq.(1)), naively a sizeable effect can result only from remarkable coherence of all these different channels. In the high $p_\perp$ region single transverse asymmetries are strongly suppressed in lowest order of perturbative QCD as a consequence of eq.(4) for all inelastic channels. There are many experimental results on this asymmetry for inclusive baryon production and the fact that Λ hyperons produced by unpolarized protons are strongly polarized is known for ten years. The Λ transverse asymmetry has the following properties[20]:

- It is *negative* with respect to the direction $\vec{p}_{inc} \times \vec{p}_\Lambda$.

- For $p_\perp$ below 0.8 GeV/c, its magnitude is approximately linear on $p_\perp$ with a slope increasing with x.

- For $p_\perp$ above 1 GeV/c, its magnitude is independent of $p_\perp$ up to $p_\perp \sim 3.5$ GeV/c and approximately linear on x.

- It is almost independent of energy for an incident energy ranging from 12 GeV/c to 2000 GeV/c.

This last property shows that a European Hadron Facility can be used to improve greatly our knowledge in this area. An example of this is shown in fig.9b which exhibits a fair consistency on hydrogen between the 28.5 GeV/c data for 0.25<x<0.6 from ref.21 (open circles) and the 1500 GeV/c data for 0.4<x<0.9 from ref.22 (filled circles). Fig. 9a shows the comparison between 28.5 GeV/c (filled circles) and 400 GeV/c (open squares) on a beryllium target, suggesting perhaps in this case a decreasing effect as the energy increases. In the FNAL energy range the measured asymmetries of Ξ^0 and Ξ^- are the same as that of Λ and opposite in sign for Σ^+ and Σ^-. In addition one finds that, in the proton beam fragmentation region, p and $\bar{\Lambda}$ are produced unpolarized(*). In the meson induced reactions[20,23], the Λ are produced unpolarized in $\pi^- p$, γp and $K^+ p$ collisions, whereas there is a large *positive* asymmetry for the hyperons produced in $K^- p \to \Lambda X$, $K^- p \to \Xi^- X$ and $K^+ p \to \Lambda X$. For illustration we show in fig. 10 the result of a recent

---

(*) A recent result[22] from $\bar{p}p$ collisions at ISR shows that $\bar{\Lambda}$'s produced in the $\bar{p}$ beam fragmentation have the same asymmetry as that of Λ's produced in the p beam fragmentation.

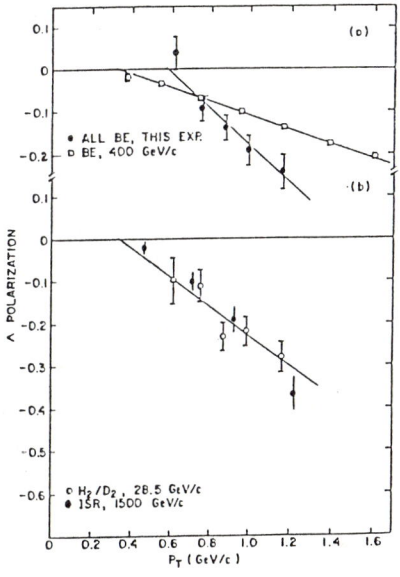
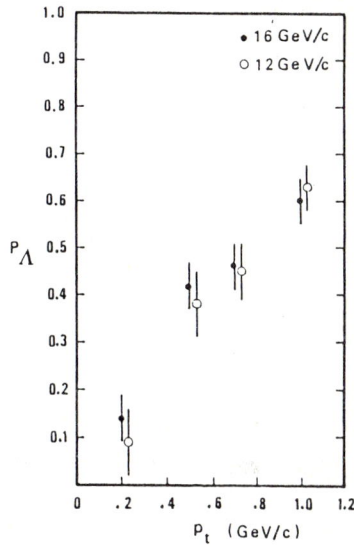

Fig.9 - Lambda asymmetry as a function of the transverse momentum of Λ.

Fig.10 - Lambda asymmetry for x>0.6 as a function of the transverse momentum of Λ at 12 and 16 GeV/c from ref. 24.

measurement at CERN [24]. In terms of the constituent quarks, proton fragmentation into a lambda with $p_\perp \neq 0$ corresponds to the replacement of a valence u quark in the projectile by a strange quark s coming from the sea which must be accelerated and must get a non zero transverse momentum. By assuming a SU(6) wave function, the (ud) system of the lambda is in a singlet state, so the polarization of the lambda is that of the strange quark. In the Lund semi classical fragmentation model[25], the confined linear colour field is stretched and the strange quark needed to make the final lambda with $p_\perp \neq 0$ is produced by a $\bar{s}s$ pair whose orbital angular momentum must be balanced by the spin of the strange quark. It results from this mechanism a negative polarization of the lambda increasing with $p_\perp$. The x dependence almost linear of the

lambda polarization is claimed to be consistent with the model[26], but the size of the polarization is difficult to predict. It is also possible to understand from the
structure of the wave functions the sign of the polarization of the other hyperons, e.g. the sigma should have a positive polarization opposite to that of the strange quark because in this case the (ud) system is in a triplet state. However the model does not apply to K⁻p→ΛX where a fast strange quark, needed to built the outgoing lambda, is already available in the incoming beam. In the recombination model[27] one assumes that a slow quark recombines with spin *down* (with respect to the normal to the production plane) whereas a fast quark recombines with spin *up*. This assumption is compatible with the observation that the lambda polarizations are opposite for pp→ΛX and K⁻p→ΛX(*). In the recombination mechanism, allowing the two possible spin states j=0 and j=1 for the spectator diquark, one is able to calculate the polarizations of the other hyperons in various reactions in terms of two parameters ε and δ. The consistency of this scheme requires ε=δ which does not seem to be supported by the data because δ is related to the size of the polarization π→Λ or K⁺→Λ as opposed to ε related to the polarization in p→Λ. A fair number of predictions have been made for spin-spin correlation effects using a polarized target or a polarized proton/hyperon beam which are of direct inte- rest for an experimental program in progress at BNL. In perturbative QCD left handed and right handed states are decoupled as a consequence of

---

(*) Comparing figs. 9 and 10, the effect is larger for K⁻→Λ but it occurs at higher x which might partly explain the discrepancy.

helicity conservation and one expects large-helicity correlations i.e. sizeable double asymmetries A_{LL} in inclusive production at large $p_\perp$. Of course the smallness of the cross section make these tests rather difficult.

4. CONCLUDING REMARKS

Large unexpected spin effects have been observed in several areas of medium and high energy physics and more remain to be discovered. They occur for exclusive reactions at short distances and for inclusive reactions mainly in the fragmentation region. They teach us that spin properties of hadron constituents are important and that spin is very useful to test some specific aspects of gauge theories of fundamental interactions. We have shown on several examples that intense polarized beams could play an essential role to uncover uniquely basic properties of mesons and baryons.

ACKNOWLEDGEMENTS

It is a pleasure to thank Tullio Bressani for organizing this stimulating school in such a pleasant atmosphere.

REFERENCES

1) For earlier comprehensive reviews see :
   C. Bourrely, E. Leader and J. Soffer, Phys. Rep. 59 (1980) 95.
   N.S. Craigie, K. Hidaka, M. Jacob and F.M. Renard, Phys. Rep. 99 (1983) 69.

2) Proceedings of the 6th International Symposium on High Energy Spin Physics, Marseille Sept. 1984, Journal de Physique, 46 C2 (1985) (editor J. Soffer).

3) Proceedings of the Workshop on Polarized Beams at SSC, University of Michigan, Ann Arbor, June 10-15 (1985), to be published.

4) J. Soffer, preprint CPT-85/P.1799 (to appear in ref.3).

5) Y. Makdisi et al., Phys. Rev. Lett. 45 (1980) 1529.

6) R.C. Fernow and A.D. Krisch, Ann. Rev. Nucl. Sci. 31 (1981) 107.

7) P.R. Cameron et al., Phys. Rev. D32 (1985) 3070.
G. Court et al., preprint UM HE 86-03 (1986).

8) S.J. Brodsky and G.P. Lepage, Phys. Rev. D24 (1981) 2848.

9) J.V. Allaby et al., Nucl. Phys. B52 (1973) 316.

10) C. Bourrely, J. Soffer and T.T. Wu, Phys. Rev. D19 (1979) 3249.

11) C. Bourrely, J. Soffer and T.T. Wu, Nucl. Phys. B247 (1984) 15 and Phys. Rev. Lett. 54 (1985) 757.

12) G. Nardulli, G. Preparata and J. Soffer, Nuovo Cimento 83A (1984) 361.

13) G. Cocconi et al., Phys. Rev. 138B (1965) 165.

14) C. Bourrely and J. Soffer, Phys. Rev. Lett. 54 (1985) 760 and preprint CPT-86/P.1899

15) Z. Asa'd et al., Nucl. Phys. B255 (1985) 273.

16) G. Nardulli, G. Preparata and J. Soffer, Phys. Rev. D31 (1985) 626.

17) G. Preparata and J. Soffer, preprint CPT-86/P.1875. (Phys. Lett. B to appear).

18) G.S. Blazy et al., Phys. Rev. Lett. 55 (1985) 1820.

19) S. Heppelmann et al., Phys. Rev. Lett. 55 (1985) 1824.

20) For a complete review on hyperon inclusive polarization see
    K. Heller in ref. 2, p.121

21) K. Raychaudhuri et al., Phys. Lett. 90B (1980) 319.

22) P. Chauvat et al., Phys. Lett. 163B (1985) 273.

23) J.R. Bensinger, AIP Conference Proceedings, 95, 77 (editor
    G. Bunce).

24) T.A. Armstrong et al., Nucl. Phys. B262 (1985) 356.

25) B. Andersson, G. Gustafson and G. Ingelman, Phys. Lett. 85B,
    (1979) 417.

26) B. Andersson, G. Gustafson, G. Ingelman and T. Sjostrand,
    Phys. Rep. 97 (1983) 31.

27) T. Degrand and H. Miettinen, Phys. Rev. D24 (1981) 2419.
    T. Degrand, J. Markkanen and H. Miettinen, Phys. Rev. D32
    (1985) 2445.

HADRONIC PHYSICS AT INTERMEDIATE ENERGY
T. Bressani, R.A. Ricci (editors)
© Elsevier Science Publishers B.V., 1986

ON THE EMC EFFECT

E. PREDAZZI

Dipartimento di Fisica Teorica - Università di Torino, Torino, Italy
and
Istituto Nazionale di Fisica Nucleare - Sezione di Torino.

I. INTRODUCTION

It was at the 1982 Paris Conference that the EMC collaboration reported[1] about a new nuclear effect showing that the quark distributions in matter is very different from the quark distribution in free nucleons. The finding can be summarized by saying that in heavy nuclei there is a depletion of large momenta partons as compared to light nuclei and that the opposite occurs at low momenta. This effect, confirmed shortly after by a reanalysis of old SLAC data[2] has been the subject of an intensive experimental and theoretical investigation. In a rapporteur talk in 1982 one could hear the speaker state that"... one can hardly think of any mechanism which could produce such effect". Two years later, in the fall 1984, the present author had a personal count of about 120 theoretical papers dealing with this problem and suggesting a large variety of theoretical explanations to it (actually, one theoretical paper[3] had **predicted** this effect).

In spite of the enormous amount of work done, the situation, to date, is still quite controversial both from the theoretical as well as from the experimental point of view, as we shall discuss shortly and we cannot but wait for the new data that are presently being collected by the New European Muon Collaboration (NEMC) at CERN and by another Collaboration working at SLAC. Hopefully, these data will settle at least the experimental controversy.

II. GENERAL INTRODUCTION TO THE PROBLEM.

As it is well known[4], the deep inelastic collision of an incoming unpolarized high energy lepton producing an outgoing lepton and nuclear debris created by the energy transferred through a virtual photon (of momentum Q) to the unpolarized nuclear target of atomic number A, is described in the parton language by two structure functions $F_i^A(x,Q^2)$ (i=1,2) where x is the Bjorken variable

(II.1) $$x = \frac{Q^2}{2m_N \nu} \qquad \begin{array}{l} 0 \leq x \leq 1 \text{ for a nucleon} \\ 0 \leq x \leq A \text{ for a nucleus A} \end{array}$$

In (II.1) ν is the energy released by the incoming lepton (of energy E) to the virtual photon of momentum Q

(II.2) $$\nu = E - E'$$

where E' is the energy of the final lepton. The afore mentioned structure functions $F_i^A(x)$ appear in the generalization of the Rosenbluth formula

(II.3) $$\frac{d^2\sigma^{(A)}}{dQ^2 dx} = \frac{8\pi\alpha^2}{Q^4 x} \left[1-y + \frac{y^2}{2(1+R)} + \frac{Q^2}{2E^2}\left(\frac{1}{1+R} - \frac{1}{2}\right) \right] F_2^A.$$

where y is the fraction of energy carried away by the virtual photon

(II.4) $$y = \frac{E - E'}{E} = \frac{\nu}{E}$$

and we have trated the structure functions F_1^A for the ratio R^A defined as

(II.5) $$R^A = \frac{\sigma_L^A}{\sigma_T^A}$$

In (II.5) σ_L and σ_T are the longitudinal and transverse virtual photon cross sections in terms of which the structure functions F_i^A are defined by

(II.6) $$F_1^A = \frac{K}{4\pi\alpha^2} \sigma_T^A$$
$$F_2^A = \frac{K}{4\pi\alpha^2} \frac{Q^2}{Q^2+\nu^2} (\sigma_T^A + \sigma_L^A)$$

and K is the flux of incoming virtual photons. For real photons, $R^A = 0$, in which case from (II.1,6) in the limit $\nu^2 \gg Q^2$ we recover the well known Callan Gros relation of the parton model

(II.7) $$2 x F_1 = F_2.$$

R^A is expected to vanish at very high Q^2 but it is still an experimentally controversial point whether or not this is actually the case. Also, the A and x dependence of R^A are quite important and largely unknown.

The reason we insist on the properties of R^A so much is that its scarce experimental knowledge is, basically, the source of most of the ambiguities connected with the empirical definition itself of the EMC effect, as we shall see. To complete our presentation, we introduce the ratios of structure functions

(II.8) $$\rho^A = \frac{F_2^A}{F_2^D}$$

(where D stays for deuterium and A is any nucleus) and of cross sections

(II.9) $$\rho_\sigma^A = d^2\sigma^A / d^2\sigma^D .$$

By (II.3) ρ^A and ρ_σ^A are related by

(II.10) $$\rho_\sigma^A = T(x,Q^2,E)\, \rho^A(x)$$

where

(II.11) $$T(x,Q^2,E) = \frac{1+R^D}{1+R^A} \frac{1+\eta(x,Q^2,E)\,R^A}{1+\eta(x,Q^2,E)\,R^D}$$

and, when $Q^2 \ll 4E^2$

(II.12) $$\eta(x,Q^2,E) = \frac{2x(x-Q^2/2mE)}{x^2+(x-Q^2/2mE)} .$$

As one sees from (II.10-12), the ratio of structure functions and of cross sections is equal only if $R^A = R^D$. We shall come back to this point later on.

An important point to stress is that the nuclear structure functions F_i^A can only be different from zero in the range $0 \le x \le A$ for a nucleus A. Thus the ratio of the nuclear to the nucleon structure function becomes infinite at x=1 where the denominator vanishes. If a nucleus behaved like a collection of free nucleons and if we take into account the known difference between the proton and the neutron structure functions, $\rho^A(x)$ should be identically equal to one.

As far as "known" nuclear effects are concerned, the only expected deviations from unity to the ratio ρ^A should come from: i) shadowing, i.e. the screening effect that the various partons and nucleons produce on one the other and ii) the Fermi motion due to the various partons and nucleons transferring stochastically part (or all) of their momenta from one to the other.

Experimentally, shadowing is observed only at very low $Q^2 (\le 1\ \text{GeV}^2)$ and very small x (≤ 0.05) so that is should be irrelevant at the Q^2's considered[1]

($10 \lesssim Q^2 \lesssim 200$ GeV^2). Theoretically, shadowing is very little known and quite difficult to take into account properly but does not seem to play an important role.

The Fermi motion effect is expected to produce strong enhancement of ρ^A at large x ($x \simeq 1$) where the denominator vanishes. The theoretical estimates[5] of this effect predicted ρ^A to start to deviate considerably from unit and to rise quickly above $x \simeq 0.5$.

The ambiguities in these calculations[5] are, however, many and varied but can be summed up in short as follows: i) who knows the actual distribution of partons in a nucleon (much less in a bound nucleon)?; ii) who knows the actual distribution of nucleons in a nucleus?; iii) what is the effect of the choice of recoyl system, of the c.m. system of the nucleus, of the interparton and internucleon correlations and so and and so forth? We shall came back briefly to this point later but a careful investigation of this effect is on the way.

III. REVIEW OF THE DATA.

The data reported in ref. 1 concern deep inelastic collisions of incoming muons of energy E between 120 to 280 GeV ($9 \lesssim Q^2 \lesssim 180$ GeV^2) impinging on an iron and on a deuterium target separately.

The result for the ratio of the iron over the deuterium structure function,[1] apart from a slight Q^2 dependance (which is expected as the result of scaling violation) exhibits a totally unexpected x behavior (see Fig. 1 where the Q^2 average data are plotted; the error bars are statistical only and the shaded area represents the systematic uncertainties) and can be summarized as follows:
i) there is a depletion of the parton distribution in iron as compared with free nucleons in the large x region (contrary to the naive expectation of the Fermi motion, at $x \simeq 0.6$ the data are still decreasing);
ii) there is an increase of parton distribution in iron as compared with free nucleons in the small x region ($x \lesssim 0.3$).

Whatever the theoretical reasons for this effect (see §.IV) it is curious that
iii) the cross over to $\rho^A = 1$ occurs at $x \simeq 0.3$ i.e. at the typical momentum of a valence parton. This point seems to have received no attention in the literature.

The unexpected behavior of $\rho^A(x)$ has come to be known as the "EMC effect".

A reanalysis of old SLAC data^2 (at much lower Q^2, $1.6 \lesssim Q^2 \lesssim 20$ GeV^2) shows agreement with the EMC findings but, extending to larger x values, shows the turnover expected from the Fermi motion effect although this occurs at considerably larger values of x than anticipated^5 (see Fig. 2 where the EMC data are also shown).

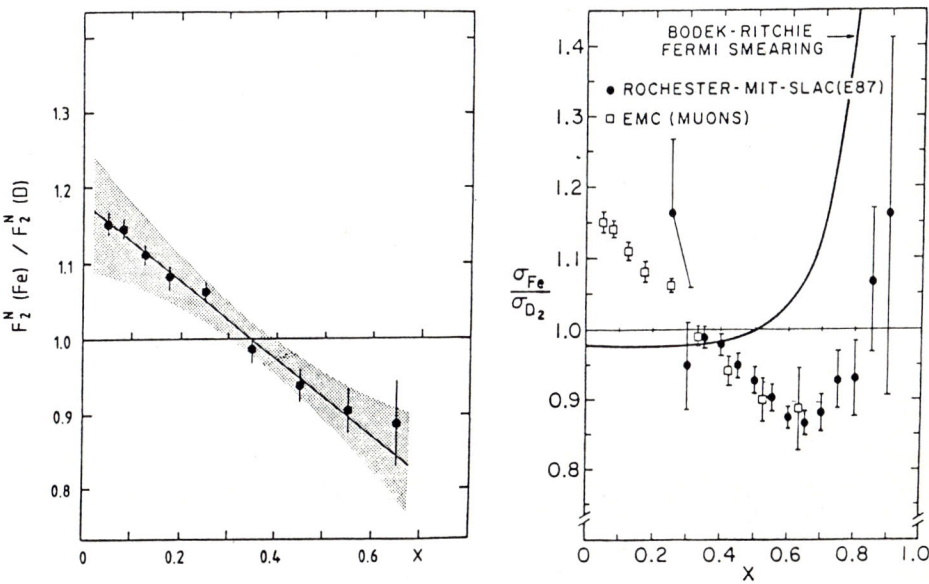

FIGURE 1
The Q^2 averaged data of the EMC collaboration (ref.1), for $\rho^{Fe}(x)$, the ratio of iron and deuterium structure functions.

FIGURE 2
The Q^2 averaged data of refs. 1 and 2 compared with the Fermi motions predictions (ref. 5).

The results of a new dedicated experiment performed at SLAC with a variety of nuclear targets with incoming electrons and $8 \lesssim Q^2 \lesssim 24.5$ GeV^2 are shown^6 in Fig. 3. While the basic trend of the EMC effect is confirmed at large x, at low x a discrepancy seems to occur. The SLAC data for Fe exhibit a substantially lower value in the domain $x \lesssim 0.25$ than those of Fig. 2. This point has raised the (still unsettled) question of what really happens in this x domain and of whether the two sets of data (refs. 1 and 6) are mutually compatible.

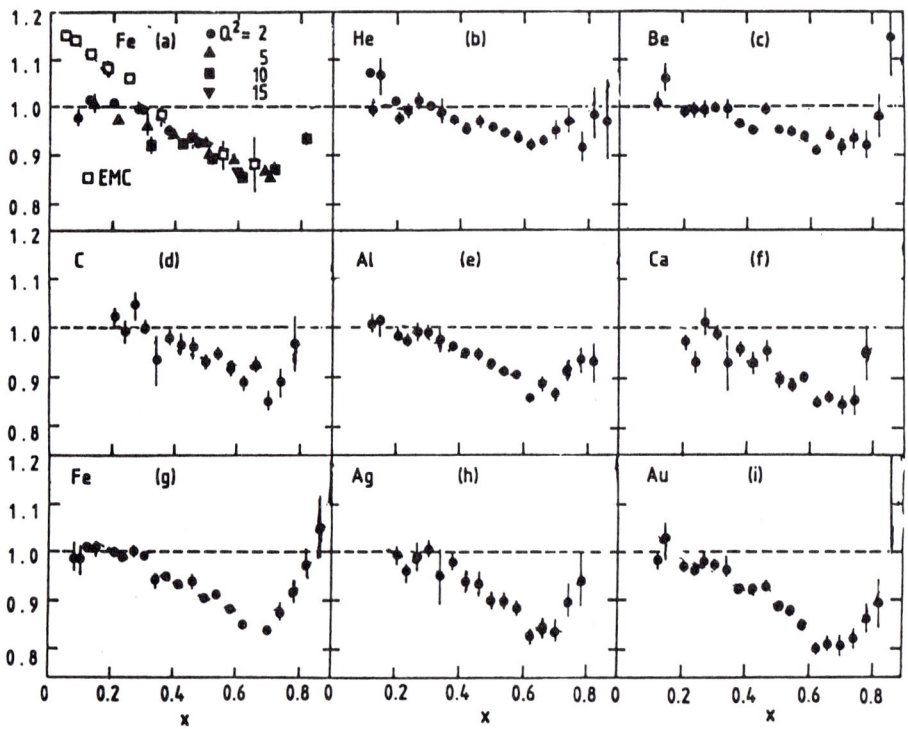

FIGURE 3
The Q^2 averaged ratios of cross sections $\rho_\sigma^A(x)$ as functions of x for various nuclei plotted vs. x.

It has been noticed, however, that whereas the EMC data[1] refer to the ratio of structure functions $\rho^A(x)$ since the coefficient R^A (II.5) measured in the kinematic domain of ref. 1 is basically zero, the ratios plotted in Fig. 3 are ratios of cross sections $\rho_\sigma^A(x)$ given that the averaged R^A factor is now not zero and different from R^D. It has therefore been suggested[7] that one should take the measured[6] R^A values and perform the conversion from $\rho^A(x)$ to $\rho_\sigma^A(x)$ before one can make a meaningful comparison between these two sets of data.

The inspection of eqs. (II.10-12) shows that, indeed, at low x (and just at low x), a small deviation from $R^D = R^A$, gives a conspicuous effect which brings very close the two sets of data^{1,6} when the comparison is performed among the some quantities i.e. when all the data are referred to $\rho^A(x)$. We shall come back later to this point which is probably the one that deserves the most accurate attention given the extremely scarce experimental knowledge of R^A in all the various possible variables Q^2, x and A. While, in fact, there is a general consensus that the measurements are largely Q^2 independent in both kinematical domains so far explored^{1,6}, in the latter (much lower energies and Q^2's) R^A is substantially non-zero and its variation with A quite non-negligible ($R^A \neq R^D$).

The dedicated experiments that have been approved and are on the floor both at CERN and at SLAC plan a careful measurement of both small x structure functions and of a large A span of R^A(II.5) to settle the above controversy.

Needless to say, if our experimental knowledge in the small x-domain is limited (to say the least), the situation is almost desperate from the theoretical point of view given that small x is the least understood kinematical domain of deep inelastic collisions since it is presumably dominated by sea quarks. The two sides, experimental and theoretical, of the difficulty to handle the low -x domain are, clearly related to the highly non-perturbative nature of the effects that govern it.

IV. BRIEF REVIEW OF THEORETICAL IDEAS.

As we have already pointed out, it is essentially impossible to review all the theoretical attempts to give account of the EMC effect. We shall therefore limit ourselves to an arbitrary classification not so much of the models as of the ideas that have been proposed to explain the EMC effect and to a few general comments.

The theoretical mechanisms that have been made responsible of the experimental deviation from 1 of $\rho^A(x)$ include

i) change of the nucleon's properties (effective radius, effective mass etc.) when it is bound in a nucleus as compared to when it is free. These changes, well known empirically in nuclear physics, include the change in the scale of confinement[8];

ii) formation of multiquark and/or multinucleon bags in a nucleus;

iii) formation of heavy isobars (Δ(1236) and contamination from the pionic cloud;

iv) additional sea contribution and impossibility to distinguish between virtual particles emitted and reabsorbed by the same or by a different nucleon in a nucleus[3];

v) formation of diquark structures in nuclei.

The above classification is, as already stated, totally arbitrary and widely different models could belong to one or more of the above entries i) to v).

Without entering into the details of anyone of the above models, let us just point out that whereas the dynamical mechanism invoked may be totally different from one model to the other and so can be the mathematical tools used to implement a given idea, one (and perhaps the only one) common denominator to all models of all categories i) to v) is that of making the effective radius of bound nucleons look larger thant that of free nucleons.

A careful analysis of the points in common of all these models has never been made and is being planned.

Another general consideration is that there has been a huge proliferation of theoretical papers attempting to give account of the original EMC data[1] but relatively few authors[8,9,10] have taken up the more complex problem of analyzing the full A dependence and only in ref. 10 an attempt is given to discuss the low x experimental discrepancies previously discussed along the lines outlined at the end of §.III.

V. A THEORETICAL MODEL AND CONCLUSIONS.

A simple approach to the EMC problem is given by a theoretical model of nuclei as two-fold composite structures[10,11] made of nucleons made in turn of partons where one assumes an explicit gaussian c.m. nuclear distribution of the form

$$(V.1) \quad |\Psi(\vec{r}_{Ii})|^2 = N^2 \prod_{I=1}^{A} \exp[-(\vec{R}_I - \vec{R})^2/r_A^2] \prod_{i=1}^{a} \exp[-(\vec{r}_{Ii} - \vec{R}_I)^2/r_b^2]$$

where A (a) is the number of nucleons (partons) whose positions are denoted by $\vec{r}_{Ii}$ ($\vec{R}_I = \frac{1}{a}\sum_{i=1}^{a} \vec{r}_{Ii}$) and $\vec{R}$ is the total c.m.

$$(V.2) \quad \vec{R} = \frac{1}{A}\sum_{I=1}^{A} \vec{R}_I = \frac{1}{Aa}\sum_{I=1}^{A}\sum_{i=1}^{a} \vec{r}_{Ii}.$$

r_A is the nuclear radius, r_b the radius of the (bound) nucleon and N is the normalization factor.

It is well known that (V.1) itself is not a truly acceptable nuclear density but that it could be made realistic by multiplying it with a suitable polynomial. We shall not worry with these details and use (V.1) as a representative of a model in which a certain number of partons semiconfined (by a gaussian) in their own nuclei make up nucleons which in turn are "bound" to form a nucleus.

The major characterization of a model of this kind is what we could call "loose colour" in that partons are not strictly bound to one particular nucleon but leakage from one nucleon to the other is only depressed by a gaussian confinement.

The model (V.1) was originally motivated as a tool to explore high energy nuclear collisions[12] and, as such, has been fairly successful in those reactions in which it has been used[13].

The nuclear density (V.1) depends, a priori, on two parameters, r_b and R_A the latter of which, however, can be taken of the usual form $r_A = 1.2 \ A^{1/3}$ fm. In addition, it depends on the effective number of partons "a_b" in a bound nucleon.

If we begin by ignoring the problem of Fermi motion, the following expression[11] obtains for $\rho^A(x)$

$$(V.3) \qquad \rho^A = N_A \exp - [(\delta_A - \delta)x^2]$$

where δ and δ_A (and, in particular, its A dependence), are explicitly known

$$(V.4) \qquad \delta = r_f^2 \, m_f^2 \, a_f / (a_f - 1)$$

$$(V.5) \qquad \delta_A = A \, a^2 \, r_A^2 \, r_b^2 \, m_b^2 / [(A-1) \, r_b^2 + A a \, (a-1) \, r_A^2]$$

N_A is some normalization parameter.

In (V.4) the index "f" denotes the various quantities referred to "free nucleons" (in contrast to the index "b" for "bound nucleons" in (V.5)).

The role of Fermi motion can be estimated by introducing one extra factor which insures us the blow up of ρ^A at x=1. This is done, for simplicity, by modifying (V.3) in the following most economic way[10]

$$(V.6) \qquad \rho^A(x) = N_A \exp [-(\delta_A - \delta)x^2] \, (1-x)^{-\Delta_A}$$

but the full analysis of the role of Fermi motion is almost completed[14] and confirms entirely the approximate validity of the **ad hoc** modification leading from (V.3) to (V.6).

The major limitation of the above approach lies in the unified treatment of valence and sea quarks which makes it impossible to extend it to really low values of x.

The parameter Δ_A is choosen so that (V.6) has an A-independent minimum as the data[6] seem to demand (Fig. 3). This gives

$$(V.7) \qquad \Delta_A = (1 - x_{min}^2)^{1/2} (\delta_A - \delta) \simeq 0.58 \, (\delta_A - \delta)$$

The result of calculating the ratio of structure functions with eqs. (V.6,7) is shown by the continuous curve of Fig. 4 for the case of Fe and of Fig. 5 for the data of Ref. 6. Remember, however, that the latter data refer to the ratio of cross sections rather than of structure functions and that the conversion obtains by means of eqs. (II.10-12). The result of performing these conversions is exhibited by the dashed curves of Figs. 4,5 and shows excellent agreement with both the EMC and SLAC data. The same agreement is shown in the comparison of SLAC data with those of the BCDMS Collaboration working at CERN[15] (Fig. 6) with muons scattering off N^{14} in the Q^2 range $26 < Q^2 < 200 \, (GeV/c)^2$.

The results shown in Figs. 4-6 are obtained with the following choice of parameters in (V.4-6)

$$(V.8) \qquad \begin{array}{l} a = a_f = 2 \\ r_b = 1.16 \, r_f \\ r_f = 0.75 \, f_m \\ m_b = 0.91 \, m_f \\ m_f = 940 \text{ MeV} \end{array}$$

furthermore, following previous authors[16], we have parametrized R^A defined in (II.5) by

$$(V.9) \qquad R^A = CA^{1/3}$$

with $C \simeq 0.07$. This choice seems to reproduce fairly well the few experimental

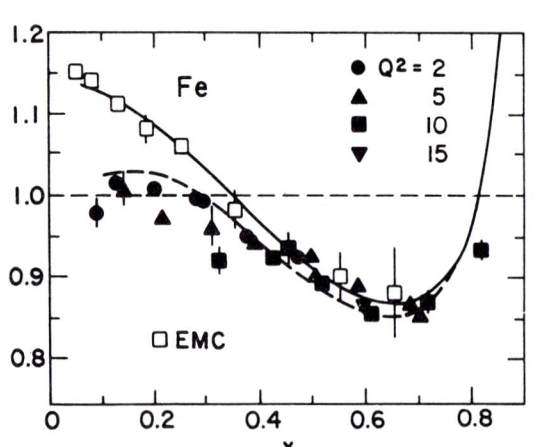

FIGURE 4
The data [1] for $\rho^{Fe}(x)$ and [6] for $\rho_\sigma^{Fe}(x)$ plotted are to be compared with the solid and dashed lines respectively as obtained from eqs. (II.10-12). See text for details.

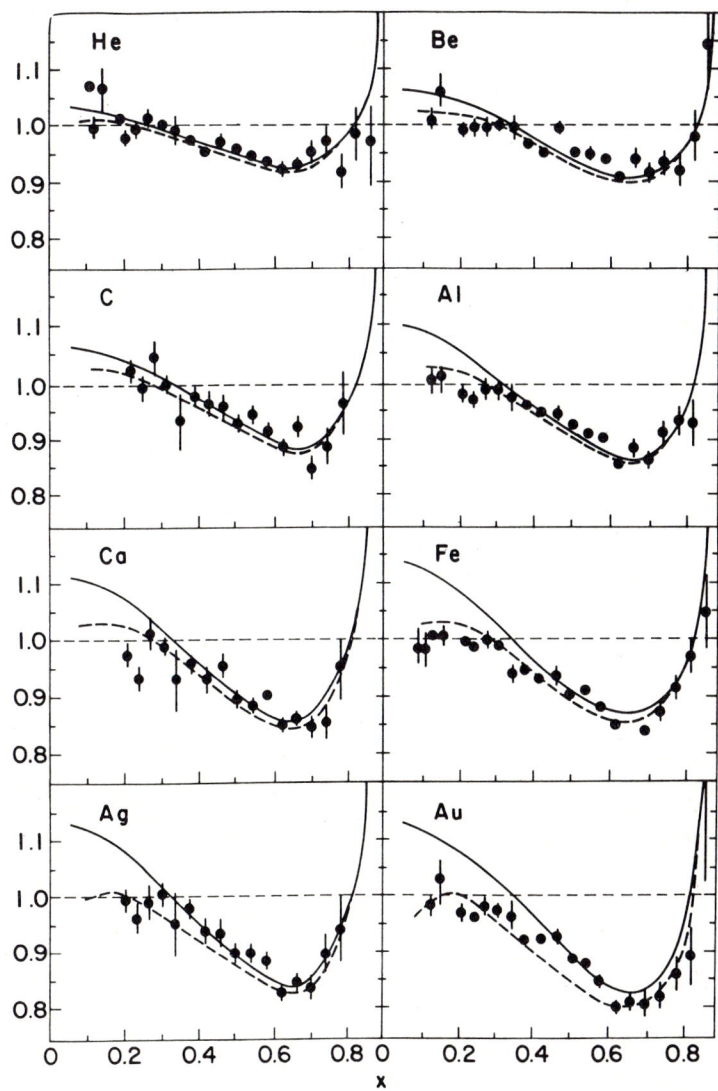

FIGURE 5

Same as in Fig. 4 for all the various nuclei considered in Ref. 6. See text for details.

data[6,17].

Some comments are in order independent of the quality of how the data are reproduced by the model for all A.

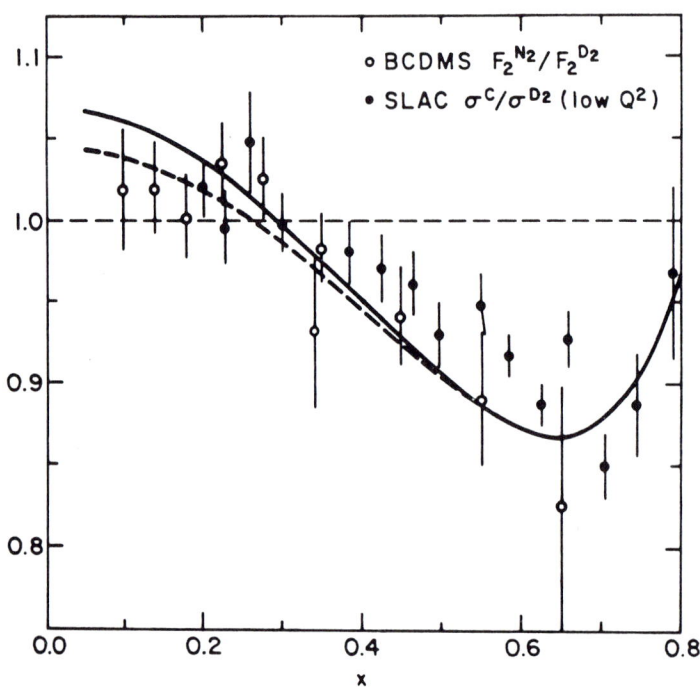

FIGURE 6
Same as in Fig. 4 for N^{14} (ref. 15); see text for details.

In the first place, we confirm the known effect of nuclear physics that bound nucleons appear somewhat larger and less massive than free nucleons. This is presumably due to the percolation of partons from one nucleon to the others within a nuclear structure. This is, in essence, why our model works so nicely i.e., because it takes into account the doubly composite structure of nuclei and it allows for partons to migrate from one nucleon to the other. This makes each bound nucleon to appear larger and less massive.

Another interesting feature is the choice a=2 which we have been led to make in order to reproduce the data. Taken at face value, this seems to imply that for a considerable fraction of time, two quarks appear clustered into a diquark which behaves as a single parton. This is not at all surprising as we know from other sets of data[18].

A final comment concerns the comparison among the two sets of curves, continuous for ρ^A and dashed for ρ^A_σ, of Fig. 4-6. The comparison between the data for iron of Fig. 4 seems to support the analysis[7] of §.II hinting that the discrepancy between the CERN data[1] and the SLAC data[6] is due to the fact that in these two experiments one measures different things: ratios of structure functions in one case[1] and of cross sections in the other[6] and that this is due to R^A being zero[1] or not[6] in the various energy and Q^2 ranges explored.

While Fig. 5 cannot teach us anything on this point, the admittedly not conclusive data of Fig. 6 seem to cast a doubt on this analysis. Working, in fact, at the same energies and Q^2's of the EMC group[1], the small x results (x ≲ 0.2) reported by the BCDMS Collaboration[15] lies lower than our theoretical prediction (dashed curve of Fig. 6) although this is well within the error bars.

Better data will come soon from both CERN and SLAC and will, hopefully, clarify the situation.

REFERENCES

1) The EMC Collaboration: J.J. Aubert et al.: Phys. Lett. 123B 275 (1983).

2) A. Bodek et al.: Phys. Rev. Lett. 50 1431 (1983); 51 534 (1983).

3) A. Krzwicki: Phys. Rev. D14 132 (1976).

4) For a complete definition and treatment of the problem, see, for instance: E. Leader and E. Predazzi: "An Introduction to Gauge Theories and the New Physics", Cambridge Univ. Press (Cambridge 1982).

5) A. Bodek and J.L. Ritchie: Phys. Rev. D23 1070 (1981), D24 1400 (1981); L.L. Frankfurt and M.I. Strickmann: Nucl. Phys. B181 22 (1981); G. Berlaud et al.: Phys. Rev. D22 1547 (1980).

6) R.G. Arnold et al.: Phys. Rev. Lett. 52 724 (1984).

7) J. Sacquin: Proc. 19th Rencontre de Moriond; La Plagne, France (March 1984).

8) F.E. Close, R.L. Jaffe, R.G. Roberts and G.G. Ross: Phys. Rev. D31 1004 (1985); Phys. Lett. 134B 449 (1984).

9) J. Dias de Deus, M. Pimenta nd J. Varela: Phys. Rev. D30 697 (1984); J. Cleymans and R.L. Thews: Phys. Rev. D31 1019 (1985); C.A. Garcia-Canal, E.M. Santangelo and H. Vucetich: Phys. Rev. Lett. 53 1430 (1984).

10) A.W. Hendry, D.B. Lichtenberg and E. Predazzi: Indiana Univ. preprint (1985)

to be publised in Nuovo Cimento A.

11) A.W. Hendry, D.B. Lichtenberg and E. Predazzi: Phys. Lett. <u>136B</u> 433 (1984).

12) A.W. Hendry, D.B. Lichtenberg and E. Predazzi: Nuovo Cimento <u>77A</u> 147 (1982), S. Forte and E. Predazzi: Nuovo Cimento <u>88A</u> 391 (1985).

13) S. Forte: "Glauber Theory of High Energy p-He Scattering with Composite Nucleons" - Univ. of Torino preprint 1985.

14) A.W. Hendry, D.B. Lichtenberg, E. Predazzi and F. Truc: in preparation.

15) BCDMS Collaboration; CERN-EP 85-132 (1985).

16) I.A. Savin and G.I. Smirnov: Phys. Lett. <u>145B</u> 438 (1984).

17) See A. Bodek: Experimental Results from Lepton-Nucleon Scatterign Experiments, Structure Functions and Nuclear Effects; Univ. of Rochester report (UR 884) COO-3065-389.

18) See, for instance, E. Predazzi: "Hadron Spectroscopy - A Relativistic Quark-Diquark Model"; These Proceedings and references therein.

II

SUBNUCLEAR PHYSICS WITH HADRONS AT INTERMEDIATE ENERGY

HADRON SPECTROSCOPY FROM $\bar{p}p$ ANNIHILATION

P. DALPIAZ

Dipartimento di Fisica, Università di Ferrara, Italy
INFN, Gruppo di Ferrara

1. INTRODUCTION

The spectroscopy is traditionally one of the most important tools for the understanding of microscopic structure. We can say: when a structure does exist we have also a peculiar spectroscopy.

It is well known the role, played by the black body spectroscopy, at the end of the last century in the introduction of the quantization of electromagnetic radiation.

The quantum mechanics and the present model of atomic structure was a direct consequence of the study of the spectra of the hydrogen atom, and the study of its hyperfine structure, Lamb shift, offered the precise test of the QeD (Quantum electro Dynamics) theory.

What we know of the structure of nuclei is mainly based on the study of nuclear levels i.e. the nuclear spectroscopy.

The SU_3 symmetry and the quark structure of hadrons came from the big work done in the 50^{ties} and 60^{ties} on the spectroscopy of mesons and baryons.

The so called 4th spectroscopy is recent story. It is the J/ψ and T spectroscopy. These spectacular and unexpected results not only introduce the "flavours" of quarks but give also important elements of the puzzle. The scenario has been changed and let the formulation of QCD (Quantum Cromo-Dynamics), the theory of the hadron structure.

Presently the hadron spectroscopy could have many applications. We limit our interest on the study of $\bar{p}p$ annihilation because we can obtain results not reachable by other means and also because we can profit of excellent antiproton beams constructed at CERN and at Fermilab for the $\bar{p}p$ colliders, and the ones expected from EHF (European Hadron Facility)[1].

The main fields where we can expect important results are related:
- To the explanation of the quark confinement
- To qualitative test of QCD; search of:
  - Dibaryons (qqqqqq)
  - Baryonium ($qqq\bar{qqq}$, $qq\bar{qq}$)

- Glueballs (gg)
- Exotic States ($\bar{q}qg$)
- To quantitative test of QCD:
  - Spectroscopy of Charmonium
  - Spectroscopy of Bottonium.
- To test of fundamental laws with heavy flavours.
- To the search of Higgs Bosons.

2. QUANTUM CROMO DYNAMICS

The Quantum Cromo Dynamics (QCD) is the most recent theory of hadron structure and dynamics.

The base of QCD is a block of six quarks with spin 1/2

$$\binom{u}{d}\binom{s}{c}\binom{t}{b}$$

with 1/3 and 2/3 electric charge. Each of these quarks is differentiated from the others by a quantity called "flavour". Each flavour exists in 3 types of colours. The force related to the colours is the binding force between quarks. To complete the scheme we need a coloured spin one octet called gluon. The mediator of the interactions between the quarks is the gluon.

In spite of the present big number of fundamental elements on which QCD is based, Quantum Cromo Dynamics is a very promising theory because it is deduced from fundamental principles and is qualitatively verified because the predicted structures of mesons ($q\bar{q}$) and nucleon (qqq) is confirmed very well by the spectroscopy of hadrons.

Quantitatively QCD is tested in some dynamics reactions like for example ν,e or μ High Energy Deep Inelastic Scattering of Nucleons. The QCD tests are related only to asimptotic processes and the verification is done quantitatively within a factor two, and that obviously means an error of 100%.

In spite of the evident success in the hadron structure predictions, other qualitative predictions for other states like (qqqqqq) Dybaryons, ($qq\bar{q}\bar{q}$ and $\bar{q}\bar{q}\bar{q}qqq$) Baryonium, (gg) Glueballs and ($q\bar{q}g$) Exotic States are not experimentally confirmed. An important work on the spectroscopy, related to those not confirmed predictions is largely justified.

Considering the quantitative tests of QCD, it is an evident success respect to other precedent hadron theories which predictions are out by orders of magnitudes, but obviously a good theory needs quantitative tests better than 1%. Such tests are possible for QCD with the spectroscopy of Charmonium and Bottonium in particular from $\bar{p}p$ annihilation.

Another important problem not justified or deduced by QCD is the confinements of the quarks in the hadrons.

With this brief description of QCD we can see that remains a lot of important problems, where the hadron spectroscopy could give important contributions.

3. LIGHT QUARKS SPECTROSCOPY

The search of Dibaryons[2] and Baryonium[3] was very popular some years ago. The unique signature is the strong coupling with nucleons. The negative and uncertain results decreased strongly the interest for the experimentation.

The search of Dibaryons must be performed with kaon beams or polarized proton beams. Presently the availability of such beams is scarce, but with EHF we can have a lot of possibilities[1].

The search of Baryonium states was mainly done for narrow width resonances in the $\bar{p}p$ total cross section. After alternating positive and negative results, related to the low statistics and the bad quality of antiproton beams, from the LEAR experiments of CERN the answer is negative.

To continue the search of Baryonium states we must study exclusive reaction and, for large width resonances, the phase shift analysis on data with polarized beams and targets.

Some work of this type was initiated at LEAR and will be continued with the new LEAR experiments. The polarized $\bar{p}$ beams from EHF could give a lot of possibility on this search.

4. GLUEBALLS AND EXOTIC STATES OF LIGHT QUARKS

In the results of Hadron Spectroscopy of Light Quarks, several particles like for example[4] $E^0(1400)$ never found a clear classification in SU_3 symmetry.

Recently[5] data from MARK III of SLAC, on the radiative decay of J/ψ shows the evidence of 3 resonances decaying into $K\bar{K}$, i.e. $\iota(1400)$, $\theta(1700)$ and $\xi(2200)$. The fact that this states decay from heavy flavours makes this particles interesting. The ι probably coincide with the E^0. For that resonances, not yet completly established, several hypothesis for their classification are formulated, one of the most popular is the attribution of a nature of Glueballs. The solution for the puzzle is related to the measurement of all the quantum numbers to understood the nature of the particle.

At LEAR, experiments like ASTERIX[6] are working on E^0/ι and PS170[7], PS185[8] study the $\xi(2200)$. In few month we expect the results.

The work will be continued at LEAR[6] in several experiments, with apparata with complete reconstruction of charged and neutral particles.

The existence of Exotic States ($q\bar{q}g$) is of fundamental importance as a qualitative test of QCD. It could provide a direct evidence for the gluonic degree of freedom and more insight into the structure of bound states in QCD.

Till now, not a lot of work has been done, because the identification of such states is not easy. We must search for states with quantum numbers not accessible to the $\bar{q}q$ system like for example $J^{PC}=1^{-+}$. Such states decay for example into π^0, so they are very difficult to be reconstructed for the presence of many neutrals in the final state. Detectors able to reconstruct such states are expected for the next LEAR phase, at that point the search for Exotic States could start.

5. QUANTITATIVE QCD TEST

The precise measurement of levels and widths of quarkonium $\bar{q}q$ constituted by heavy quarks like $c\bar{c}$ or better $\bar{b}b$ could be considered a unique quantitative test of QCD[9,10,11].

The $\bar{q}q$ system is the simplest case to be treated in QCD. In fact the attraction of $\bar{q}q$ can be described by the coupling α_s depending on the separation r

$$\alpha_s(r) = \frac{12\pi}{33-n_f} \frac{1}{\Lambda^2 \ln 1/r^2} \tag{1}$$

where the n_f is the number of effective flavours and $\Lambda \sim 300$ MeV is the QCD scale parameter $r \sim 1/\Lambda \sim 0.7$ fm is the confinement radius.

The quark masses are m_u and $m_d < 10$ MeV, $m_s \sim 500$ MeV, $m_c \sim 1500$ MeV and $m_b \sim 4800$ MeV.

Bound states of heavy quarks ($m_q \gg \Lambda$, that is c, b and t) can be treated in QCD as non relativistic two body system. Heavier are the masses of quarks better is the QCD approximation.

In analogy with QeD calculation for the positronium (e^+e^-) system, QCD can give precise predictions, without free parameters, of the value of the masses and widths, for the $q\bar{q}$ system constituted of heavy quarks like $\bar{c}c$, $\bar{b}b$ and $\bar{t}t$.

We can conclude that precise measurements, of the masses and widths of $\bar{c}c$ and better of $\bar{b}b$ states give a unique quantitative test of QCD, comparable for the $\bar{b}b$ with the Lamb shift for QeD[10].

As it is well known the e^+e^- colliding beams produce excellent results on $J^P=1^-$ states, when the machine resolution is sufficient, tuning the machine energy on the particle. For $J^P \neq 1^-$ states the detection of particles is performed only with spectrometers and their typical resolution do not allow the determination of several parameters like the width and frequently the existence of the state itself. In the case of $b\bar{b}$, 18 states have not yet been discovered and no width has been measured.

The $\bar{p}p$ annihilation is not dominated by one photon exchange: for this reason, a huge background with respect to e^+e^- channels, (when we look for a particular channel), is expected. However the states with $J^P \neq 1^-$ are produced directly, and we can profit of the resolution of the p and $\bar{p}$ beams. Due to the high background we must select a well signed channel. This method allows for the direct measurement of the widths of all the states[12,13,14]. It looks also very promising to discover the missing states, detecting decay modes with particular signatures like for example

$$\bar{p}p \to \chi \to J/\psi + \gamma \qquad (2)$$
$$\qquad\qquad\qquad \hookrightarrow e^+e^-$$

$$\bar{p}p \to \eta_c \to \gamma\gamma \qquad (3)$$

This method has been tested at the ISR of CERN with one $\bar{p}$ beam impinging an H_2 gas jet-target[13,14,15]. The results show an excellent resolution with very low background in spite of the poor statistics[16]. Unfortunately the ISR experiment was not been completed: running time has been shorted down before the ISR of CERN closure.

From an experimental point of view, the $q\bar{q}$ spectroscopy requires an energetic resolution of about 100 KeV to 1 MeV, in all the possible J^P final states; the reason for it being twofold: from one side narrow new states can be discovered, from the other the measurement of the width of the states of heavy quarkonium is of primary importance to test quantitatively QCD[9,10].

In analogy with $c\bar{c}$ study we can propose the same study for the bottonium family[13,14].

The predictions for the rates of bottonium states formed in $\bar{p}p$ annihilation are not obvious because the $(\bar{p}p \to T_{STATES})$ are not yet measured, and a complete dynamical theory does not still exist.

To make a guess we must rely on a QCD model that gives the dependence on the mass of the $p\bar{p}$ coupling to the $b\bar{b}$ states with respect to the already measured $c\bar{c}$ states[17].

The model predicts the dependence of $\sigma_{b\bar{b}} = \sigma_{c\bar{c}} (m_c/m_b)^{-n}$ for S, P and D states n=8 and for singlet states n=10.

These dependences are not tested but do not seem optimistic. For example, it was not taken into account that heavy flavours seem favoured in $\bar{p}p$ annihilation, as indicated by the high production of strange particles in $\bar{p}p$ interactions. If we consider the cross section integrated over the whole resonance widths we obtain:

$\sigma_{\bar{c}c}$ measured	$\sigma_{\bar{b}b} = \sigma_{\bar{c}c}(m_c/m_b)^{-n}$
$(\bar{p}p \to J/\psi) \sim 3\ \mu b$	$(\bar{p}p \to \Upsilon) \sim 1\ nb$
	$(\bar{p}p \to \chi_b^2) \sim 0.1\ nb\ (\Gamma\chi_b^2 \sim 180\ KeV)$
$(\bar{p}p \to \chi_c^2) \sim 20\ nb\ (\Gamma\chi_c^2 = 2.6\ MeV)$	$(\bar{p}p \to \eta_b) \sim 0.01\ nb$

with a luminosity of $L=(10^{32}-10^{33})cm^{-2}s^{-1}$ we can obtain:

$$\bar{p}p \to \eta \to 10^4-10^5\ events/day$$

$$\bar{p}p \to \chi_b^2 \to 10^3-10^4\ events/day$$

$$\bar{p}p \to \eta_c \to 10^2-10^3\ events/day$$

These rates are calculated considering all the decay channels. If we look at the e^+e^- signature like in reaction (2) and (3) for bottomium, we loose another factor 100 in the rates. The ratio signal e^+e^- over the background of hadrons should be $\sim 10^{-10}-10^{-11}$. It is possible to overcome this background[7] because the kinematical constraints reduce it by a factor of 10^3-10^4. But in any case the rates are very low, if we consider a realistic experiment.

If we perform the experiment with a 14 GeV $\bar{p}p$-collider, one consideration taken from results of UA1 and UA2, is that at large angle, very few energetic particles are founded. This large p_t events are originated in UA1 and UA2 by the decays of heavy mass states, or from hadronic jets. In the energy range of 10 GeV with the 14 GeV $\bar{p}p$-collider the jet production is negligible. In the case of Υ STATES formation, all the energy is used to form the states; while in the case of UA1 and UA2 experiments $m_{W\pm} \sim 540 GeV/6$. We can put a high momentum cut, at wide angles (large p_t). All the particles formed with some quarks, existing in proton or in the antiproton, due to leading effect, goes to small angle.

With a 4π, magnetic and very fine grained detector, triggered over one or two high momentum particles at large angle, the selection of hadronic decays of bottomium states seems possible, tuning the 14 GeV $\bar{p}p$-collider over the resonance mass. With such a trigger the event rates are sufficient to perform the bottonium spectroscopy[18].

6. BEAUTY AND CHARM SPECTROSCOPY, AND VIOLATION OF CONSERVATION LAWS LIKE PARITY, CP AND CPT

As it is well known the study of strange particles, after the first period when their properties are measured, gives important information both on conservation of fundamental laws like CP, parity conservation, and on weak interactions.

For charmed and beauty particles the measurements are presently focused on the discovery of the different states and on the measurement of their properties. It is not clear how much the study of these particles can contribute practically to the study of the fundamental laws.

Theoretically, it is expected a strong interference effect between $B^0\bar{B}^0$ mesons, that can play a role similar to the $K^0\bar{K}^0$, but with heavy flavours. Not many other predictions exist but the experimentation could give many surprise if we are able to produce heavy flavours particles with sufficient statistics.

If we limit our speculation to the charmed particles we can observe that they are produced in the different interactions with the following rates:

$\nu p \to C^*+X$ with a rate 10^{-1} i.e. $\sigma \sim 10^{-38} cm^2$

$\gamma p(e^+e^-) \to C^*+X$ with a rate 10^{-2} i.e. $\sigma \sim 10^{-31} cm^2$

$hp \to C^*+X$ with a rate 10^{-3} i.e. $\sigma \sim 10^{-29} cm^2$

The cross section for charmed particles is favoured by order of magnitudes in strong interactions, but the background is very high. To discriminate from background we must have 4π detector very sophisticated. The possibility to perform the trigger is favoured working with leptons in final states at large p_t[18]. The same argument could be applied also at beauty particles.

P. Pistilli present a precise proposal[19] based on $\bar{p}p$-collider with maximum $\sqrt{s}=20$ GeV (i.e. $p_{\bar{p}}=10$ GeV/c) with 4π detector with silicon strip microvertex to have a further discrimination to background. With an integrated luminosity of 5.10^{38} cm^{-2}, because $\sigma(\bar{p}p \to c\bar{c})/\sigma_{TOT} \sim 10^{-3}$ and $\sigma(\bar{p}p \to b\bar{b})/\sigma_{TOT} \sim 10^{-7}$ with a collider of $\sqrt{s}=20$ GeV it is possible to collect 10^{10} $D\bar{D}$ and 10^6 $B\bar{B}$.

It is clear that no other way exists to produce so many beauty particles. In this field the EHF could be unique if it is coupled with a $\bar{p}p$ collider of $\sqrt{s}=20$ GeV.

7. A $\sqrt{s}$=9-20 GeV $p\bar{p}$-COLLIDER

We have already shown how important is to couple EHF with a $p\bar{p}$-Collider working in the $\sqrt{s}$ range of 9 to 20 GeV.

At the 3th LEAR Workshop was presented[20] one ring $p\bar{p}$-Collider called SUPER-LEAR, filled with 10^{12} $\bar{p}$ from ACOL the new CERN $\bar{p}$-source that could reach a luminosity of $\sim 3 \cdot 10^{30}$ cm^{-2}s^{-1}. As was pointed out in the same workshop[21] a $p\bar{p}$-Collider constructed with two separate rings filled one, with a unique long bunch of 10^{12} $\bar{p}$ and the other with 10^{12} p, with electron cooling at the appropriate energy on both beams and with low β section a luminosity of L=10^{32} cm^{-2}s^{-1} could be reached. If the machine is filled with 10^{13} $\bar{p}$, i.e., less than 1/10 of the EHF $\bar{p}$/day production[1], a luminosity at L=10^{33} cm^{-2}s^{-1} and $\Delta p/p \sim 5 \times 10^{-5}$ could be reached with e-cooling at working energy[22]. With such a machine coupled with EHF the physics of chapters 5 and 6 becomes possible.

8. EXOTIC STATES OF HEAVY QUARKS AND $\mu^{\pm}e^{\mp}$ SPECTROSCOPY

8.1. Spectroscopy of Exotic States of heavy quark

The QCD predicts several extra states like $c\bar{c}g$ 4.3 < $m_{c\bar{c}g}$ < 5.8 GeV, and $b\bar{b}g$, 10.7 < $m_{b\bar{b}g}$ < 11.6 GeV.

The $p\bar{p}$ collider proposed in chapter 7 covers the energy range of such states. The $c\bar{c}g$ could be investigated with a H_2 gas jet target on the ring of $\bar{p}$ and the $b\bar{b}g$ with the collider itself. The difficulty in the search of $q\bar{q}g$ state is the clear signature. The only way is the search of states $J^{PC}=1^{-+}$ not permitted in $q\bar{q}$ system for exemple with $J^{PC}=1^{-+}$. As a trigger we can look for reactions $c\bar{c}g \to J/\psi + ...$ and $b\bar{b}g \to \Upsilon + ...$ with large solid angle detector.

8.2. $\mu^{\pm}e^{\mp}$ Spectroscopy

The $\mu^{\pm}e^{\mp}$ spectroscopy is responsible for the discovery of the $\tau^{\pm}$. Performing such a spectroscopy is equivalent to search for $\tau^{+}\tau^{-}$ decays.

One big question remains to be confirmed in the unified electroweak theory, i.e. the existence of the scalar Higgs (H^0) boson, expected with $J^P=0^+$, width of 100 KeV and strongly coupled to the heavy fundamental fermions and in particular with $\tau^{+}\tau^{-}$. The present limit of the mass is $M_{H^0} > 7$ GeV. There is a precise prediction for a Higgs boson at 10.4 GeV[23].

With 14 GeV $p\bar{p}$-Collider we can search for such Higgs boson using the following reaction:

$$\bar{p}p \to H^0 \to \tau^{\pm}\tau^{\mp} \qquad (4)$$
$$\hookrightarrow e^{\pm}\nu\bar{\nu}$$
$$\hookrightarrow \mu^{\pm}\nu\bar{\nu}$$

Tuning the $p\bar{p}$-collider and searching for a signature of the type ($\pi^{\pm}e^{\mp}$),

($\mu^+\mu^-$), (e^+e^-), in the correct proportion, we can perform a search for Higgs bosons, or particles decaying in $e^\pm\mu^\mp$.

The cross section of formation of Higgs boson in $\bar{p}p$ annihilation was guessed in $\sigma(\bar{p}p \to H^0) \sim 10pb$, in analogy with the Υ singlets states.

The ratio[24] of H^0 in $\tau^+\tau^-$ is BR=$\Gamma(H^0 \to \tau^+\tau^-)/\Gamma(H^0 \to all)$=0.25-0.5 and the BR=$\Gamma(\tau^\pm \to e^\pm(\mu^\pm)\nu\bar{\nu})/\Gamma(\tau^\pm \to all)$=0.18.

Computing all these numbers, we deduce a rate of 20 events/day in the $\mu^\pm e^\mp$ channel with a luminosity of 10^{33} cm^{-2}s^{-1}. For the search of such an important resonance as the one of the Higgs boson, this is not a low rate; in any case, however, an accurate prediction of the mass is very important, since the scanning over 100 KeV needs a lot of time.

9. CONCLUSION

As we have seen before the spectroscopy of hadrons could give fundamental contribution to the physics if performed with adequate apparata and beams and if the experiments are choosed with particular care.

REFERENCES

1) F. Bradamante, on this book.

2) E. Predazzi, on this book.

3) P. Schiavon, on this book.

4) J. Baillon et al., Nuovo Cimento 50A (993) 1967.
   R. Armenteros et al., CERN EP 80-1.

5) W. Toki et al., SLAC - PUB - 3111 (1983).
   T. Barnes, like ref. 10, pg 191.

6) S. Ahmad et al., like ref. 9, p. 253 and like ref. 11, p. 347.

7) G. Bardin et al., like ref. 9, p. 347.
   G. Bardin et al., like ref. 11, p. 517.

8) P. D. Barnes, like ref. 11, p. 283.

9) E. Remiddi, "Physics at LEAR with low energy cooled antiproton beams", ed. U. Gastaldi and R. Kapish, PLENUM N. Y. and London 1984, p. 711.

10) W. Büchmüller, Quarkonium Spectroscopy "Fundamental Interactions in Low-Energy Systems", Ed. P. Dalpiaz, G. Fiorentini and G. Torelli, PLENUM Press N.Y., 1985, p. 233 and references therein.

11) A. Martin, "Physics with antiprotons at LEAR in the ACOL hera" Ed. by U. Gastaldi. R. Klapisch, J.M. Richard, L Van Than Van. Edition Frontiere,

1985, p. 321 and References therein.

12) P. Dalpiaz, Electromagnetic Annihilation in Low Energy $\bar{p}p$ Colliding beam, CERN-$\bar{p}p$ note 06 (March 1977), $\bar{p}p$ First Study Week, CERN.

13) P. Dalpiaz, Charmonium and Other Onia at Minimum Energy, K.f.K. 2836. 111 (1979).

14) P. Dalpiaz, Proceedings 5th European Symposium on Nucleon-Antinucleon Interactions, ed. CLEUP, Padova 1980, p. 711.
P. Dalpiaz, Research Program at LEAR, LA 8775C, 300 (1981).

15) P. Dalpiaz, V. Gracco and M. Macri CERN/ISRCI 79-23.

16) M. Macri, like ref. 11, p.423.

17) W. Büchmüller, like ref. 11, p. 327.

18) P. Dalpiaz, like ref. 1, p. 441.

19) P. Pistilli, on this book.

20) D.Möhl et al., like ref. 11, p. 83.

21) L. Tecchio, like ref. 11, p. 135.
U. Bizzarri, M. Conte, P. Dalpiaz, L. Tecchio and A. Vignati, "Physics at LEAR with low energy cooled antiproton beams", ed. G. Gastaldi and R. Klapish, PLENUM N.Y. and London 1984, p. 729; and references therein.

22) L. Tecchio, Private communication.

23) S. Coleman and E. Weinberg, Phys. Rev. D7 (1973) 1982.

24) Ellis, M.K. Gaillard and Nanopoulos, CERN-TH 2634 (1979).

RARE DECAYS OF K MESONS

Rosanna CESTER

Istituto Nazionale di Fisica Nucleare - Sezione di Torino

K mesons have played an important role in the understanding of the structure of electro-weak interactions. This is not accidental since the two iso-spin doublets:

$$\begin{pmatrix} K^+ \\ K^\circ \end{pmatrix} \qquad \begin{pmatrix} \bar{K}^\circ \\ K^- \end{pmatrix}$$

a) are produced abundantly in strong interactions at intermediate energies;
b) are sufficiently long lived to allow for the construction of high quality secondary beams and
c) have a mass large enough to yield a variety of decays with up to 4 particles in the final state.

Moreover the $K^\circ, \bar{K}^\circ$ system has unique properties. The two states, of strangeness +1 and -1 respectively, are mixed by weak interactions. As a consequence the mass eigenstates K°_S, K°_L are linear superposition of the two strangeness eigenstates. The interplay of strong and weak interactions gives rise to a variety of interesting experimental situations that have been exploited to probe, to a high degree of sensitivity, the structure of weak interactions (see for example N. Paver paper in these proceedings).

Three fundamental discoveries can be traced back to the experimental study of K decays:

1) the discovery of maximal parity violation in weak interactions from the observation of K decay states, $2\pi^+ + \pi^-$ and $\pi^+ + \pi^-$, of opposite parity;

2) the discovery of CP violation through the observation of the forbidden decay $K^\circ_L \to \pi^+ + \pi^-$ and

3) the discovery of a suppression mechanism (which implied the existence of a fourth quark flavour (Charm)) to explain the measured rate difference between strangeness changing decays mediated respectively by charged and neutral currents:

$$K^+ \to \mu^+ + \nu = .63 \quad (K^+ \to \text{ALL}) \quad^{(1)} \qquad \text{1a)}$$

$$K^0_L \to \mu^+ + \mu^- = (9.1 \pm 1.9)\ 10^{-9} \quad (K^0_L \to \text{ALL}) \quad^{(1)} \qquad \text{1b)}$$

Glashow, Iliopolus and Maiani observed that the addition of a new term:

$$\bar{c} \gamma_\mu (1-\gamma_5)\ (s\ \cos\theta_c - d\ \sin\theta_c)$$

to the weak current:

$$J_\mu = \bar{u}\ \gamma_\mu (1-\gamma_5)\ (d\ \cos\theta_c + s\ \sin\theta_c)$$

cancels the first order contribution to the decay 1b). Its rate must then be explained by second order effects. The graphs that are expected to yield the larger contributions are the second order weak

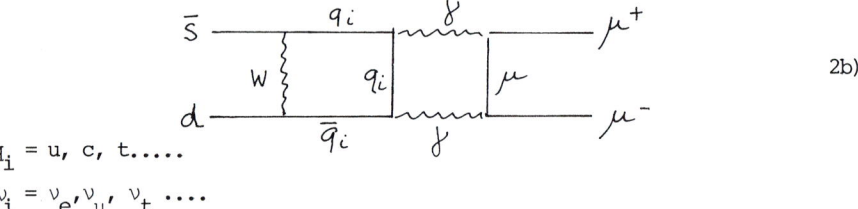

2a)

and the weak-electromagnetic one:

2b)

$q_i = u, c, t, \ldots$
$\nu_i = \nu_e, \nu_\mu, \nu_t \ldots$

From the measured branching ratio:

$$BR(K_L \to \gamma\gamma) = (4.9 \pm .4)\ 10^{-4} \quad^{(1)}$$

and calculating with QED the electromagnetic part of the graph, one can estimate the contribution of 2b) to BR $(K_L \to \mu\mu)$ from γ's on the mass shell. This turns out to be:

$$(4.9 \pm .4)\ 10^{-4} \times 1.2\ 10^{-5} \simeq 6\ 10^{-9}$$

Comparing this number to the experimental value for BR$(K_L \to \mu\mu)$ one concludes that the dominant contribution to 1b) comes from graph 2b).

The study of suppressed (forbidden at first order) weak processes is of great interest because a) they provide a test ground for weak interaction calculations at higher order, which receive, through the inner loop of the graphs, contributions from all the fermion spectrum and b) occuring at a very low level they might be one of the best places to look for new effects that will violate the predictions of the standard model.

As we just saw the decay $K_L \to \mu\mu$ is not a good candidate for this studies (at least the rate measurement is not) since electromagnetic effects dominate. A better decay channel is:

$$K^+ \to \pi^+ + \nu + \bar{\nu} \qquad 3)$$

where electromagnetic contributions are absent.

The main contributions to 3) come, in the standard model from graphs of the type:

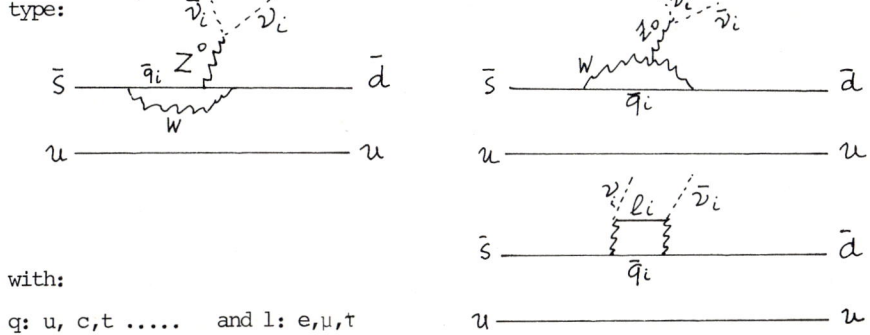

with:

q: u, c, t ..... and l: e, μ, τ

If only the three established generations are taken into account one can explicity compute the branching ratio for 3) as a function of the parameters of the theory (fermion masses and mixing angles):

$$BR(K^+ \to \pi^+ + \nu + \bar{\nu}) \simeq 0.61 \ 10^{-6} \times$$
$$\times \{D(\chi_c) + S_2(S_2 + S_3 e^{i\delta})D(\chi_t)\}^2 \qquad 4)$$

where D is a function of $\chi_q = (m_q/m_w)$, $S_K = \sin\theta_K$

with θ_K (K=1,2,3) the quark mixing angles and δ the CP violating phase. If one uses all the available experimental information to constrain the elements of the Kobajashi Maskaya (K-M) matrix and assumes for the mass of the top quark the value: $m_t \sim 40$ GeV/c^2 the bounds:

$$2 \ 10^{-11} > \sum_i BR(K^+ \to \pi^+ \nu_i + \bar{\nu}_i) < 9 \ 10^{-11}$$

with i=e, μ τ

are obtained, to be compared with the 90% C.L. limit from recent experiments of $1.4 \ 10^{-7}$ 1).

Since the outcoming ν's are not detected, this type of experiment must be designed to select events of the type:

$$K^+ \to \pi^+ + \text{missing energy}$$

The signature is therefore not particularly strong and requires high quality identification of the K and π, momentum measurement of these particles and, above all, a powerfull hermetic veto system to reject all events with other electro-magnetically or strongly interacting particles originating from the decay vertex.

Assuming 100% rejection of unwanted background the experiment will yield a measurement of the π momentum spectrum and a rate estimate.

The π momentum spectrum might signal the presence of two body decays through the superposition of narrow lines to the broad 3 body continuum. The appearance of such lines would be extremely interesting even though their interpretation would not be unique. Possible decay channels to explain such lines are:

$$K^+ \to \pi^+ + H^0 \quad (H^0 \text{ a low mass Higgs})$$
$$K^+ \to \pi^+ + \text{Axion}$$
$$K^+ \to \pi^+ + f \quad (f \text{ stands for familon, the Goldstone boson present in spontaneously broken family simmetries})$$
$$K^+ \to \pi^+ + \pi^0 \quad (\pi^0 \to \nu\bar{\nu})$$

for this last decay mode the momentum of the π^+ is known (205 MeV/c` in the K rest system); the decay $\pi^0 \to \nu\bar{\nu}$ is allowed only if neutrinos are massive or in models with more then one Higgs multiplet.

The continuous three body spectrum can also carry important informations: when a high statistics sample of data will be available the study of the π^+ spectrum will allow for a precise estimate of the upper limit for the mass of the ν_τ (to $\sim$50 MeV). Moreover the shape of the spectrum will depend on whether the ν_τ is a Dirac or a Maiorana particle.

As for the absolute rate, if it falls within the predicted bounds, it will only add a constrain to the values of the parameters appearing in equation 4) and provide an additional input to a multidimensional fit to determine the elements of the K-M matrix. If on the other side the rate turns out to be much higher then expected this will be a signal for new physics.

Possible explanations for a rate increase are:
1) the existence of more then three generations. This could however also give rise to a decrease in rate if the new quark amplitudes should interfere negatively with those of the three established generations;
2) contributions from decays with supersymmetric particles in the final state (in some models photinos have low mass) and
3) contributions from $\pi^+ + \nu + \nu'$ final states which may occur in flavour changing theories.

An experiment to detect $K^+ \to \pi^+$ + missing energy is in the construction stage at the AGS and should reach a sensitivity of $2 \cdot 10^{-10}$. The experiment will be run at a low energy beam, stopping the K's inside an active target. The pion identification is based on momentum/range and energy/range measurements and on the detection of the $\pi \to \mu \to e$ decay chain from stopping π's. The time sequence of this decay chain provides a powerful signature for the pion. A complete description of the experiment and an estimate of the dominant backgrounds can be found in the AGS proposal 787 (see also in this volume the paper presented by L. Littenberg at the Mainz conference).

Based on the previous discussion we observe that a definitive experiment to study $K^+ \to \pi^+ \nu \bar{\nu}$ should reach a sensitivity of $10^{-12} - 10^{-13}$ (branching ratio equivalent) per event.

The requirement of excellent particle identification and momentum measurement favours low energy experiments. Indeed all experiments of this kind either completed or in progress, study the decay of stopped K's. If we want to reach the indicated level of sensitivity we need to stop $(10^{12} - 10^{13})/\varepsilon$ K's, where ε is the detection efficency which will be arbitrarely set at 30%. To perform the experiment in a reasonable time interval (the canonical 1000 hours) one then requires $10^6 - 10^7$ K stops/second. This is to be compared to an available flux of K stops at the AGS of $\sim 1.5 \cdot 10^5$/second (per $5 \cdot 10^{12}$ protons on target). With the high intensity facilities now under discussion (EHF, LAMPFII and TRIUMF) which are designed for 10^{15} circulating protons, separated low energy beams yielding 10^6, 10^7 K^+ stops/second should be easily obtainable.

An other group of very interesting K-decay experiments are searches for decay modes that separately violate muon and electron number conservation.

We restrict the discussion to the two channels which are more interesting from the experimental point of view:

$$K^0 \to e^{\pm} + \bar{\mu}^{\mp} \quad \text{a}$$
$$K^+ \to \mu^- + e^+ + \pi^+ \quad \text{b} \quad \quad 5)$$

The present experimental limits on these decay channels are respectively 2 10^{-9} [2] for a) and 7 10^{-9} for b) [1].

In the $SU(2)_L \times U(1)$ gauge model of the electroweak interactions there are no flavour changing couplings of the neutral gauge bosons to fermions. μ (or e) number non conservation can in principle still occur through the mixing of massive neutrinos. The experimental limit on ν masses are however constraining the upper values of these decay modes to:

$$BR(K^0 \to e^{\pm} + \bar{\mu}^{\mp}) < 5 \; 10^{-16}$$
$$BR(K^+ \to \mu^- + e^+ + \pi^+) < 6 \; 10^{-18} \quad \quad 6)$$

a range of values which is not accessible to experiments of the present and not even the next generation of accelerators. If a signal should be found above these levels it would definetly be evidence for new physics.

The replication of fermion generations, each with the same $SU(2)_L \times U(1) \times SU(3)_C$ structure, suggests the existence of a new symmetry (horizontal) which relates the different generations. This may be a local gauge symmetry with associated gauge bosons (Y). The exchange of these bosons will lead to flavour changing interactions. In this scheme the decays 5) are allowed and described by graphs of the type:

The simplest gauge group that can be added to the electroweak group is $U(1)$. The Lagrangian describing the coupling of the Y to the particles of the first two generations (d,s,e,μ) is in this case

$$L_H = \tilde{g}(\beta_V \bar{s}\gamma_\lambda d + \beta_A \bar{s}\gamma_\lambda \gamma_5 d + \xi_V \bar{d}\gamma_\lambda d + \xi_A \bar{d}\gamma_\lambda \gamma_5 d$$
$$+ \zeta_V \bar{s}\gamma_\lambda s + \zeta_A \bar{s}\gamma_\lambda \gamma_5 s + \sigma_V \bar{\mu}\gamma_\lambda e + \sigma_A \bar{\mu}\gamma_\lambda \gamma_5 e + \rho_V \bar{e}\gamma_\lambda e + \rho_A \bar{e}\gamma_\lambda \gamma_5 e$$
$$+ \kappa_V \bar{\mu}\gamma_\lambda \mu + \kappa_A \bar{\mu}\gamma_\lambda \gamma_5 \mu)Y^\lambda + \text{H.c.},$$

where $\tilde{g}$ is the horizontal gauge group coupling constant and the parameters β_V

β_A.... are all real if CP is to be conserved. It is then possible with standard techniques to calculate the branching ratios for the decays 5) in term of the parameters of the Lagrangian:

$$BR(K_L \to \mu e) \sim (4 \; 10)^3 \; (m_w/m_y)^4 \; (\tilde{g}/g)^4 \; \beta_A^2 \; (\sigma_V^2 + \sigma_A^2)$$

$$BR(K_L^+ \to \mu e_\pi) \sim 40 \; (m_w/m_y)^4 \; (\tilde{g}/g)^4 \; \beta_V^2 \; (\sigma_V^2 + \sigma_A^2)$$

where m_y, is the Y mass and $\tilde{g}$ the electroweak coupling constant.

It should be noted that the two channels are sensitive to different parameters (β_V and β_A respectively) and yield therefore complementary information. We can rewrite the first relation as:

$$BR(K_L \to \mu e) \sim (0.94 \; \text{TeV}^4) \; (\tilde{g}/m_y)^4 \times k$$

where k is a factor which depends only on the fermion mixing angles.

It follows that an experimental limit on the branching ratio implies a bound on the ratio of $m_y/\tilde{g}$. If we set $\tilde{g} \sim g$ and k∿the 4 th power of the Cabibblo angle we can establish the sensitivity of such an experiment in term of m_y. One obtains:

$$m_y > 15 \; \text{TeV} \; \text{for} \quad BR(K_L \to \mu e) < 10^{-8} \; \text{and}$$

$$m_y > 150 \; \text{TeV} \; \text{for} \quad BR(K_L \to \mu e) < 10^{-12}$$

The assumptions on the value of $\tilde{g}$ and k are completly arbitrary and this must be understood as a way to characterize the relative sensitivity of experiments in term of only one parameter.

The question then arises wether it is possible to estimate the values of this branching ratio. Indeed this is the case: the K_L, K_S mass difference $\Delta m_{L,S}$ has been determined with great accuracy from the study of the time evolution of the K_L, K_S system. The measured value is: $\Delta m_{L,S} = 3.5 \; 10^{-6}$ eV.[1]

Since Y exchange would add a contribution to $\Delta m_{L,S}$, the measured value allows to bound the parameters of this model. From the extreme hypothesis that the K_L, K_S mass difference is entirely due to horizontal interactions of the type described, a lower bound:

$$m_y > (1.5 \; 10^3) \; (\tilde{g}/g) |\beta| \; \text{TeV}$$

is derived,

For BR ($K_L \to \mu e$) this implies:

$$BR(K_L \to \mu e) < (3 \cdot 10^{-14}) \, (\sigma_V^2 + \sigma_A^2)/\beta_A^2$$

The prediction on the decay branching ratio is then a function only of the ratio between the strength of the coupling of the Y boson to (sd) and to (eµ) respectively. For equal strengths:

$$BR(K_L \to \mu e) \sim 2 \cdot 10^{-14}$$

for a strength of the (eµ) 10^3 times that of the (sd) coupling:

$$BR(K_L \to \mu e) \sim 10^{-8}$$

This is only an example of the interplay between different experimental measurements and the parameters of a theory. Other horizontal models based on non-abelian Gauge groups and/or incorporating CP violation will lead to different predictions. Moreover, the exchange of Y bosons is not the only mechanism through which flavour changing reactions can occur. In recent years other extensions of the standard model have been proposed that can also lead to muon number non conservation. These include models with extended Higgs sectors, extended technicolor or involving the exchange of leptoquark gauge bosons. For a complete discussion and an extended reference list on earlier work see P. Herczeg. Proceedings of the "Workshop on Nuclear and Particle Physics at Energies up to 31 GeV" pag. 58 Los Alamos National Laboratory, Jan. 5-8, 1981.

I will conclude by briefly reviewing the status of experiments to detect muon number non conserving K decays:
the quoted limit on BR ($K_L \to \mu e$) ($\sim 10^{-8}$) is controversial since it comes from an old experiment which failed to detect the decay $K_L \to \mu\mu$. For this reason the "Particle properties data" booklet gives a much less stringent upper limit of $6 \cdot 10^{-6}$. There are two experiments in preparation at the AGS[3] and one at KEK[4] to search for the $K_L \to \mu e$ decay with a claimed sensitivity in the range of 10^{-11}, 10^{-12}. Again at the AGS an experiment has been approuved to search for $K^+ \to \pi \mu e$ [5] at a sensitivity level of 10^{-11}. Since a detailed description of these experiments can be found in L. Littenberg contribution to the Mainz Conference, I will only address one general question relevant to the design of the next generation of experiments to detect the decay $K_L \to \mu e$: does the limiting factor in present experiments come from the fact that K

fluxis are too low or from the difficulty of rejecting background events? Typical fluxis of K_L's at the AGS are $4.5 \; 10^6$ per 10^{12} protons on target, within a solid angle acceptance of 3.5 μsterad. These K_L are distributed over a wide band of momenta (centered at ~ 10 GeV/c) and are accompanied by a flux of ~1.5 10^8 neutrons. With an average decay length of τcp/m ~ 300 meters, only ~ 2% of the K_L will decay in a 6 meters decay region, yielding ~ 10^5 K_L per pulse or, with a 2.5 sec repetition time, $1.4 \; 10^{11}$ K_L decays in 1000 hours. Of course the K_L flux can be increased by increasing the number of protons on target (by no more then a factor 5 at the AGS) and increasing the beam angular acceptance (easily by a factor 5) to a level of $3.5 \; 10^{12}$ in 1000 hours. The problem is that the flux of neutrons increases by roughly the same ratio to ~ $4 \; 10^9$ per pulse, a level which will impose strong constrains on the design of beam pipe and detector. With a much higher flux of protons available, as would be the case at K factories, a cleaner K_L beam could be obtained filtering out neutrons with an absorber, exploiting the fact that K_L interact less then neutrons. Also, to remove completely the neutron component, the possibility of building K_L beam out of secondary high intensity π beams has been discussed.

The worse background to $K_L \to \mu e$ comes from the aboundant (38%) 3 body decay:

$$K_L \to \pi + e + \nu$$

where the π decays in flight to $\mu + \nu$. If the π decay is not detected, the event will appear as:

$$K_L \to \mu + e + \text{missing energy}$$

and a discrimination against this background will have to rely only on a precise measurement of the charged particles momenta.

Since the momentum of the K_L is not known there are only two constraints to be applied in the event reconstruction: a) the invariant mass of the two charged particles:

$$(m_e^2 + m_\mu^2 + 2 E_e E_\mu - 2 p_e p_\mu \cos\theta_{e\mu})^{1/2}$$
$$\sim \{m_\mu^2 + 2 E_e (E_\mu - p_\mu \cos\theta_{e\mu})\}^{1/2}$$

must be equal, within errors, to the mass of the K^0 and b) the vector sum of the two particles (μe) momenta must point back to the K^0 production target.

If in the K_L center of mass system both the ν_e and the ν_μ from the K_{e3} and π decay chain had vanishingly small energy, the background event would be indistinguishable from a $K_L \rightarrow \mu e$ decay event. However the π from K_{e3} decay has a maximum velocity $\beta = .854$ and, as a consequence the ν_μ has a minimum energy of $p^*\gamma (1.-\beta) = 8.4$ MeV. The reconstructed K_L mass will then be always < 489.3 MeV even if the ν_e energy vanishes. Moreover the V-A matrix element for the K_{e3} decay goes to zero when the neutrino energy goes to zero and, as a consequence, the probability for the ν_e to have an energy E less then E_0 is a steep function of E_0:

$$P (E<E_0) \sim .04 \; (E_0/50 \text{ MeV})^3.$$

It can however happen, if the π decays within the volume of the analyzing magnet with the decay plane nearly coinciding with the bend plane, that the μ

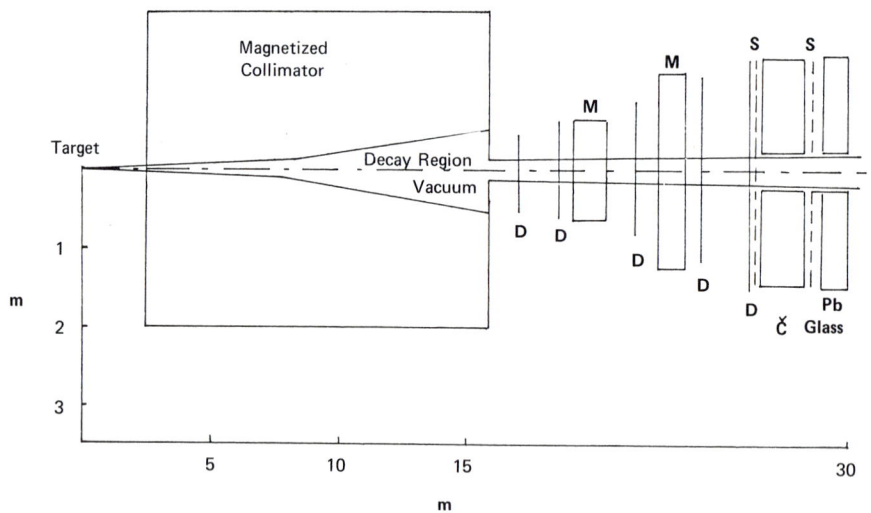

Figure 1

A schematic diagram of the apparatus for an experiment to search for $K_L \rightarrow \mu e$. The bending magnets are denoted "M", the drift chambers "D", and "C" is a hydrogen Cerenkov counter.

momentum is uncorrectly measured. To avoid this possible source of systematic error one of the experiments at the AGS proposes the use of a spectrometer with two magnets, so that each particle is measured twice. Events in which the two measurements on one track do not agree will be rejected. Fig. 1 shows the design of an experiment for LAMPF II[6] which also makes use of this technique to suppress background. The proponents claim that with the new generation of beams available at K factories a 10^{-13} sensitivity level could be reached on the $K_L \to \mu e$ branching ratio.

The conclusion then appears to be that the sensitivity is at present limited by the beam requirement. This must be taken with some caution, however, since new and unexpected sources of background could show up limiting the success of the present generation of experiments and suggesting the introduction of new concepts in detector design.

REFERENCES

1) Review of Particles Properties, Rev. Mod. Phys. Vol. 56, N.2, Part. II, April 1984.

2) Clark et al., Phys. Rev. Lett. 26, 1667 (1971).

3) M. Schmidt et al. AGS Proposal N. 780 (1981)

4) T. Inagaki et al., KEK Proposal N.   (1985)

5) P. S. Cooper et al., AGS Proposal N. 777 (1982)

6) D. Bryman et al., Search for the decay $K_L \to \mu e$ LAMPF II Protoproposal (1986).

CHARM AND BEAUTY PHYSICS AT $\sqrt{s}$ = 20 GeV WITH A $p\bar{p}$ COLLIDER
OF HIGH LUMINOSITY

Pio PISTILLI
Dipartimento di Fisica "G. Marconi", Roma, INFN, Roma

1. INTRODUCTION

In this talk I shall try to discuss the possibility to study production and decay of heavy (charm and beauty) particles at $\sqrt{s} \sim$ 10-20 GeV by using a $p\bar{p}$ collider of very high luminosity that can be achieved in a future European Hadron facility.

I shall try to point out the most interesting items in the heavy flavours physics that may be studied with such a machine.

Some consideration on the experimental problems as well as a comparison with e^+e^- rings will be done.

2. HEAVY FLAVOUR HADROPRODUCTION

The main interest of $p\bar{p}$ interactions with respect to charm and beauty physics deals with the possibility to create $c\bar{c}$ and $b\bar{b}$ resonant states in the energy region $\sqrt{s} \sim$ 3-4 GeV and $\sqrt{s} \sim$ 10-11 GeV respectively.

In particular $p\bar{p}$ collisions directly produce J=0 resonant states, not allowed in e^+e^- annihilation with great advantage for the study of the $c\bar{c}$ and $b\bar{b}$ spectroscopy.

At higher energy however processes like

$$p\bar{p} \rightarrow D\bar{D}\ (\Lambda_c \bar{\Lambda}_c) + X$$
$$p\bar{p} \rightarrow B\bar{B}\ (\Lambda_b \bar{\Lambda}_b) + X$$

1)

take place.

In these processes the study of the production mechanisms of charmed (beauty) mesons and baryons can be done.

Moreover with a machine of high luminosity processes like 1) become a rather intense source of heavy particles useful for the experimental investigation of the spectroscopy of such states and for the study of their decay characteristics.

Up to now the experimental information on beauty hadroproduction is very poor.

One event has been identified in emulsion by WA75 collaboration at CERN[1] as a BB pair produced in a π^- interaction at 350 GeV. Another experiment (WA78) looking at events with three muons in the final state produced in π^-U interaction at 320 GeV gives for the total BB cross section a value of few nanobarn[2].

Concerning charm production the behaviour of the total cross-section as a function of $\sqrt{s}$ is reported in Fig.1.

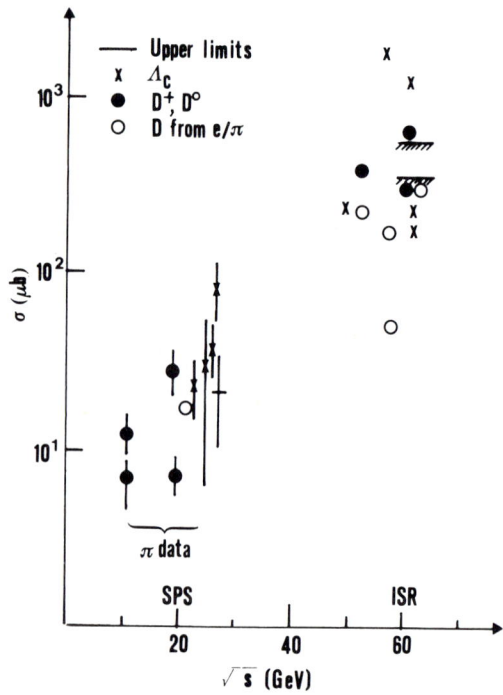

Fig.1 Summary of data on charm production cross-section

Various experiments[3] at CERN have at present collected few hundreds of reconstructed charmed particles. The goal of the experimental program at CERN in the next years is to collect $\sim 10^4$ reconstructed charm produced in pion and proton beams in order to achieve a detailed knowledge of the production mechanism and of the decay characteristics.

From a theoretical point of view the heavy flavour production is usually described as the sum of two kind of processes:
a) the so called flavour creation (Fig. 2a) in which the heavy quark pair is produced via the gluon-gluon, gluon-quark, quark-antiquark interaction.
b) the flavour exitation (Fig. 2b) in which the heavy quark pair is extracted from the "sea" of the incident particles.

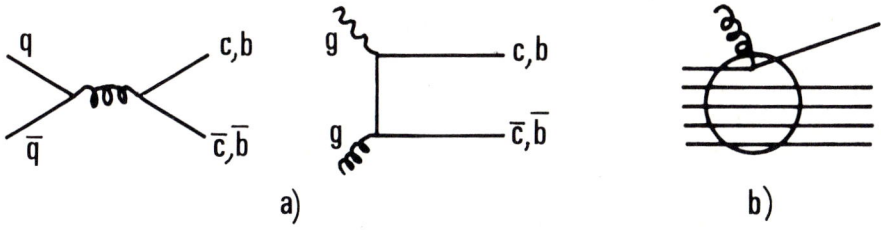

Fig.2 Mechanisms for charm and beauty production
a) flavour creation, b) flavour excitation

The contribution of flavour creation is computable with Q.C.D. and is given by the covolution of the structure functions $F(x,Q^2)$ of the incident particles multiplied by the elementary cross sections $\sigma_{gg} \rightarrow c\bar{c},(b\bar{b})$, $\sigma_{q\bar{q}} \rightarrow c\bar{c}(b\bar{b})$.

A theoretical estimate[4] of the beauty cross section is given in Fig. 3. As one can see the difference between pp and $\bar{p}p$ interactions turns out to be of several order of magnitude in the energy range $\sqrt{s} \simeq$ 15-20 GeV. This is due to the fact that in low energy region the $q\bar{q}$ annihilation is the dominant contribution.

In general the comparison of the cross section behaviour among π, p, $\bar{p}$ expecially at energies not too much larger than the threshold is a very powerful tool to understand the production mechanism.

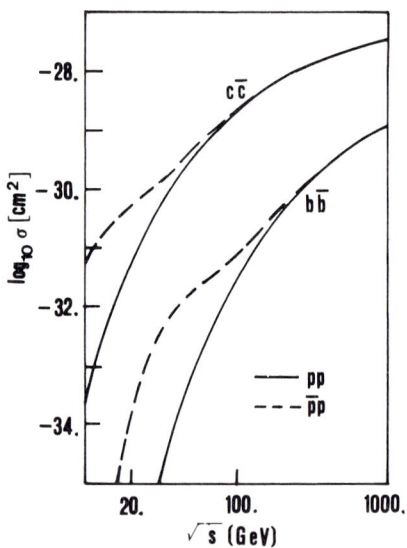

Fig.3 Associated production of charm and beauty via the flavour-creation mechanisms

As one can see from Fig.3 the $B\bar{B}$ cross section falls down very fastly a low energy so that the measure is perhaps possible only in the energy region $\sqrt{s} \gtrsim 15$ GeV.

If we consider a $p\bar{p}$ ring with $\sqrt{s} = 20$ GeV with a luminosity of $5 \cdot 10^{31}$ cm^{-2} sec^{-1} in about two years of operation one may achieve an integrated luminosity of $\sim 10^{39}$ cm^{-2}. Under these conditions $\sim 10^{11}$ $D\bar{D}$ and $\sim 10^7$ $B\bar{B}$ are produced.

These numbers should be compared with the $\sim 10^5$ $D\bar{D}$ and $\sim 2 \cdot 10^4$ $B\bar{B}$ produced in e^+e^- annihilation at $\sqrt{s} \sim 30$ GeV with the luminosity presently integrated at PETRA and at PEP of ~ 1000 pb^{-1}.

The comparison shows that a $p\bar{p}$ ring of the kind we are considering is a very powerful source of heavy mesons.

It has to be remarked however that while in e^+e^- interactions the $c\bar{c}$ and $b\bar{b}$ cross sections correspond to about 30% and 10% of the total in $p\bar{p}$ collisions in the considered energy range one has:

$$\frac{\sigma_{D\overline{D}}}{\sigma_{tot}} \simeq 10^{-3} \qquad \frac{\sigma_{B\overline{B}}}{\sigma_{tot}} \simeq 10^{-6} - 10^{-7}$$

Therefore the identification of the heavy flavour state expecially in the $b\overline{b}$ case is an extremely difficult experimental task that will need sofisticated detectors.

3. HEAVY FLAVOUR DECAY

The study of the decay characteristics of the heavy flavour states is one of the most important items in elementary particles physics.

Several informations on the weak mixing angles of the K.M. matrix are obtained by the experimental values of the lifetimes and the branching ratios of charm and beauty decay. The present situation[5] for charm decay is summarized in Table I.

TABLE I
Recent measurements of $\tau(D^+)$ and $\tau(D^\circ)$ in units of 10^{-12} sec.

GROUP	(D^+)	(D°)
NA 27	$11.6 \genfrac{}{}{0pt}{}{+3.1}{-2.2}$	$3.9 \genfrac{}{}{0pt}{}{+0.9}{-0.7}$
SLAC HYB	$8.6 \pm 1.3 \genfrac{}{}{0pt}{}{+0.7}{-0.3}$	$6.1 \pm 0.9 \pm 0.3$
MARK II	$8.9 \genfrac{}{}{0pt}{}{+3.8}{-2.7} \pm 1.3$	$4.7 \genfrac{}{}{0pt}{}{+0.9}{-0.8} \pm 0.5$
TASSO	-	$4.6 \genfrac{}{}{0pt}{}{+2.9}{-1.7} \genfrac{}{}{0pt}{}{+1.0}{-1.3}$
NA 11	$10.6 \genfrac{}{}{0pt}{}{+3.6}{-2.4}$	$3.7 \genfrac{}{}{0pt}{}{+1.0}{-0.7}$
CLEO	-	$4.1 \pm 1.0 \pm 0.7$
DELCO	-	$4.6 \pm 1.5 \genfrac{}{}{0pt}{}{+0.7}{-0.6}$
HRS	-	$4.5 \pm 1.4 \pm 0.8$

The most striking features is the large difference between the lifetime of $D^{\pm}$ and D°.

A precise determination of the various lifetimes will tell the relative importance of the diagrams responsible for the charm decay.

In fact while the spectator diagram (Fig. 4a) is present in the decay of each particle and is the only mechanism allowed in the $D^{\pm}$ decay the exchange diagram (Fig. 4b) contributes to D° decay.

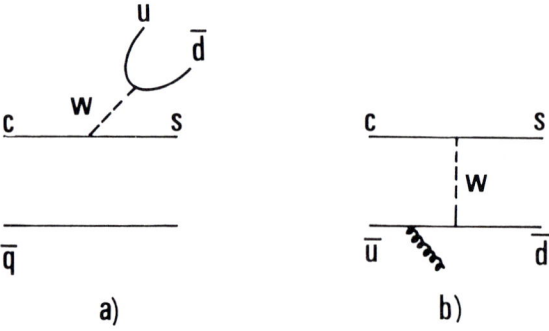

Fig.4 Mechanisms for charm decay

Concerning the beauty the present status of the measurements of the meanlife is summarized in Fig. 5.

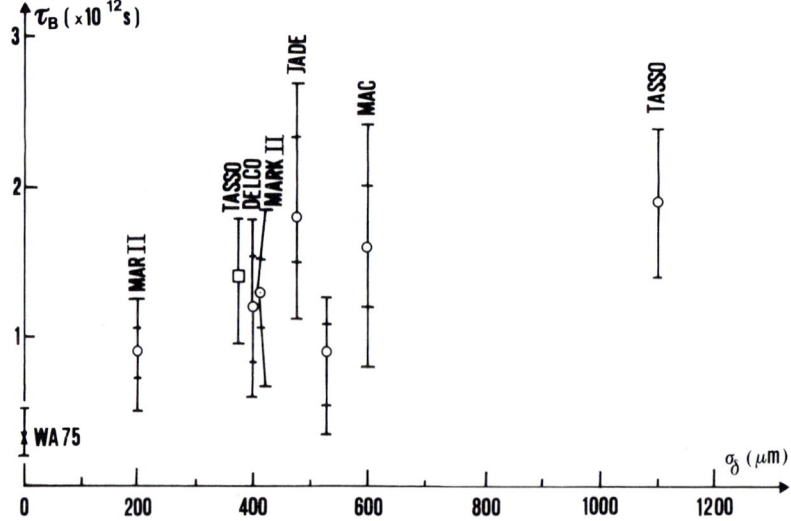

Fig.5 Plot of the beauty hadron lifetime as a function of the experimental resolution

The measures are still affected by large uncertainties and more precise and detailed data are needed.

From this point of view the very large number of BB pairs available in an high luminosity $p\bar{p}$ ring may be very useful.

It is worth to mention in the study of the rare decay of the heavy flavor the possibility to detect the very interesting phenomena of the $\Delta Q = 0$, $\Delta C, \Delta B = 2$ transitions.

Similarly of the $K_0 \bar{K}_0$ system a mixing between neutral D and B meson are expected with an oscillation between $D_0 \bar{D}_0$, $B_0 \bar{B}_0$, $B_s - \bar{B}_s$ states. The diagrams responsible of these effect are represented in Fig.6; the oscillation probability is dependent on the weak mixing angles and from the heavy quark (included top) masses.

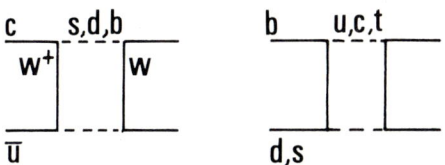

Fig.6 Mechanisms for mixing

4. EXPERIMENTAL PROBLEMS

The detection of the BB pairs over a background $10^6 - 10^7$ times greater than the signal is obviously an experimental task of great complexity.

It is obviously impossible to discuss in detail the problem for a machine still not fully defined and for a nonexisting apparatus.

I will do only some considerations on general characteristic of the signal that can be useful in the background rejection.

a) For $\sqrt{s} \sim 20$ GeV the BB pair is produced with a slow in the $p\bar{p}$ system.

This means that its decay products will have rather high energies (1-2 GeV) at large angle.

The background is mainly due to forward low p_t particles.

This different pattern may be very useful for the selection of the signal both at trigger level than in the final off-line reconstruction.

b) The B mean life is about 10^{-12} sec; the mean decay path turns out to be of the order of few hundreds μm for the mostly of the produced pairs.

With the fastly depeloping technology of the vertex detectors the identification of the distance between production and decay points can be done with an accuracy of the order of ten μm.

Therefore the finite decay path will become a very strong signature of the heavy particles.

c) The B and D have a large (~ 20%) semileptonic branching ratio. The identification of high p_t muons and electrons will substantially improve the signal/background ratio.

5. CONCLUSIONS

A $p\bar{p}$ ring of high luminosity at $\sqrt{s} \simeq$ 15-20 GeV is a very intense source of heavy particles. The behaviour of the cross section in that energy range is of particular interest for the understanding of the production mechanism.

The high number of produced charmed and beauty particles can be used to study the detail of the decay, but expecially in the BB case one has to overcome severe experimental problems due to the presence of a large background.

REFERENCES

1) J.P.Albanese et al. P.L. 158B, (1985) 186.

2) WA78 coll. private communication.

3) For a review of charm program at SPS see: L.Foà, Workshop on SPS Fixed-Target Physics, Cern 83-02.

4) B.L.Cambridge N.P. B151 (1979) 429.

5) For a recent review see: A.J.Buras HEP 85 proceedings Bari 1985 and bibliography quoted therein.

6) For a recent review see J.Sacton proceeding International Conference SSNTD Rome (Sep. 1985) and bibliography therein.

INTRODUCTION TO SPIN PHENOMENA IN HIGH ENERGY PARTICLE PHYSICS

Paolo SCHIAVON
Dipartimento di Fisica, Università di Trieste, Italy and
Istituto Nazionale di Fisica Nucleare, Sezione di Trieste, Italy

1. INTRODUCTION

This lecture is intended to give an elementary introduction to the spin phenomena in the field of intermediate and high energy particle physics. In particular, only hadrons of spin 1/2 will be considered in the following.

In the past 15 years, the technology on which the experimentation with spin is based has made great advances. Namely polarized targets of improved quality and efficiency have been built and new techniques of producing polarized beams have been developed, making possible much more refined measurements and rising a general interest for the spin physics.

It is clear that the possibilities offered by new accelerators of very high intensity proposed at intermediate energies, around 40 GeV, will be of the greatest importance for the future of the spin physics.[2,3,4]

2. LARGE SPIN EFFECTS

I don't intend to make a review of the spin experiments (there are many of them around: see, e.g., the Proceedings of the Marseille Symposium[1]). I will mention only a few large and unexpected spin effects discovered in the recent past.

A common feature is their occurence at large momentum transfer values which is generally believed to give the indication of an interaction between constituents. QCD provides definite predictions and in particular, helicity conservation along the quark lines[5], suggests that double-spin longitudinal asymmetries should be large, whereas transverse spin effects are expected to be strongly suppressed.

A number of experimental results are in contradiction with these expectations and their theoretical explanation seems to be not yet fully established.

Large polarization parameters have been measured in:

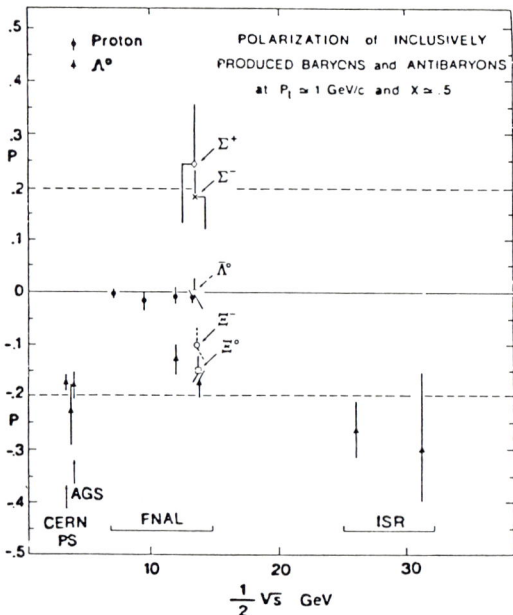

Fig. 1. Compilation of Polarization data of inclusively produced baryons and antibaryons vs energy. The effect seems to be energy independent over a large energy range.

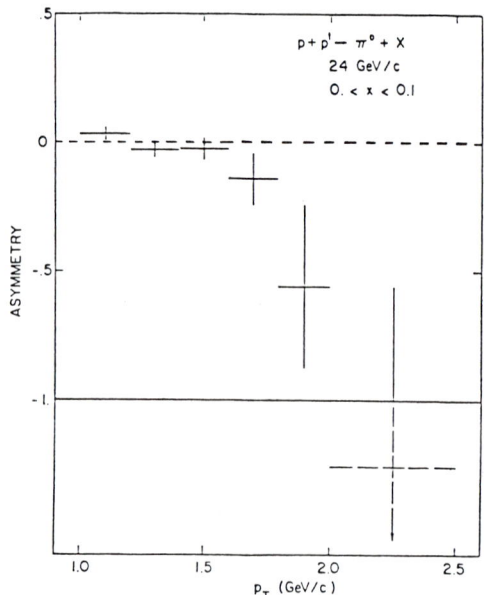

Fig. 2. Asymmetry in the inclusive production of π^0 in pp scattering at 24 GeV/c vs transverse momentum.

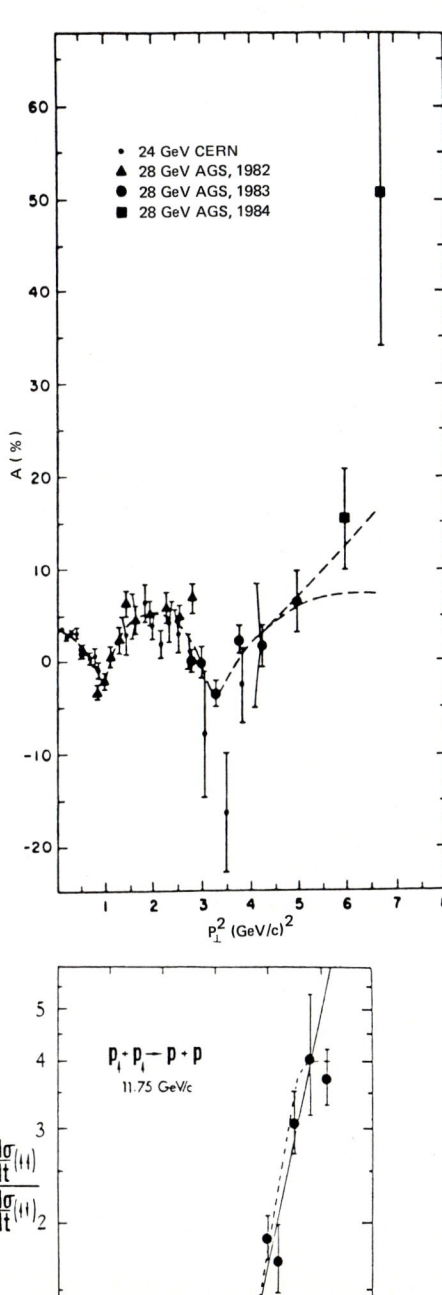

Fig. 3.
Polarization Parameter in elastic p-p scattering, showing the latest results from AGS.

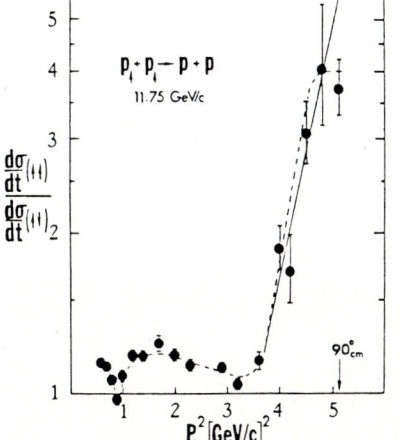

Fig. 4.
Ratio of differential cross-section in elastic p-p scattering in pure spin states.

i) the inclusive production of hyperons in pp and p-Nucleus;[6] (Fig. 1).
ii) the inclusive production of π^0 in pp scattering[7] (Fig. 2),
iii) the pp elastic scattering at 28 GeV and large p_T^2 (~ 6 GeV^2)[8] (Fig. 3) and at 150 GeV/c[9] and of even greater interest,
iv) the large values of the parameter A_{oonn} measured in pp elestic scattering at 11.75 GeV/c at the ZGS[10].

In this experiment a transversely polarized proton beam is made to collide with a transversely polarized proton target, both when spins are parallel and antiparallel. The ratio of the differential elastic cross-sections shows a very large deviation from unity (up to a value of 4) in the region of p_T^2 above 3.5 GeV^2 (Fig. 4). This evident change in the spin-spin interaction forces, has not yet a full explanation in terms of QCD and in the framework of the conventional quark model it is in contradiction with the assumption of quark independence[8].

It is clear that a complete understanding of these phenomena can lead to a better picture of the fundamental dynamics and that, on the other hand, it is of the greatest importance to perform new and more precise spin measurements.

3. FORMALISM OF THE NUCLEON-NUCLEON SCATTERING

The general form for the nucleon-nucleon elastic scattering matrix (taking into account parity and time reversal invariance) is the following:

$$M(\vec{k}_f, \vec{k}_i) = \frac{1}{2}\{(a + b) + (a - b)(\vec{\sigma}_1 \cdot \vec{n}) \otimes (\vec{\sigma}_2 \cdot \vec{n}) +$$
$$+ (c + d)(\vec{\sigma}_1 \cdot \vec{m}) \otimes (\vec{\sigma}_2 \cdot \vec{m}) +$$
$$+ (c + d)(\vec{\sigma}_1 \cdot \vec{\ell}) \otimes (\vec{\sigma}_2 \cdot \vec{\ell}) +$$
$$+ e [(\vec{\sigma}_1 \otimes I_2 + I_1 \otimes \vec{\sigma}_2) \cdot \vec{n}] +$$
$$+ f [(\vec{\sigma}_1 \otimes I_2 - I_1 \otimes \vec{\sigma}_2) \cdot \vec{n}]\}$$

where:

k_i, k_f are unit vectors in the direction of the initial and final particle momenta in the centre of mass system (c.m.s.) of the reaction. $\vec{\sigma}_1$, $\vec{\sigma}_2$ are the Pauli spin matrices in the spin space of particle 1 or 2. $\vec{\ell}$, $\vec{m}$, $\vec{n}$ are three unit vectors defining a reference system:

$$\vec{\ell} = \frac{\vec{k}_f + \vec{k}_i}{|\vec{k}_f + \vec{k}_i|} \qquad \vec{m} = \frac{\vec{k}_f - \vec{k}_i}{|\vec{k}_f + \vec{k}_i|} \qquad \vec{n} = \frac{\vec{k}_f \times \vec{k}_i}{|\vec{k}_f \times \vec{k}_i|}$$

a, b, c, d, e and f are complex amplitudes, functions of two variables, e.g. the c.m.s. energy E and the scattering angle ϑ. If the scattering is between identical particles, f is zero, leaving 9 (plus one phase) real quantities to be determined.

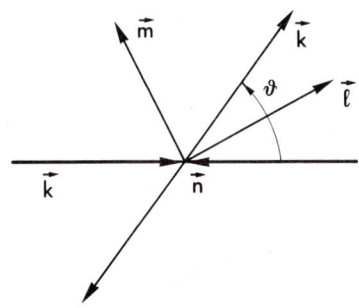

Fig. 5 . Definition of basis vectors $\vec{\ell}$, $\vec{m}$, $\vec{n}$ in the c.m.s..

Assuming isotopics spin invariance, the scattering matrices for pp, nn and np scattering can be expressed as

$$M(\vec{k}_f,\vec{k}_i) = M_0 \left[\frac{1-(\tau_1 \cdot \tau_2)}{4} \right] + M_1 \left[\frac{3+(\tau_1 \cdot \tau_2)}{4} \right]$$

where τ_1 and τ_2 are the nucleon isospin matrices and M_0 and M_1 are isosinglet and isotriplet scattering matrices. We have

$$M(pp \to pp) = M(nn \to nn) = M_1$$
$$M(np \to np) = M(pn \to pn) = (M_1 + M_0)/2$$
$$M(np \to pn) = M(pn \to np) = (M_1 - M_0)/2$$

The amplitudes of the scattering matrix can be written in many forms; a complete review can be found in ref. 11.

I will recall here only the helicity amplitudes related to the helicity formalism introduced first by Jacob and Wick and used by many other authors.

The helicity amplitudes can be written as $\langle \lambda_3 \lambda_4 | M | \lambda_1 \lambda_2 \rangle$ where $\lambda_{1,2,3,4}$ are the helicities (the projections, $\pm 1/2$ in units of $\hbar$, of the spin along the momentum direction) of the four nucleons:

$$M_1 = \langle ++|M|++\rangle = \langle --|M|--\rangle$$
$$M_2 = \langle ++|M|--\rangle = \langle --|M|++\rangle$$
$$M_3 = \langle +-|M|+-\rangle = \langle -+|M|-+\rangle$$
$$M_4 = \langle +-|M|-+\rangle = \langle -+|M|+-\rangle$$
$$M_5 = \langle ++|M|+-\rangle = \langle -+|M|--\rangle = \langle --|M|+-\rangle = \ldots\ldots$$

In the scattering of spin 1/2 particles experimental quantities are usually defined, in the c.m.s., in the form of tensors X_{pqik} where p,q refer to the final state polarization of scattered and recoil particle respectively and i, k to the initial state polarization of beam and target. A "pure experiment" is one involving only spin projections on basis vectors which in the c.m.s. are $\vec{\ell}, \vec{m}, \vec{n}$ as defined above. Then in general 256 such experiments can be defined as summarized in the following table:

Polarized beam	no	yes	no	yes
Polarized target	no	no	yes	no
Differential cross section	I_{oooo}	A_{ooio}	A_{oook}	A_{ooik}
Polarization of scattered particles	P_{pooo}	D_{poio}	K_{pook}	M_{poik}
Polarization of recoil particles	P_{oqoo}	K_{oqio}	D_{oqok}	N_{oqik}
Correlation of polarization	C_{pqoo}	C_{pqoo}	C_{pqok}	C_{pqik}

The various quantities can be expressed in terms of the scattering matrix, e.g.:

$$I_{oooo} = \sigma = \tfrac{1}{4} \text{Tr}(MM^+) \qquad \text{unpolarized differential cross section}$$

$$\sigma P_{pooo} = \tfrac{1}{4} \text{Tr}(\sigma_{1p} MM^+) \qquad \text{polarization of scattered particle}$$

...

$$\sigma A_{oook} = \tfrac{1}{4} \text{Tr}(M \sigma_{1k} M^+) \qquad \text{asymetry in cross section due to polarized target}$$

...

$$\sigma D_{poio} = \tfrac{1}{4} \text{Tr}(\sigma_{1p} M \sigma_{2i} M^+) \qquad \text{depolarization tensor for polarized beam}$$

...

$$\sigma C_{pqik} = \tfrac{1}{4} \text{Tr}(\sigma_{1p}\, \sigma_{2q} M \sigma_{1i}\, \sigma_{2k} M^+) \qquad \begin{array}{l}\text{contribution to polarization correlation}\\ \text{from polarized beam and target.}\end{array}$$

Taking into account parity conservation, the generalized Pauli principle and time reversal invariance, the number of independent experiments reduces to 25, namely

$$\sigma = I_{oooo}$$
$$P = P_{nooo} = P_{onoo} = A_{oono} = A_{ooon}$$
12 two component tensors
9 three component tensors
2 four component tensors.

Similar experimental quantities are defined in the laboratory system (l.s.); in this case the choice of the basis vectors depend on the particle considered and on the other experimental constraints, as the measurement of the final state particle polarization in presence of magnetic fields. A typical choice of the basis vectors is

$$\vec{n} = \frac{\vec{k} \times \vec{k}'}{|\vec{k} \times \vec{k}'|} \qquad \text{(the same as in the c.m.s.)}$$

$$\vec{s} = \vec{n} \times \vec{k} \qquad \vec{s}' = \vec{n} \times \vec{k} \qquad \vec{s}'' = \vec{n} \times \vec{k}''$$

where $\vec{k}$, $\vec{k}'$ and $\vec{k}''$ are unit vectors in the direction of the momentum of incident, scattered and recoil particle respectively.

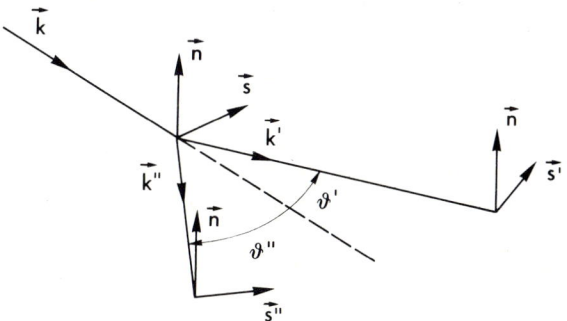

Fig. 6. Definition of basis vectors $\vec{n}$, $\vec{s}$, $\vec{s}'$, $\vec{s}''$ in the l.s..

A parallel formalism can be developed for the l.s. quantities, as well as the relations between l.s. and c.m.s. parameters. As an example, I will mention the useful choice of the well-known Wolfenstein parameters, in the case in which, using a polarized beam, the polarization of the scattered particle is analized

$$D = D_{nono}$$
$$R = D_{s'oso}$$
$$A = D_{s'oko}$$

(the same as in the c.m.s.)
$$R' = D_{k'oso}$$
$$A' = D_{k'oko}$$

4. EXPERIMENTAL METHODS

What in most cases is measured, is an asymmetry in the angular distribution of the particles scatterd by a target or a polarimeter; this means that a difference between two angular distributions has to be measured. A high statistics, one or two order of magnitude larger than in the case of a differential cross-section measurement, is therefore needed. The role of a high intensity accelerator becomes of primary importance in this kind of experiments.

Let me discuss some elementary examples of polarization experiments

a) $pp\uparrow \to pp$ elastic scattering.

This type of measurement is characterized by the use of a polarized proton target (PT). Most PT used consist of a solid sample rich in hydrogen atoms whose protons are polarized in a fixed direction, namely the vertical in the l.s.

The measured parameter is $P = A_{ooon}$ (polarization parameter of the reaction). The scattering matrix reduces to the form

$$M = a + e\, \vec{\sigma}\cdot\vec{n}$$

$$\frac{d\sigma}{d\Omega} = Tr(\rho_f) = Tr(M\,\rho_i\,M^+) \qquad \rho_i = \frac{1}{2}(1+\vec{P}_T\cdot\vec{\sigma})$$

where ρ_i and ρ_f are the density matrices of the initial and final states respectively. Combining the above relations

$$\frac{d\sigma}{d\Omega} = \frac{d\sigma}{d\Omega}\bigg|_0 (1+P\,\vec{P}_T\cdot\vec{n}) \sim (1+P\,P_T\cos\phi)$$

$$\frac{d\sigma}{d\Omega}\bigg|_0 = \frac{1}{2}(|a|^2+|e|^2)$$

$$P = \frac{Re(a^*e)}{|a|^2+|e|^2}$$

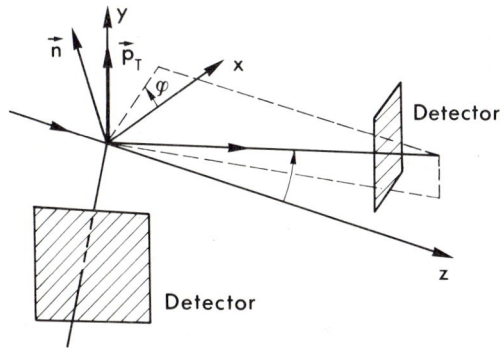

Fig. 7. Schematic view of the experimental set-up.

$\vec{P}_T$ is the target (PT) polarization vector, $\vec{n}$ is the normal to the scattering plane. The detectors for scattered and recoil protons are shown, asymmetric with respect to the beam axis. ϕ is the angle between the scattering plane and the horizontal one. The scattering is detected mainly in the horizontal plane ($\cos\phi \simeq \pm 1$); then to evaluate the asymmetry, we count the particles scattered to the left ($N^L(\theta)$) and to the right ($N^R(\theta)$), as a function of the scattering angle. Because the apparatus does not allow to measure to the left and to the right, the target polarization is revised instead (P_T^+ and P_T^-). Then

$$N^L(\theta) \sim (1+PP_T^+)$$

$$N^R(\theta) \sim (1-PP_T^-)$$

$$P(\theta) = \frac{N^L - N^R}{N^L P_T^- + N^R P_T^+} \simeq \frac{N^L - N^R}{N^L + N^R} \quad \text{if} \quad P_T^+ \simeq P_T^-$$

The fact that the apparatus is exactly the same for the left and right measurements, results in a garantee that no spurious asymmetries are introduced due to experimental biases, a crucial point in this kind of measurements.

A solid polarized proton target is a rather complicated device requiring a lot of advanced technology in the fields of very intense and very uniform magnetic fields, extreme cryogeny (down to a few m°k), power microwave handling. It is obvious that its development can be done only in few large research laboratories and by very specialized teams. Here I want to discuss the experimental constraints due to the use of such a target
i)   A magnetic field up to 2.5 Tesla is needed to keep the polarization of the

sample. This field has to be uniform down to 10^{-4} in the region of the target; as a consequence the target has to be small in dimensions and the counting rate is correspondingly low. Moreover the PT is surrounded by a large region of highly non-uniform fringing field which complicates the detector geometry and the analysis of the data.

ii) The PT consists of a material in which the hydrogen content of its molecules (typically organic compounds) is low with respect to other elements (C, O, N,...); this means that the ratio of the free protons (which can be polarized) to the unpolarized bound nucleons is small ($\sim 1 : 10$). As a consequence, if the scattering is exclusive (elastic, e.g.), the reaction has to be fully measured and kinematically reconstructed in order to separate the wanted process form the background due to the scattering off the bound nucleons. This is possible because the Fermi motion of the bound nucleons spreads the corresponding kinematical distributions. An efficient method to improve the background subtraction is to collect "pure" background data using a carbon "dummy" target, with no free protons, and then compare the two samples.

In the case of inclusive scattering, this background cannot be rejected and we must account for a dilution of the effective polarization of the same factor as the ratio of bound to free nucleons.

The following table shows the relevant properties of the mostly used materials for PT:

		free pr/bound nucl.	polarization	temperature
Propanediol	$C_3H_8O_2$	12%		
Butanol	$C_4H_{10}O$	16%	$\leq 90\%$	$\leq 0.5 °k$
Ammonia	NH_3	21%		

The error with which the PT polarization is known is of the order of 1% or less.

Typical polarization reversal time: $10 \div 30$ min.

A different kind of solid PT is the "frozen spin" target (FST), based on very long (~ 10 days) spin relaxation times. After a polarization procedure in a high magnetic field ($\simeq 2.5$ Tesla) and at very low temperature ($\simeq 50$ m°k), the FST can be used in the experiment in a much lower holding magnetic field ($0.3 \div 0.5$ Tesla). Another big advantage of the FST is the possibility of rotating the polarization vector direction, rotating the direction of the holding magnetic field (typical rotation time: 30 min/90°).

A more recent type of proton polarized target is the jet target (PJT), consisting of a jet of gaseous atomic or molecular hydrogen form a polarized source. This PJT is intended to be used directly inside the vacuum pipe of the accelerator, which has to be adapted to this particular mode of operation. A low density of target protons (and then a low counting rate) is well compensated by many advantages like the possibility of detecting very low momentum recoil particles, the absence of background from bound nucleons, the possibility of giving any useful direction to the target proton polarization.

b) p↑p→pp elastic scattering

This kind of experiment is much the same as that discussed above. The main difference is the use of a polarized proton beam instead of a PT.

The measured parameter is $A = A_{oono}$ i.e. the left-right asymmetry in the angular distribution in the pp elastic scattering.

Scattering matrix and angular distribution have the same expression as in the previous example; then

$$A(\theta) = \frac{\pi}{2} \frac{N^+ - N^-}{N^+ + N^-}$$

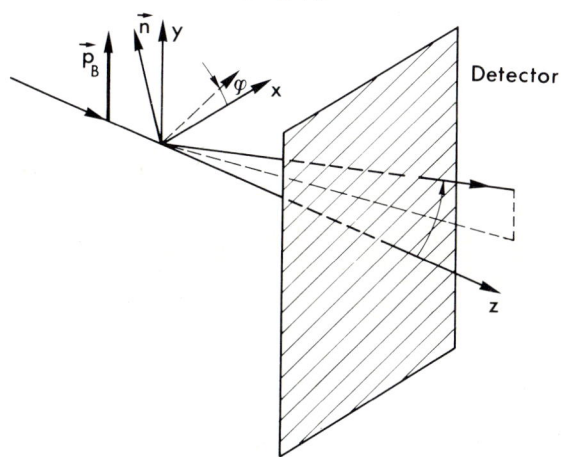

Fig. 8. Schematic view of the experimental set up. $\vec{P}_B$ is the beam polarization vector (vertical in the l.s.). The detector for small angle scattered protons is symmetric around the beam direction; only the energy of the recoil protons is measured (they do not escape the solid target material).

where
$N^+ = N^L{\uparrow} + N^R{\downarrow}$
$N^- = N^l{\downarrow} + NR{\uparrow}$

L, R mean scattered to the left or to the right and the arrows point according to the beam polarization.

The beam polarization is reversed every accelerator pulse (∿1 sec) and has a typical value of 55%. This value should be continuously monitored with a suitable polarimeter; the error on the polarization is then a function of the precision of this measurement.

Polarized beams can be obtained accelerating polarized particles (protons, deuterons,...). This is the best technique, but many technical problems, connected mainly with the depolarization of the beam particles during the acceleration, have to be solved. For this reason, a particle accelerator has to be suitably designed or the existing ones adapted; most of them (SIN, Saturne II, AGS, KEK,...) are at present able to produce polarized beams or will soon be. The relevant advantages are a high intensity and a good polarization value (50 ÷ 90%).

Polarized proton beams are obtained also by elastic scattering of protons by nuclei (pC, pBe,...). This procedure is simpler than the previous one, but the intensity of the polarized beam is much lower, typical of a secondary beam, and the polarization value don't exceeds 50%.

A further method to get polarized protons is from hyperon decay ($\Lambda^0 \to p\pi^-$, and also $\bar{\Lambda}^0 \to \bar{p}\pi^+,\ldots$). In this case the polarized beam intensity is even lower, because of the two stage production mechanism, but its value is reasonably high (40 ÷ 50%).

c) The decay of the Λ^0 hyperon: $\Lambda^0 \to p + \pi^-$

We know that in some cases hyperons are produced polarized, but what I want to discuss now is the possible polarization of the hyperon decay products. The decay (parity violating) matrix, can be written:

$$M = a + b\vec{\sigma} \cdot \vec{k}$$

where $\vec{k}$ is a unit vector in the direction of the proton momentum in the Λ^0 c.m.s.

$$\frac{d\sigma}{d\Omega} = \frac{d\sigma}{d\Omega}\bigg|_0 (1-\alpha \vec{P}_\Lambda \cdot \vec{k})$$

where $\vec{P}_\Lambda$ is the Λ^0 polarization vector

$$\frac{d\sigma}{d\Omega}\bigg|_0 = |a|^2 + |b|^2$$

$$\alpha = -\frac{2\,\mathrm{Re}\,(a^*b)}{|a|^2 + |b|^2}$$

As a consequence we expect a decay asymmetry with respect to the plane normal to $\vec{P}_\Lambda$ in the Λ^0 c.m.s..

The decay proton is also polarized; we may compute

$$\vec{P}_p = \frac{1}{1-\alpha \vec{P}_\Lambda \cdot \vec{k}} \left[(-\alpha + \vec{P}_\Lambda \cdot \vec{k})\vec{k} + \beta \vec{k} \times \vec{P}_\Lambda + \gamma \vec{k} \times \vec{P}_\Lambda \times \vec{k} \right]$$

where β and γ are coefficients in our case close to 0.

Integrating over the Λ^0 polarization, the result is

$$\vec{P}_p = -\alpha \vec{k} \quad (\Lambda^0 \text{ c.m.s.}) \qquad \text{with} \quad \alpha \simeq 0.65$$

In the decaying Λ^0 c.m.s. there is a net longitudinal polarization of the proton. This proton polarization is no more purely longitudinal in the l.s., but very close to it for small angle decay, parallel or antiparallel to the proton laboratory momentum, depending on the forward or backward emission in the Λ^0 c.m.s.. Then a proton momentum analysis can separate the two samples.

This experiment is not only interesting by itself, it also provides a good alternate method to produce high energy proton polarized beams[12] and even more interesting, polarized beams of antiprotons, for what some difficulties (compared to proton polarized beams) seem to exist[13].

REFERENCES

1) Proceedings of the 6-th International Symposium on High Energy Physics, Sept. 12-19, 1984, Marseille, Ed. By J. Soffer, Journal de Physique, Tome 46, Colloque C2, Fev. 1985.
2) LAMPF II Proposal, LA-UR-84-3982, December 1984.
3) Proceedings of Second Kaon Factory Physics Workshop, Vancouver, 1981, ed. R.M. Woloshyn and A. Strathdee, TRIUMF, TRI-81-4.
4) Proceedings of Workshop on the Future of Intermediate Energy Physics in Europe, Freibourg, 1984, ed. H. Koch and F. Scheck.
5) J. Babcock, E. Monsay and D. Sivers, Phys. Rev. $\underline{D19}$, 483 (1979).
6) K. Heller, AIP Conf. Proc. $\underline{95}$, 320 (1983).
7) J. Antille et al., Phys. Lett. $\underline{94B}$ (1980) 523.
8) A.D. Krish, Ref. (1), p. 511.
9) G. Fidecaro et al., Nucl. Phys. $\underline{B173}$, 513 (1980).
   G. Fidecaro et al., Phys. Letters $\underline{105B}$, 309 (1981).
10) D.G. Crabb etal., Phys. Rev: Lett. $\underline{41}$, 1257 (1978).
    J.R. O'Fallon et al., Phys. Rev. Lett. $\underline{39}$, 733 (1971).
11) J. Bystricky et al., Journal de Physique, Tome 39, 1 (1979).
12) P. Auer et al., Fermilab Proposal 581 (1978).
13) C. Beard et al., Phys. Letters $\underline{155B}$, 437 (1985).

COHERENT PRODUCTION

G. BELLINI and L. MORONI

Dipartimento di Fisica dell'Universita' and I.N.F.N., Milano, Italy

1. INTRODUCTION

The physics of the coherent productions started in 1963 and completed a first development period in 1983. The first experiments have been carried out using the techniques of the nuclear emulsions and bubble chamber: just in an experiment with heavy liquid bubble chamber a first evidence of the phenomenon was obtained in 1963 (see ref. 1). But in these experiments the low statistics and the poor variety of target nuclei were limiting factors to an actual understanding of the physical mechanism.

Two important steps in the understanding of the coherent productions were represented by two counter experiments, started in 1967: one at Desy[2], with a photon beam, and a second one at CERN[3], with a pion beam. In both cases several nuclear targets were used.

Before to start the description and the analysis of what we have learned from the coherent productions, we will spend few words for definitions and kinematics.

A coherent reaction is a process which maintains the nucleus in its ground state. These processes can arrive only at very small momentum transfer, because, in that case, due to the uncertainty principle, a spatial region larger than the nuclear radius is involved and the nucleus reacts as an all; namely

$$q_l \cdot R_A \geq 1 \qquad q_t \cdot R_A \geq 1$$

where q_l and q_t are the longitudinal and the transverse three-momentum transfers, respectively.

A coherent process can be either <u>elastic</u>:

$$h + A \to h + A \qquad (1)$$

or <u>inelastic</u>:

$$h + A \to Y + A \qquad (2)$$

In this second case the process is called <u>diffractive dissociation</u> or <u>coherent production</u>.

If, on the other hand, the nucleus breaks-up or it is excited, the reaction is no more coherent, but <u>incoherent</u>.

The special kinematics of a coherent production process, as in (2), allows few simplifications. The momentum transfer is very small and, due to the large nuclear mass, the energy transfer is negligible; $P_Y \simeq P_h$; the production angle θ (between Y and the beam in the L.S.) is very small; P_h and P_Y are definitively larger than m_h and M_Y.

The minimum four-momentum transfer, which allows the reaction, corresponds to $\theta \simeq 0$:

$$|t|_{min} = q_1^2 = m_h^2 + M_Y^2 - 2\sqrt{P_h^2 + m_h^2} \cdot \sqrt{P_h^2 + M_Y^2} + 2P_h^2$$

and, after a Mc Lauren expansion,:

$$q_1 \simeq \frac{M_Y^2 - m_h^2}{2P_h} \qquad (3)$$

We can define:

$$t' \equiv |t| - |t_{min}| = 2P_h P_Y (1-\cos\theta) \simeq \frac{P_h}{P_Y} q_t^2 \simeq q_t^2 \qquad (4)$$

t' is the most characteristic parameter which is studied in the coherent production analysis. Of course in the coherent elastic scattering $t' \equiv t$.

In order to explain better the physical meaning of the coherent mechanism, we mention the main features of the coherent elastic processes and we compare the results from the coherent reactions with few incoherent data.

In this frame we would like to recall the main quantities measured in the study of the multiparticle production on nuclei, as:

$$h + A \rightarrow \Sigma h' + \Sigma p + \Sigma n + A' \qquad (5)$$

$\Sigma h'$ are secondary fast particles ($\beta > 0.7$), also called <u>shower</u> particles; Σp are mostly bound protons knocked out from the target nucleus: they are called <u>gray</u> tracks and their β ranges from 0.3 to 0.7; finally Σn are evaporation (<u>black</u>) tracks with $\beta < 0.3$.

The effect of a possible cascade within the nuclear matter is reflected mostly by the shower particles, which do not include the target fragmentation products. The most significant parameters is the ratio:

$$R = \frac{\langle n_s \rangle_A}{\langle n_s \rangle_H} \qquad (6)$$

where $\langle n_s \rangle_A$ and $\langle n_s \rangle_H$ are the mean shower multiplicities on nucleus and on hydrogen, respectively. R is studied as a function of $\bar{\nu}$, that is:

$$\bar{\nu} = \frac{R}{\lambda} = \frac{A \cdot \sigma_{inel} \ (h-H)}{\sigma_{inel} \ (h-A)} \qquad (7)$$

$\bar{\nu}$ is the nuclear size, measured in mean free path units, but its meaning is also the average number of inelastic collisions through the nuclear matter.

A further meaningful behaviour is the dependence of the mean shower multiplicity on the rapidity or pseudorapidity.

2. COHERENT ELASTIC PROCESS

The coherent elastic process (1) is well described by a Glauber picture of the interaction. The main lines of the Glauber approach can be summarized as follows.

If the nucleons should be "weak scatterers", the nuclear amplitude could be obtained simply by adding <u>coherently</u> the scattering amplitudes by a single nucleon:

$$\frac{d\sigma}{d\Omega} = A^2 \cdot |f(q)|^2 \cdot S^2(q) \qquad (8)$$

where $f(q)$ is the amplitude on a single nucleon and $S(q)$ is the nuclear form factor.

The $d\sigma/d\Omega$ vs q is a very steep function, which can be roughly approached by an exponential:

$$\frac{d\sigma}{d\Omega} \div \exp[-t/t_0] \qquad (9)$$

where $t_0 = \dfrac{3}{R^2_{r.m.s.}}$

Just for comparison the incoherent scattering gives a flatter distribution, smoothly going down as

$$A \ |f(q)|^2 \ [1-S(q)] \qquad (10)$$

But the nucleons are not weak scatterers. Then multiple scattering within the nuclear matter has to be taken into account. Glauber solved this problem by developing its <u>nuclear diffraction theory</u> or <u>multiple scattering theory</u>[4].

The amplitude by a single nucleon is written following two main assumptions:
a) the momentum $\hbar k$ of the incident particle is high enough so that the wavelength k^{-1} is much smaller than the range of interaction between the particle and a nucleon;

b) the forward-peaked character of the angular distributions of diffractive scattering implies that the partial waves contribute coherently to the scattering.

The two-dimensional Fourier integral of the profile function $\Gamma(b)$ gives the amplitude on a nucleon:

$$f(\underline{q}) = \frac{ik}{2\pi} \int e^{i\underline{q}\cdot\underline{b}} \Gamma(\underline{b}) d^2b \qquad (11)$$

where $\Gamma(b) = 1-e^{2i\delta_l} = 1-e^{i\chi(b)}$. $\chi(b)$ is the phase shift function.

The collision amplitude by the nucleus is written by making use of the <u>eikonal approximation</u>, following which the phase shifts χ_j by the individual nucleons are combined <u>additively</u>:

$$\chi_A(\underline{b}, \underline{s}_1, \dots \underline{s}_A) = \sum_j \chi_j(\underline{b}-\underline{s}_j) \qquad (12)$$

and the total profile function:

$$\Gamma_A = 1 - \prod_{j=1}^{A} \left[1 - \int d^2s\, dz\, \rho(\underline{s}, z)\, \Gamma_j(\underline{b}-\underline{s}) \right] \qquad (13)$$

$\rho(s, z)$ is the density of the nuclear matter. The correlations among the bound nucleons are neglected.

To obtain the single profile function we can invert the Fourier transform (11):

$$\Gamma(b) = \frac{1}{2\pi i k} \int e^{-i\underline{q}\cdot\underline{b}} f(q) d^2q \qquad (14)$$

$f(q)$ is measured experimentally from experiments on free hydrogen.

Finally the total amplitude on the nucleus is:

$$f_A(q) = \frac{ik}{2\pi} \int d^2b \cdot e^{i\underline{q}\cdot\underline{b}} \Gamma_A(b) \qquad (15)$$

Further simplifications can be introduced for the heavy nuclei ($A \gg 1$), where the range of the high energy interactions is much smaller than the nuclear radius. The total profile function (13) becomes:

$$\Gamma_A = 1 - \prod_{j=1}^{A} \left[1 - \frac{1}{A} f(0) T(b) \right] \qquad (16)$$

$T(b)$ is the thickness function, which corresponds to the integral of the nuclear density along a straight-line path:

$$T(\underline{b}) = A \int_{-\infty}^{+\infty} \rho(\underline{b}, z) \, dz \tag{17}$$

To the heavy nuclei limit the nuclear diffraction theory gives the same cross sections as the optical model:

$$\sigma_{el}^A = \int d^2b \left(1 - e^{-\sigma_1/2 \, T(b)}\right)^2$$

$$\sigma_{tot}^A = 2 \int d^2b \left(1 - e^{-\sigma_1/2 \, T(b)}\right) \tag{18}$$

where σ_1 is the collision cross section by a single nucleon (=free hydrogen).

The nuclear diffraction theory works very well up to 30-40 GeV incident energy. It reproduces, without free parameters, the differential and total cross sections with an excellent agreement[5].

At incident energy larger than 40 GeV the total cross section calculated with the Glauber theory is larger than the experimental data[6]. This discrepancy can be solved if we introduce the presence of inelastic intermediate states[7,8]. Collisions of the incident hadron h within the nuclear matter can open inelastic channels with production of intermediate states h^*; these states are combined again into the hadron h before leaving the nucleus. Even if the same cross section on nucleon is assumed for h^* and h, the total cross section σ_{tot}^A is depressed because the introduction of intermediate channels enhances the double scattering term, which in the multiple scattering expansion is of opposite sign with respect to the single scattering term.

The reduction of the total cross section is expected to be more important at higher-energy because in the inelastic double scattering diagram the outgoing hadron could still maintain almost the same phase as the incoming one.

3. COHERENT PRODUCTION

The inelastic coherent reactions as the process (2) can be treated with the nuclear diffraction theory, where the transition from the incident particle h to the produced state Y is introduced. Because experimentally both the coherent and incoherent productions are present in the data, the differential cross section can be written simply as a sum of the coherent and incoherent contributions:

$$\frac{d\sigma}{dt'} = \frac{d\sigma_C}{dt'} + \frac{d\sigma_I}{dt'} \tag{19}$$

For the coherent contribution the formula of Kolbig and Margolis[9] assumes that the production be point like and instantaneous and takes into account the probability of absorption in the nuclear matter of the incident and the outgoing states. The coherent cross section is written in the following way:

$$\frac{d\sigma_C}{dt'} = C_o(M) \cdot A^2 \cdot |F(t', M)|^2 \qquad (20)$$

where $C_o(M)$ is the intercept, at t=0, of the collision cross section by a single nucleon and $F(t', M)$ is the nuclear form factor, which includes the absorption effects:

$$F(t', M) = 2\pi \int_{-\infty}^{+\infty} dz \cdot \int db \cdot b \cdot \exp\left[i \frac{M_Y^2 - m_h^2}{2P_h} \cdot z\right] \cdot J_o(\sqrt{t'}\, b) \cdot \rho(b, z) \cdot$$

$$\cdot e^{i\chi_{coul}(b)} \cdot \exp\left[-(1-i\alpha_1)\frac{1}{2}\sigma_1 T_1(b, z)\right] \cdot \exp\left[-(1-i\alpha_2)\frac{1}{2}\sigma_2 T_2(b, z)\right] \qquad (21)$$

T_1 and T_2 are the profile functions:

$$T_1(b, z) = \int_{-\infty}^{z} A \cdot \rho(b, z) dz' \qquad T_2(b, z) = \int_{z}^{+\infty} A \cdot \rho(b, z') dz' \qquad (22)$$

The nuclear density, $\rho(b, z')$ is normally written following the Wood-Saxon distribution.

For the Coulomb phase, the following formula can be used:

$$\chi_{coul}(b) = \frac{27}{137}\left\{\ln(.P_h b) + 4\pi\int_0^\infty \rho(r) \cdot \left[\ln\left(\frac{r}{b} + \sqrt{\frac{r^2}{b^2} - 1}\right) - \sqrt{1 - \frac{b^2}{r^2}}\right] r^2 dr\right\} \qquad (23)$$

In the nuclear form factor the two exponentials take into account the absorption effects: σ_1 and α_1 are the cross section and the ratio of the real to the imaginary part of the forward scattering on nucleon for the incident particle h; σ_2 and α_2 have the same meaning for the produced state Y.

For σ_1 and α_1 we normally use the measured values by free hydrogen: σ_2 and α_2 are free parameters to be fitted on the experimental distributions.

The incoherent cross section can be approximated simply by an exponential:

$$\frac{d\sigma_I}{dt'} = I_o(M, A) \cdot e^{-B(M) \cdot t'} \qquad (24)$$

$B(M)$ is the slope of the distribution by a single nucleon and $I_o(M, A)$ is a normalization factor. $B(M)$ and $I_o(M, A)$ are obtained separately by fitting with the formula (24) the differential cross section at large t' (as $t' > 0.1(GeV/c)^2$), where the coherent production does not contribute (see Fig.2).

Then the total cross section is fitted with the expression (19) taking three parameters free: $C_o(M)$, σ_2, α_2. For correlation problems between α_2 and σ_2 we prefer to do the fit for various α_2, with <u>$C_o(M)$ and σ_2 only as free parameters</u>. Anyway different values of α_2 normally do not affect the σ_2 best fit values.

Before to discuss the experimental results we have to mention that the coherent mechanism implies precise selection rules for the produced state Y.

The collision amplitudes by a single nucleon consist of a term, no-spin and no-isospin dependent, which is the same for all the nucleons, and of terms, spin and/or isospin dependent, which are different in sign according to the type (n or p) and to the spin (up or down) of the knocked nucleon. Because the amplitude by a nucleus is obtained by adding in phase the amplitudes by single nucleons, it turns out that the contribution of the terms spin-isospin dependent is at most of the order of A^{-1}.

It follows that:

i) the Isospin I_Y of the produced state Y must be the same as the isospin I_h of the incident particle, because T≠0 exchanges are forbidden;

ii) the helicity of Y must be the same as the helicity of h, for forward production at very small angles;

iii) if h is a spinless particle, the parity of Y must be $P_Y=(-1)^J P_h$, because the only term which can give an "unnatural" parity exchange is $q_h \times q_Y$, which is =0 in forward direction.

Then, if the incident particle is a 0^- state as π or k, only systems belonging to the $(-1)^{J+1}$ spin parity series are allowed. In addition the transition π→2π is forbidden for spin-parity and g-parity conservation, and k→kπ is forbidden for spin-parity conservation.

As a further example, a coherent interactions induced by an incident π^- can produce the resonance $A_1(J^P I^G=1^+1^-)$ and the resonance $A_3(2^-1^-)$, while the production of A_2 (2^+1^-) is forbidden.

Of course this is true strictly in forward direction; at larger t' the ω exchange contribution, even if depressed with respect to the Pomeron exchange, can give violations of these selection rules.

4. ABSORPTION OF UNSTABLE STATES PRODUCED WITHIN THE NUCLEUS.

The cross section σ_2 for collision of unstable states Y by bound nucleons has been studied at various energies.

The coherent production of ρ and φ by 5-7 GeV incident photon allowed to obtain the cross sections (ρ-n) and (φ-n); they turned out to be 26.7±2 mb for ρ and 12±4 mb for ϕ[2,10].

The cross section σ_2 has been studied also for baryonic resonances as (pπ°) and (pπ⁺π⁻) produced by 23 GeV protons; the [(pπ°)-n] and the [(pπ⁺π⁻)-n] cross sections range from 18±6 to 33±7mb[11].

Systematic measurements of σ_2 have been carried out for 3π and 5π production by incident π^- at four incident energies: 8.9 and 15.1 at CERN[3], 22.5 at Brookhaven[12], 40.0 at Serpukhov[13,14]. These experiments used various nuclear targets, from Be to U, and analyzed a wide mass spectrum.

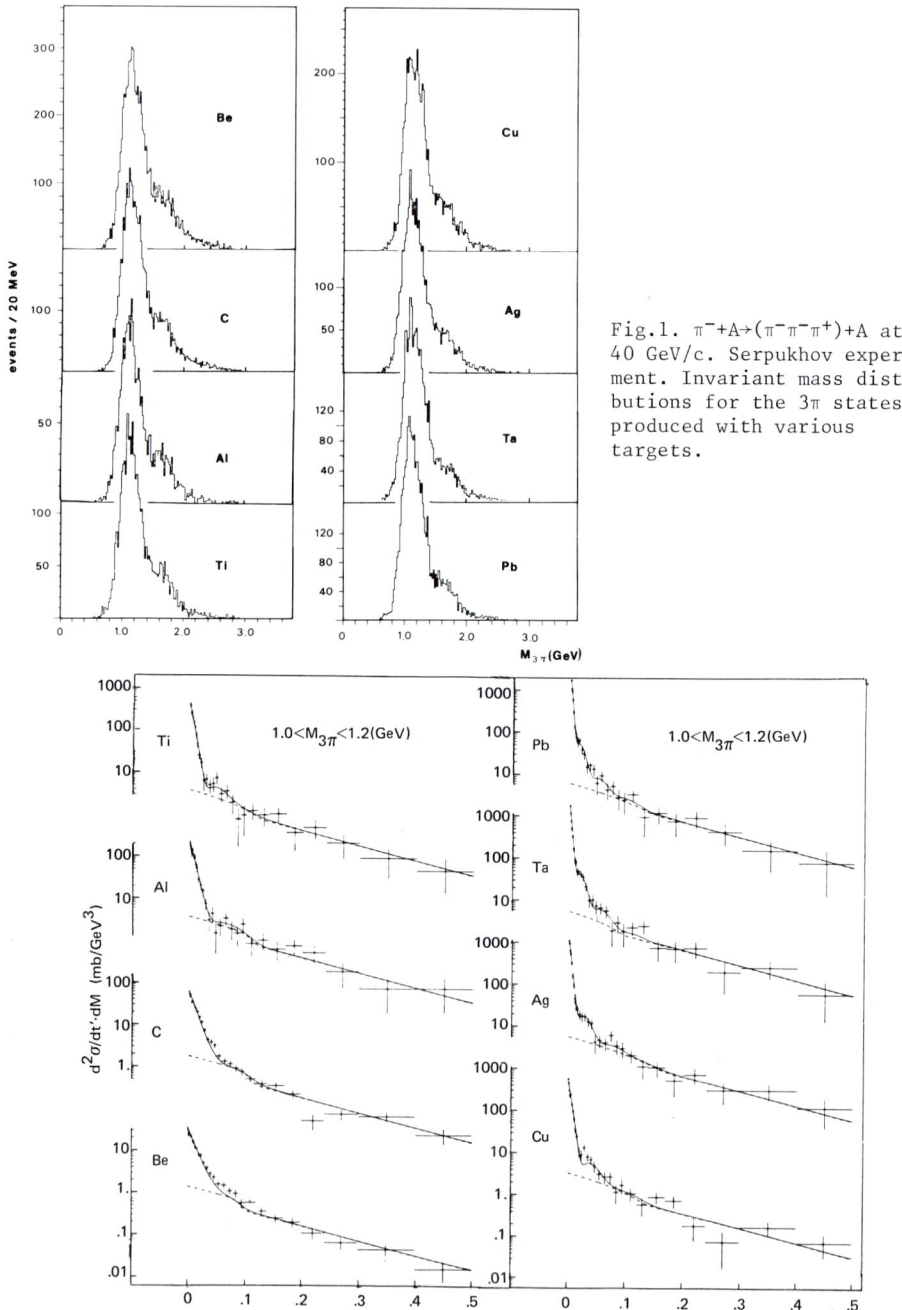

Fig.1. $\pi^- + A \to (\pi^-\pi^-\pi^+) + A$ at 40 GeV/c. Serpukhov experiment. Invariant mass distributions for the 3π states produced with various targets.

Fig.2. Serpukhov data. Small crosses: experimental data. Full line: fit with the formula (19) ($\alpha_\rho = 0$ and $\sigma_2 = 16.9 \pm 0.6$ mb. Dotted line: extrapolation of the incoherent part only; the deviation from the exponential shape is due to experimental biases.

Just as an example we report in Fig. 1 the 3π mass spectra obtained in the Serpukhov experiment with various targets; as it can be easily observe the nuclear form factor is restricting more and more the large mass tail with increasing A. The t' distributions are displayed in Fig. 2: the coherent slope, more and more steeper with growing A, and the flatter incoherent exponential are easily identified. The σ_2 best fit value is obtained by fitting with (19) the surface $d\sigma/dt'$ vs t' vs A for different regions of invariant 3π mass, following the procedure described in the previous paragraph.

As alternative method σ_2 can be obtained by fitting the integrated coherent cross section as function of A (see Fig. 3).

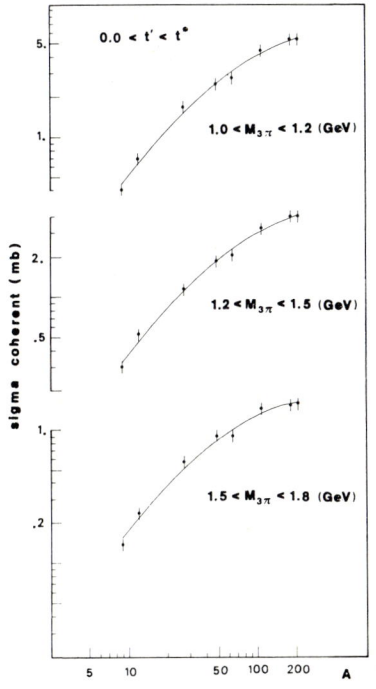

Fig.3. Serpukhov data. σ_c vs A. Black points: experimental $d\sigma_c/dt'$ integrated up to the first diffraction minimum, once subtracted the incoherent background. Full line: fit with the formula (20) integrated up to the first diffraction minimum. The σ_2 best fit values are: 16.6±0.7 mb for the $M_{3\pi}$ interval 1.0÷1.2 GeV; 15.4±1.9 mb for 1.2÷1.5 GeV; 17.4±1.9 mb for 1.5÷1.8 GeV.

The 3π states have been analyzed in partial-waves. σ_2 for the various waves contributions have been studied by fitting the total coherent cross sections, corresponding to fixed spin-parity states as function of $A^{15,16,17,3}$. This analysis was very detailed for the 1^+ and 0^- states, which gave a very large contribution.

A summary of the σ_2 best fit values for 3π and 5π states is shown in Table I. An overall inspection of the results reveals striking behaviours, which can hardly agree with an interpretation of σ_2 as an actual cross section for collision by bound nucleons. The main properties od σ_2 can be summarized as follows:

i) σ_2 seems independent from the resonant behaviour of the 3π system. Its values for the various mass regions are consistent into the errors: they are of the order of π-(free hydrogen) cross section, even where the three pions would be produced as uncorrelated particles. The only exception is the value for the 0^- state in the mass region from 0.9 to 1.2 GeV, at 15.1 GeV/c.

ii) σ_2 for the 5π states[18,19] is even smaller than the cross section of the 3π states. σ_2 seems to decrease with increasing multiplicity.

iii) σ_2 tends to decrease with increasing invariant mass and incident energy: nevertheless these general rules show many exceptions.

But the more general and striking conclusion is that σ_2 is <u>very small</u>, if interpreted as an actual cross section. This trasparency of the nuclear matter is confirmed also by the results obtained on the multiparticle production on nuclei, studied in inclusive way, e.i. with no selection of channels.

The average multiplicity of shower particles is very small and no large differences exist from this point of view between Hydrogen and Uranium targets. A good interpolation gives the relation:

$$R \sim 0.4 + 0.6\,\bar{\nu}$$

just as an example, R is ~ 2.5 for a lead target at ~100 GeV.

In addition, if we take into account only the events with $\Sigma n=0$, e.i. without evaporation tracks and therefore with small or negligible nuclear excitation, R tuns out to be $\simeq 1$. The multiparticle production with $\Sigma n=0$ are not so different from the coherent productions, at least from the energetic point of view. Therefore the very small multiplicity (R≃1) and the small value of σ_2 should be referred to the same physical behaviour.

The basic property of the interaction mechanism, which has to be taken into account, is that the production is neither instantaneous nor localized, as assumed by the Glauber-Kolbig-Margolis approach. Since the time when the interaction takes place up to the time when the produced system reaches a fixed state a formation interval and therefore a formation length l_f is needed. For fast secondaries l_f can be even comparable to the nuclear radius.

Several approaches have been used to describe the space-time development of the collision. Feinberg, Van-Hove, Gottfried and other theoreticians developed interesting models which include, more or less directly, this idea of a formation length. In terms of quarks we can also say that in a collision only bare quarks are emitted and they reach usual properties only when they are dressed by gluons.

A more recent approach[20,21] has been developed, which is connected, in a certain sense, to the intermediate state idea of Gribov, Karmanov and

Kondratyuk. In this approach the hadron travelling through the nucleus is represented in terms of eigenstates of t-matrix in nuclear matter: they include states with very small absorption (passive states). Due to the presence of these passive states, the hadron absorption in nuclear matter turns out to be lower than we can expect from a Glauber approach.

We would end this paragraph by concluding that σ_2 obtained from the experimental data by means of the Kolbig-Margolis formula is not anymore a meaningful parameter.

TABLE I

σ_2 (mb)

3π

$M_{3\pi}$ (GeV)	P_{inc} (GeV/c)	8.9 CERN	15.1 CERN	22.5 BNL	60.0 SPK
T O T A L	0.9-1.2	26±2.	21±1.1	26±2.	16.6±0.7
	1.2-1.5	–	17±2.2	22±2.	15.4±1.9
	1.5-1.8	–	20±3.5	19±3.	17.4±1.9
1^+	0.9-1.2	–	17±1.2	25.4±1.	14.1±1.
	1.2-1.5	–	–	23±2.	11. ±1.5
	1.5-1.8	–	–	–	14. ±2.
0^-	0.9-1.2	–	49±8.	–	23. ±3.
	1.2-1.5	–	–	30. ±5.	28. ±5.
	1.5-1.8	–	–	–	16. ±7.

5π

$M_{5\pi}$	P_{inc} (GeV/c)	15.1	40.0
	1.6-2.2	17.±5.	8.8±0.5.
	2.2-3.0	–	5.5±0.6

5. THE NUCLEUS AS A SELECTOR OF RESONANT STATES

Even if we reached the conclusion that the Kolbig-Margolis formula does not give the cross section on the nucleons of the states produced within the nucleus, nevertheless information on the absorption in nuclear matter can be obtained in a model independent way. In Table II the contributions of the two partial waves, which give most contribution, 0^-s and 1^+s are shown for four samples in which the data on the various targets of the Serpukhov experiment[14] were grouped. From these results it is clear that the 0^-s contribution is going down with increasing A, while the 1^+s is growing up. This can be clearly interpreted as a selective behaviour of the nuclear absorption: the 0^-s states are more absorbed than the 1^+s.

TABLE II

Partial wave contribution (%)

	target $M_{3\pi}$ (GeV)	Be+C	Al+Si	Ti+Cu	Ag+Ta+Pb
0^-s	0.9-1.2	12.5±1.2	12.5±0.7	11.2±1.6	9.6±1.0
	1.2-1.5	17.3±2.2	14.4±1.0	13.5±2.2	9.4±1.3
1^+s	0.9-1.2	64.4±3.4	64.8±1.5	68.8±3.6	66.9±2.3
	1.2-1.5	38.3±3.1	37.5±1.3	43.7±3.1	50.9±2.0

The partial waves analysis was carried out for both the CERN[22] and the Serpukhov[23] experiment using the Illinois group program. In order to select samples enriched in coherent events, a t' cut has been introduced at the first diffractive minimum. This cut reduces the incoherent background less than 1.5% for the Pb target and less than 10% for Be. The background for the other targets is of course intermediate between these two numbers. The statistics is not depressed anyway, because most of the coherent events are produced in the first diffractive peak.

We would like to stress that a very large part of the data consists of events at very small t': just as an example, more than 1/2 of the events produced with the Pb target are in the t' range 0.-0.003 $(GeV/c)^2$.

The partial waves analysis of coherent samples are very simplified and more reliable with respect to the data on hydrogen because:
i) the number of waves which give contribution is reduced (all spin-flip are suppressed) as a consequence of the selection rules of the coherent processes;
ii) the degree of coherence is $\simeq 1$: this means that the differences between the phases of two amplitudes are really measurable.

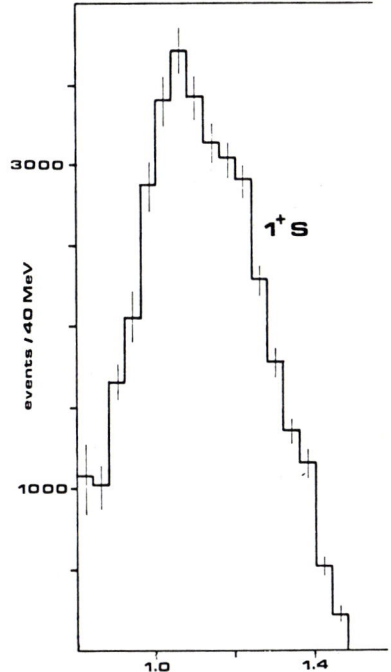

Fig. 4. The 1^+S amplitude vs $M_{3\pi}$(GeV) Serpukhov data.

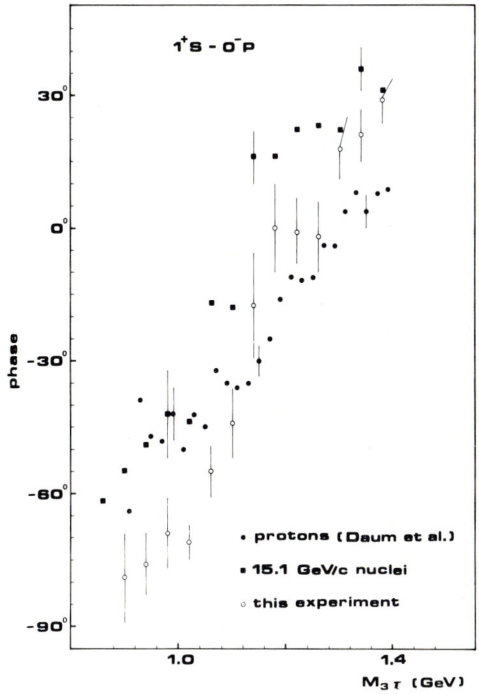

Fig. 5. Relative phase between 1^+S and 0^-P vs $M_{3\pi}$. Empty points: 40 GeV/c, Serpukhov experiment, nuclear targets. Filled squares: 15.1 GeV/c, CERN exp., nuclear targets. Filled points: hydrogen data.

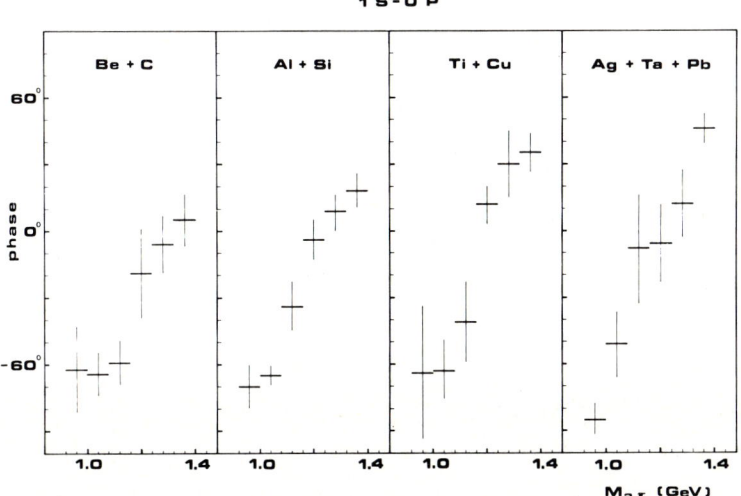

Fig. 6. Relative phase between 1^+S-0^-P for the $M_{3\pi}$ interval 0.9÷1.4 GeV and for groups of nuclear targets. Serpukhov data.

The 1^+S wave was carefully analyzed: it exhibits a very clear enhancement in the region of A_1 and, what is more important, shows a fast change in phase through the peak. In Fig. 4 the 1^+S contribution is shown as function of $M_{3\pi}$ mass. In Fig. 5 the 1^+S-0^-S phase difference is displayed for the CERN and Serpukhov experiments on nuclei: 0^-P is chosen as reference wave because it is flat in the 3π mass region from 0.9 to 1.5 GeV[23].

The 1^+S phase is fastly moving through the A_1 mass, changing of ~110° at 40 GeV and of ~90° at 15.1 GeV/c. In the data collected with an hydrogen target a change in phase was never found in the A_1 region.

Therefore, in 1977 we were able to demonstrate, for the first time, the resonant behaviour of the A_1 structure[22]. Only in 1980 the ACCMOR collaboration obtained a 1^+S phase change of ~60°, anyway definitively smaller than in the data on nuclei [24] (Fig. 5).

The effect of the nucleus on the 1^+S phase is further demonstrated by the plots of Fig. 6[25]. The 1^+S-0^-P phase difference is studied separately for four groups of nuclear targets: Be+C, Al+Si, Ti+Cu, Ag+Ta+Pb: the data were collected at 40 GeV/c. The phase change shows systematic increase from ~60° for the lightest nuclei (similarly to what obtained by hydrogen) to ~140° for the heaviest nuclei.

As we can conclude from Fig. 7, only at very small t' this large phase moving is found: this means that this effect of enhancing the resonant 1^+S state is present only in the coherent sample.

It is known since many years that the Deck effect gives contribution to the A_1 enhancement in the 3π invariant mass distribution. The Deck effect consists of a ρ-production at the upper vertex and of a production at an intermediate vertex of a pion which gives a diffractive scattering on the nucleus. As pointed out by the ACCMOR collaboration the Deck contribution, in the hydrogen data, is much higher than the resonant contribution[24].

We have studied the contribution of the Deck mechanism and of the direct production of resonant states to the $\pi^-\pi^-\pi^+$ channel for various nuclear targets, by using the data of the Serpukhov experiment. To this purpose the two component model, as formulated by Bowler, has been used[25]: in this model the coherent Deck contribution is considered, in addition to a direct resonance production.

The Deck amplitude is written in the following way:

$$D(M_{\pi\pi}, t_\pi, t) = \frac{A_{\pi\pi}(M_{\pi\pi}, t_\pi) A_{\pi N}(S_{\pi N}, t)}{m_\pi^2 - t_\pi} \quad (25)$$

$M_{\pi\pi}$ is the l=1 $\pi\pi$ scattering amplitude through ρ meson; $A_{\pi N}$ is the π-nucleus coherent scattering amplitude: we have used the formula (15) following an

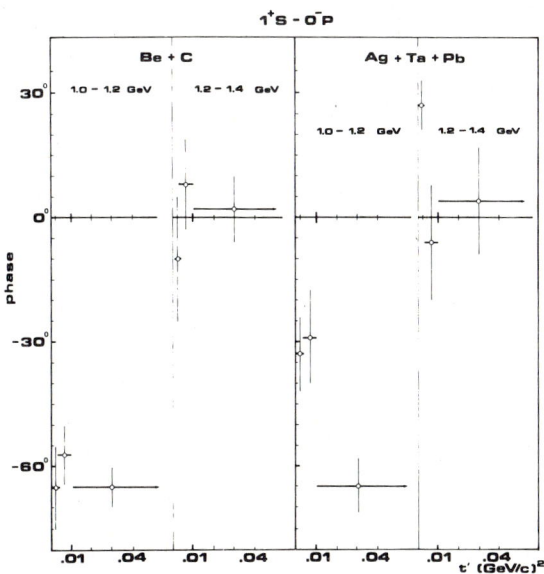

Fig. 7. Relative phase between 1^+S and 0^-P vs t' for two $M_{3\pi}$ intervals and two groups of targets: Be+C and Ag+Ta+Pb. Serpukhov data.

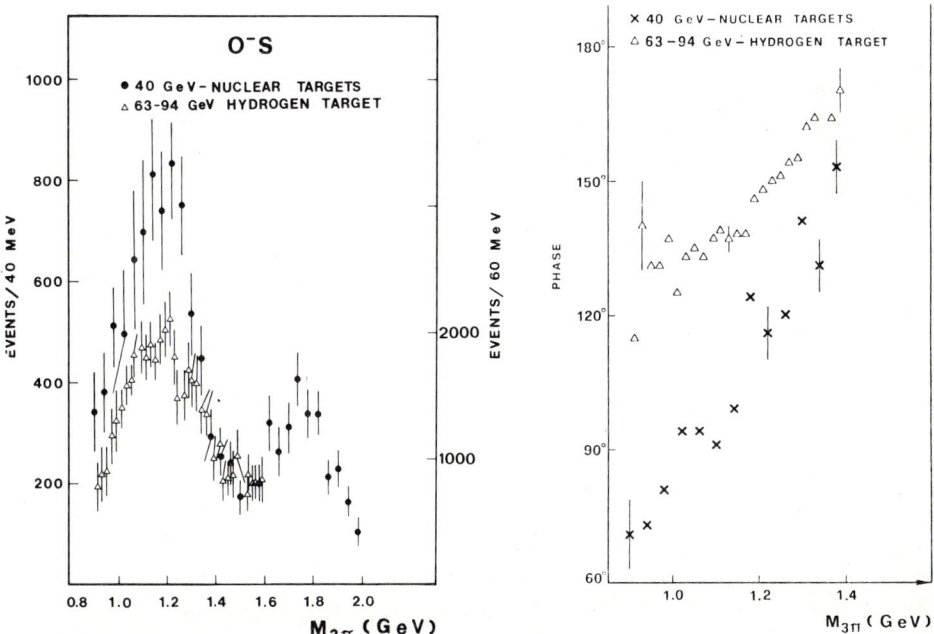

Fig. 8. Mass dependence of 0^-S. Filled squares: Serpukhov data; compilation of all nuclear targets (left y scale). Small triangles: hydrogen data (right y scale).

Fig. 9. 0^-S-0^-P relative phase. Crosses: Serpukhov data; compilation of all nuclear targets. Small triangles: hydrogen data.

approach "a' la Glauber".

The direct A_1 production is assumed to follow the Kolbig-Margolis formula as in (20).

The total amplitude is an additive combination of the Deck and the direct production terms, with a relative phase and a relative weight. With this amplitude the 1^+S intensity and the 1^+S-0^-P relative phase were fitted for the samples by the various targets, leaving as free parameters the relative weight of the resonant contribution (called resonant component) and the A_1 mass. The results show that the resonant component grows up fastly with increasing nuclear target mass: from 0.13 ± 0.04 for Be and 0.19 ± 0.05 for Si to 0.44 ± 0.11 for Pb. This means that the not-resonant background is more absorbed than the resonant signal and therefore it decreases with increasing nuclear size.

We can conclude that the cleanest results on the A_1 resonance are obtained with the heaviest nuclear targets, because only there the A_1 state is produced not fully masked by the Deck background. From the Pb target sample the A_1 mass turns out to be 1.22 ± 0.03 GeV.

This selective effect of the nuclear targets is confirmed by the analysis of the 0^-S contribution. Also in this case we refer to the Serpukhov experiment data.

The 0^-S amplitude (Fig. 8) shows two clear enhancements, the first one near $M_{3\pi}\simeq1.2$ GeV and the second one above 1.6 GeV. The 0^-S-0^-P phase difference through the first enhancement is shown in Fig. 9 and it is compared with the hydrogen data of the ACCMOR collaboration[23,25]. The data on nuclei exhibits a phase motion of ~80° and it is fully consistent with a resonant behaviour of the 0^-S wave in that region: this change in phase is much larger than in the results by hydrogen.

In the analysis of the second enhancement, the 1^+S wave has been used as reference wave instead of 0^-P, which becomes negligible above $M_{3\pi}\simeq1.6$ GeV; on the other hand 1^+S is nearly constant in this region. In addition the 1^+S wave, used as a reference for 2^-S wave, reproduces well the resonant behaviour of A_3.

The 0^-S-1^+S phase difference changes of ~100° in the 3π mass region from 1.4 to 2.0 GeV, revealing the resonant nature also of this second enhancement (Fig. 10).

Unfortunately the possible dependence of the 0^-S phase change on the target atomic weight cannot be investigated, as it was done for 1^+S, due to statistical reasons (the 0^-S contribution is definitively lower than 1^+S). Nevertheless, the effects of the nuclear targets were evident from the comparison between the data on nuclei and the data on hydrogen.

From the 0^-S results we were able to conclude that we found for the first time two radial excitations of the pion. The experimental distributions, fitted with a Breit-Wigner plus a second-order polynomial give the following

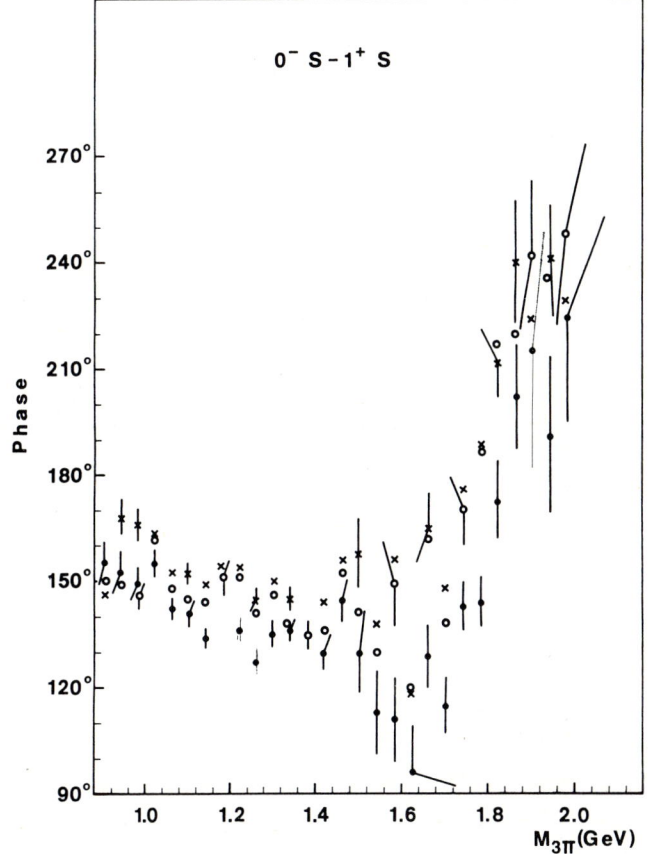

Fig.10. 0^-S-1^+S relative phase. Serpukhov data: the different symbols refer to different parametrizations used in the PW analysis. In the A_1 region the distribution is flat because both 0^-S and 1^+S show a resonance behaviour (π' and A_1).

best fit values:

$M_{R1} = 1.24 \pm 0.03$ GeV $\Gamma = 0.35 \pm 0.12$ GeV
$M_{R2} = 1.77 \pm 0.03$ GeV $\Gamma = 0.31 \pm 0.05$ GeV

6. CONCLUSIONS

The results of this large experimental program push to the conclusion that the nuclear targets are a very good tool in selecting new states. The nuclear targets are very efficient in cleaning up the resonant signals.

Two main reasons contribute to this effect:

1. The coherent mechanism requires selection rules, which allow the production of certain waves only. This makes easier and more reliable the partial wave analysis.
2. The nuclear matter absorbs more the not-resonant background that the resonant signals.

REFERENCES

1) G. Bellini et al., Nuovo Cimento 29 (1963) 896.

2) S.C.C. Ting, in "Interactions of Elementary Particles with Nuclei", ed. by G. Bellini, L. Bertocchi, S. Bonetti, Trieste 1970, p. 171.

3) P. Muhlemann et al., ETH-Milano-Imp. College Coll, Nucl. Phys. B59 (1973) 106.

4) R. Glauber, in "Lectures in Theoretical Physics, Vol.1, New York, Interscienze 1959, p.273.

5) R. Glauber and G. Matthiae, Nucl. Phys. B21 (1970) 135.

6) G. Matthiae, in "Multipartice production on nuclei at very high energies" ed. by G. Bellini, L. Bertocchi, P.G. Rancoita, ICTP Trieste 1977, IAEA-SMR-21, p. 351.

7) V.N. Gribov, Sov. Phys. JETP 29 (1969) 483.

8) A. Karmanov and L.A. Kondratyuk, JETP Lett. 18 (1973) 266.

9) K.S. Kolbig and B. Margolis, Nucl. Phys. B6 (1968) 85.

10) S.C.C. Ting, Proc. of the XIV Int. Conf. on High Energy Physics, Vienna 1968 p. 43.

11) P.R. Murthy et al., Nucl. Phys. B92 (1975) 269.

12) R.M. Edelstein et al., Phys. Rev. Lett. 38 (1977) 185.

13) G. Bellini, in "Multiparticle production on nuclei at very high energies" ed. by G. Bellini, L. Bertocchi, P.G. Rancoita, ICTP Trieste 1977, IAEA-SMR-21, p.505.

14) G. Bellini et al., Nucl. Phys. B199 (1982) 1.

15) G. Bellini, Proc. Topical Seminar on High Energy Collisions Involving Nuclei, Trieste 1974 (ed. Compositori, Bologna, 1975) p.317.

16) W. Beusch et al., CERN-ETH-Imperial College-Milano Coll., Phys. Lett. 55B (1975) 97.

17) T.J. Roberts et al., Phys. Rev. D18 (1978) 59.

18) C. Bemporad et al., CERN-Imperial College-Milano-ETH Coll., Nucl. Phys. B33 (1971) 397.

19) G. Bellini et al., Phys. Lett. 126B (1983) 140.

20) H.I. Miettinen and J. Pumplin, Phys. Rev. D18 (1978) 1696.

21) N.N. Nikolaev, Sov. Phys. JETP 54 (1981) 434.

22) J. Pernegr et al., CERN-ETH-Imperial College-Milano Coll., Nucl. Phys. B134 (1978) 436.

23) G. Bellini et al., Phys. Rev. Lett. 48 (1982) 1697.

24) C. Daum et al., Phys. Lett. 89B (1980) 281.

25) G. Bellini et al., Nuovo Cimento 79A (1984) 282.

USE OF THE EUROPEAN HADRON FACILITY COMBINED WITH AN
UNDERGROUND DETECTOR TO STUDY NEUTRINO OSCILLATIONS

M. DE VINCENZI and P. PISTILLI
Dipartimento di Fisica, Università di Roma "La Sapienza",
P. le A. Moro 2, 00185 Roma, Italy

1. INTRODUCTION

The possible existence of oscillation between neutrinos of different flavours is one of the most interesting open problems of the elementary particles physics[1].

A discovery of a finite neutrino mass and of a violation of the lepton number will have deep consequences in the understanding of the unification of the interactions and, at the same time, very important implications in several astrophysical problems.

Up to now the experiments dedicated to neutrino oscillation have been carried out with negative results using neutrinos from reactors, accelerators and cosmic rays. An experiment[2] performed by detecting solar neutrinos, has given puzzling results that may be interpreted as due to the existence of neutrino oscillation. However this interpretation is affected by several uncertainties due to the poor knowledge of the energy spectrum of the neutrinos coming from the sun.

The improvement of present limits is an experimental task that will be carried out in the future with different tecniques. The use of underground experiments to detect a ν-beam produced by accelerators in order to study ν-oscillation has been discussed since a long time[3].

In this lecture we discuss the practical feasibility of such an experiment using a very intense 30 GeV proton beam such as that foreseen in a future European Hadron facility combined with a detector placed in the Gran Sasso Laboratory.

2. PHENOMENOLOGY OF ν-OSCILLATIONS

In the more general case of three neutrinos, if the weak interaction eigenstates (ν_e, ν_μ, ν_τ) do not coincide with the mass eigenstates (ν_1, ν_2, ν_3) we can write:

$$\begin{pmatrix} \nu_e \\ \nu_\mu \\ \nu_\tau \end{pmatrix} = U \begin{pmatrix} \nu_1 \\ \nu_2 \\ \nu_3 \end{pmatrix} \qquad (1)$$

where U is a unitary (real if CP holds) matrix that connects the two sets. The state produced at the time t=0 as a weak eigenstate will evolve in time

$$\nu_\alpha(t) = \Sigma_i U_{\alpha i} e^{(-iE_i t)} \nu_i(0) \qquad (2)$$

$\alpha = e, \mu, \tau$
E = neutrino energy.

Therefore, starting with neutrinos of a given flavour α one will have a probability different from zero to find, after a time t, neutrinos of a flavour β. We restrict for the sake of simplicity our analysis to the case of two neutrinos only ν_e, ν_μ.

In this case the transformation equation will be:

$$\nu_e = \nu_1 \cos\theta + \nu_2 \sin\theta$$
$$\nu_\mu = -\nu_1 \sin\theta + \nu_2 \cos\theta \qquad (3)$$

and the probability to find a ν_μ starting from a pure ν_e state at a distance L will be given by:

$$P(\nu_e \to \nu_\mu) = P(\nu_\mu \to \nu_e) = \sin^2 2\theta \, \sin^2(\alpha \, \delta m^2 \, L/E)$$

E = neutrino energy $\qquad (4)$

$\delta m^2 = m_1^2 - m_2^2$ ($m_{1,2}$ neutrino masses)

α (constant factor) = 1.267 for L(Km), E(GeV), δm^2(eV^2)

Two kinds of experiments are usually performed to detect $\nu_\mu - \nu_e$ oscillation. In the so called "disappearance" experiments one measures the flux of neutrinos of a given flavour at different distances from the source looking for a deviation from unity. In the so called "appearance" experiments one starts with an (almost) pure beam of a given flavour and looks for the presence of another one. It is interesting to remark that in the eq.(4) the δm^2 terms is multiplied by the factor L/E.

Therefore, the sensitivity to very small value of δm^2 will be obtained by enhancing the term L/E either performing the experiment with a large distance between the source and the detector or working with low energy neutrinos.

On the other hand the oscillation probability is proportional to $\sin^2 2\theta$; therefore the sensitivity to small values of the mixing angles will be limited by the statistics. The main advantage of the use of a high intensity proton beam combined with an underground detector consists in the possibility to exploit the large distance between the detector and the accelerator while maintaing a good neutrino flux in order to perform the experiment in a reasonable time.

3. THE EXPERIMENTAL SCHEME

We have considered a 30 GeV energy extracted proton beam with an average intensity of $\sim 10^{15}$ pps placed at CERN.

The π's and K's produced in the interaction of the protons over a thick target (about two interaction lengths) are allowed to decay over a distance of about 500 m along a decay tunnel directed toward the Gran Sasso Laboratory (Fig. 1).

After the target we consider placing a magnetic horn of the Gargamelle type that enhances the neutrino flux in the forward direction by about a factor of four.

The neutrino beam will have the energy spectrum shown in Fig. 2.

If this beam is directed toward a detector of 5 Ktons of fiducial mass like ICARUS[4] placed in the Gran Sasso Laboratory at a distance of ~ 700 Km we obtain as an estimate of the ν interactions in the detector

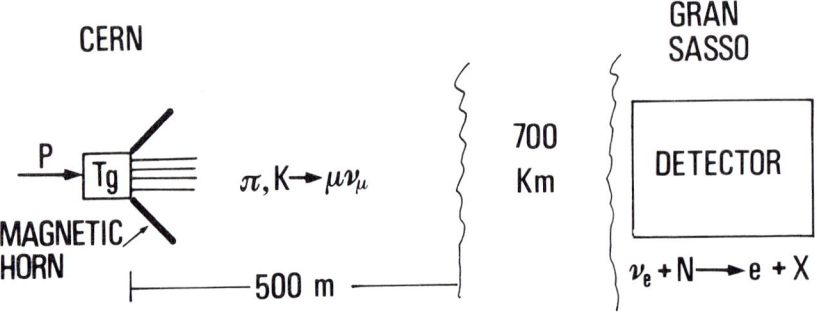

Fig.1
Scheme of the experiment

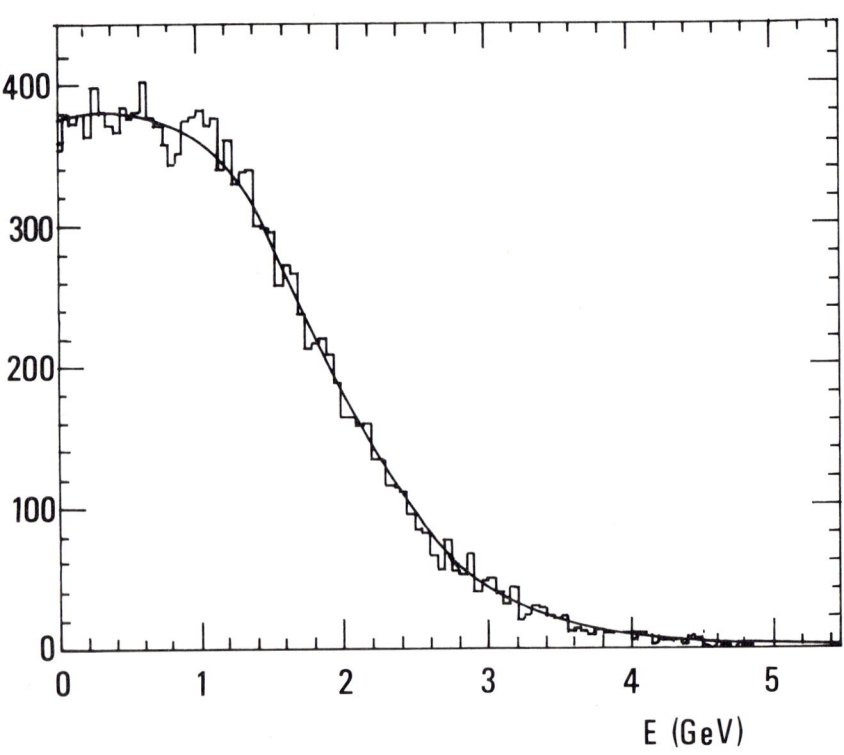

Fig.2
Neutrino energy distribution

$$N_{int} \sim 10 \text{ int/day}$$

The beam is mainly ν_μ with a contamination, at the level of a few percent, of ν_e coming from K decay. The $\nu_\mu - \nu_e$ oscillation is identified with an excess of ν_e interactions in the apparatus with an electron in the final state.

We have computed the correction due to the so called "matter effect" that will be discussed in the next part of this lecture. This effect turns out to be small in the condition we are discussing. Concerning the performance of the apparatus we have consider two cases:

a) an ideal apparatus capable of identifying electrons with full efficiency over all the energy range

b) an apparatus such as the CHARM detector used for the neutrino oscillation experiment at CERN[5].

We have assumed an uncertainty in the ν_e contamination from K decay corresponding to 1% of ν_μ. This gives the limit on the sensitivity in the case of the ideal detector. For case b) the limit is given by the capability of the apparatus to identify ν_e interactions (we have assumed a 5% upper limit for the ratio ν_e/ν_μ). The sensitivity of the experiment in $\sin^2 2\theta$, δm^2 is shown in Fig.(3) corresponding to $\sim$ 200 interactions in the apparatus equivalent to about one month of data taking. In the same figure the present limits coming from reactors[6] and accelerators[7] are shown.

4. MATTER EFFECTS

The formalism of neutrino oscillations previously discussed uses the assumption that neutrinos travel in vacuum. The presence of matter introduces a correction with a mechanism similar to the K_{os} regeneration, due to the fact that ν_e and ν_μ have different interactions with the electrons present in the matter[8].

The parameters θ and δm^2 that give the oscillation probability in eq.(4) are modified in θ_c and δm_c^2 by the matter effect according to the formulae:

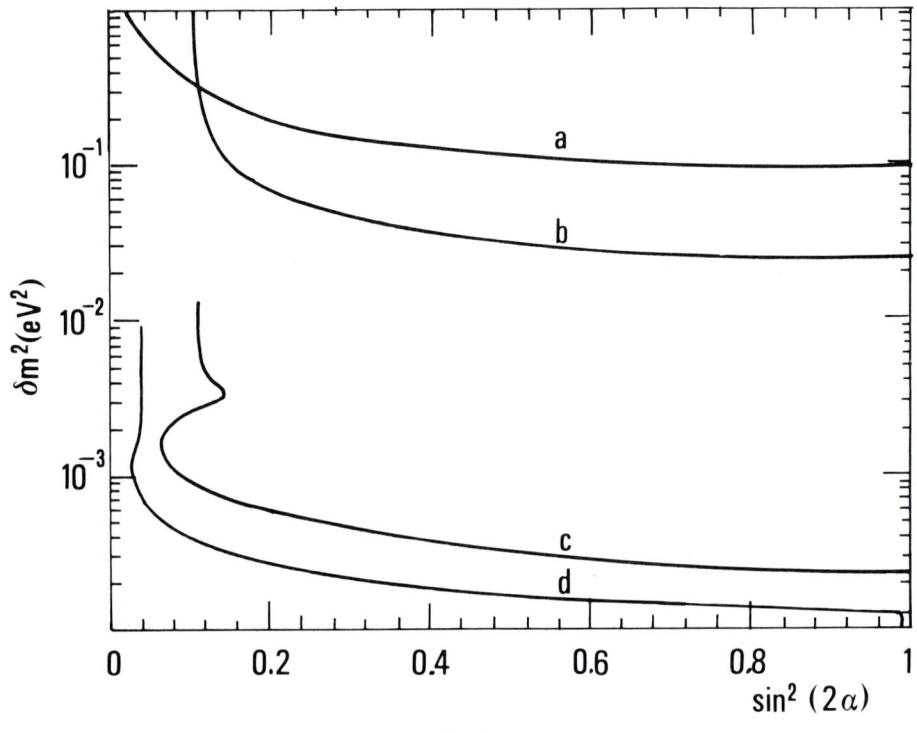

Fig.3
90% C.L. limits on the oscillation parameters
a) from accelerators
b) from reactors
c) sensitivity of the discussed experiment with a CHARM-like apparatus
d) sensitivity of the discussed experiment with an ideal apparatus.

$$\sin^2 2\theta_c = \sin^2 2\theta / (1 - 2x\cos 2\theta + x^2)$$

$$\delta m_c^2 = \delta m^2 (1 - 2x\cos 2\theta + x^2)^{1/2}$$

(5)

where

$$x = \lambda / \lambda_0$$

$$\lambda = 4\pi E / \delta m^2$$

$$\lambda_o = 2\pi/(\sqrt{2}\ G\ N_e)$$

where G is the weak coupling constant and N_e the number of electrons per unit volume.

The corrections (5) have been used in the computation of the limits presented in Fig.(3) and as stated above this effect is small in this configuration.

It should be remarked, however, that the matter effect has a resonant behaviour[9] that in particular conditions strongly enhances the oscillation probability.

Therefore, for other configurations of beam energy and of the distance between the accelerator and the detector this effect may be very important.

5. CONCLUSIONS

We have discussed a possible experiment dedicated to ν-oscillation using the intense beam of the European Hadron facility and an underground detector capable to identify electrons.

As shown in Fig.(3) an experiment of this kind will substantially improve the present limits on the oscillation parameters in a reasonable time of data taking.

- ACKNOWLEDGEMENT

We are indebted to U.Dore and D.Zanello for several clarifying discussions.

- REFERENCES

1) For a recent review of the ν-oscillation see: V.Flaminio and B.Saitta "Neutrino oscillation experiments" INFN PI/AE 85/6 to be published in "Rivista del Nuovo Cimento".
2) R.Davis et al.: Proceedings of the Conference on the Intersection between Particle and Nuclear Physics; Steamboat Springs, 1984.
3) A.Zichichi: "The Gran Sasso Project", GUD Workshop Roma 1981.

4) CERN, Harvard, Milano, Padova, Roma, Tokyo, Wisconsin Collaboration (ICARUS) INFN/AE 85/7.
5) Bergsma F. et al., P.L. B142 (1984) 103.
6) Zacek V. et al., Preprint SIN PR-85-11.
7) Katsanevas G., BEBC presented at XXI Rencontre de Moriond 1986.
8) L.Wolfestain, Phys.Rev. D, 17, 2369 (1978).
   L.Wolfestain, Phys.Rev. D, 20, 2634 (1979).
9) S.P.Mikheyev and A.Yu.Smirnov, submitted to Nuovo Cimento.

PRESENT KNOWLEDGE OF THE AXIAL-VECTOR WEAK INTERACTION COUPLING CONSTANT

A. BERTIN and A. VITALE

Dipartimento di Fisica dell'Università di Bologna, and
Istituto Nazionale di Fisica Nucleare, Sezione di Bologna, Italy

1. INTRODUCTION

One of the still open problems in the field of low-energy weak interactions is to determine the values of the different coupling constants involved in a precise and reliable way. The current attitude is to assume that a general agreement exists between the results of experiment and theory: such conclusion, which qualitatively is correct, was attained thanks to the systematic and refined experimental work carried out in more than two decades.

The aim of the present lecture is to show that, in the particular case of the axial-vector coupling constant, to state that there is an agreement between the results of experiment would be nowadays a too approximate assumption. It will also be underlined that a new generation of experiments is demanded both on neutron decay and on muon capture in hydrogen and deuterium; developments along this last direction will be presented in some detail in our second lecture[1].

To lay down some vocabulary, let us first remind that weak processes like the nuclear muon capture by a proton

$$\mu^- + p \rightarrow n + \nu_\mu \qquad (1)$$

or the neutron decay

$$n \rightarrow p + e^- + \bar{\nu}_e \qquad (2)$$

represent interaction involving quarks of the first generation (u and d). The structure of the weak charged current which describes them is defined by the V-A theory[2]. According to this theory, for instance, the matrix element (ME) for reaction (1) is of the type

$$ME = (G/\sqrt{2})\cos\theta_C \ (i \ \bar{u}(\nu) \ \gamma_\lambda (1+\gamma_5) u(\mu)) \langle n | I_\lambda(0) | p \rangle \qquad (3)$$

where $I_\lambda = V_\lambda - A_\lambda$, and

$$\langle n | V_\lambda(q^2) | p \rangle = i \langle \bar{u}(n) | \{g_V(q^2)\gamma_\lambda + (g_M(q^2)/(2M_p))\sigma_{\lambda\rho} q_\rho + \qquad (4)$$

$$+ i (g_S(q^2)/m_\mu) q_\lambda \} \tau_- | u(p) >. \qquad (4)$$

Here M_p and m_μ are the proton and muon mass, respectively; $\sigma_{\lambda\rho} = (\gamma_\lambda \gamma_\rho - \gamma_\rho \gamma_\lambda)/(2i)$ (and the γ's are the Pauli matrices), τ_- is the isospin operator ($\tau_- |p> = |n>$), and q_λ is the momentum transfer. Similarly,

$$<n|A_\lambda(q^2)|p> = i< \bar{u}(n) | \{g_A(q^2)\gamma_\lambda \gamma_5 + i(g_P(q^2)/m_\mu)q_\lambda \gamma_5 +$$
$$+ (g_T(q^2)/(2M_p))\sigma_{\lambda\rho} q_\rho \gamma_5 \} \tau_- | u(p) > . \qquad (5)$$

In Eqs. (4) and (5), g_V, g_M, g_S, g_A, g_P and g_T are six form factors, calculated at the specifical q^2 value, which are called the vector, weak magnetism, scalar, axial-vector, pseudoscalar and tensor coupling constant respectively, θ_C is the Cabibbo angle, and G is the Fermi coupling constant.

Within the framework of this theory, the decay of the neutron is determined, with good approximation, only by the coupling constants g_V and g_A. In the muon capture process, instead, other form factors play a significant role. Among these, the induced pseudoscalar term g_P can be preferentially determined by muon capture experiments, given its particular dependence on the momentum transfer. As far as regards the others, starting from the Conserved Vector Current (CVC) hypothesis[3], and using the experimental results on the proton and neutron electromagnetic form factors, one gets at the momentum transfer $q^2 = 0.88 m_\mu^2$ (corresponding to process (1))

$$g_V(0.88 m_\mu^2) = 0.975 \pm 0.001 \qquad (6)$$

$$g_M(0.88 m_\mu^2) = 3.588 \pm 0.152 \qquad (7)$$

$$g_S(0.88 m_\mu^2) = 0 \qquad (8)$$

Finally, it is generally assumed $g_T(q^2) = 0$, since no experimental evidence is available on the presence of tensor coupling term effects.

For the present, we shall concentrate our attention on the results one obtains from experiment on the axial-vector coupling constant g_A. The dependence on momentum transfer of such a form factor can be written as[4]

$$g_A(q^2) = g_A(0) \{ 1/[1-(q^2/m^2)]^2 \} \qquad (9)$$

where $m = (930 \pm 30)$ MeV[5]. One may note here, as a final comment to this Introduction, that the value of the axial-vector coupling constant (which is essential to the comprehension of the elementary weak processes) is also significant for the problem of the solar neutrino flux. Indeed, the fact that the flux of observed neutrinos in a certain energy range is significantly smaller than the predicted one could be explained by g_A values higher than the currently accepted ones.

2. METHODS TO DETERMINE g_A

In principle, the axial-vector coupling constant g_A can be determined independently by measuring different quantities, which will be shortly discussed in what follows.

(i) The free neutron decay rate Γ. Γ depends only on the axial and vector form factors $g_A(0)$ and $g_V(0)$ (see Eqs.(4) and (5)), due to the fact that the momentum transfer is vanishing in this process. The rate expression is related to the coupling constants through the relation

$$\Gamma = 0.47 G^2 \cos^2\theta_C (M_n - M_p)^5 (1/(60\pi^3)) [1 + 3 g_A^2(0)/g_V^2(0)] \quad (10)$$

where M_n is the neutron mass.

(ii) The correlation between the electron momentum and the neutron spin. For polarized neutrons, the angular distribution of the decay electrons is given by the relation

$$dW \sim 1 - \frac{2(g_A(0)/g_V(0)) [g_A(0)/g_V(0) - 1]}{1 + 3 g_A^2(0)/g_V^2(0)} (\bar{v}_e \cdot \bar{p}_n) \quad (11)$$

where $\bar{p}_n$ is the neutron polarization and $\bar{v}_e$ is the electron velocity.

(iii) The angular correlation between the electron and the neutrino. In its turn, this is given by

$$dW \sim 1 - \frac{g_A^2(0)/g_V^2(0) - 1}{1 + 3 g_A^2(0)/g_V^2(0)} (\bar{v}_e \cdot \bar{v}_\nu) \quad (12)$$

where $\bar{v}_\nu$ is the neutrino velocity.

(iv) The parameters of the hyperons semileptonic weak decays. Neglecting the mass of the leptons in the final state, the rate of these processes can be expressed as

$$(B \rightarrow B' \ell \nu) = G^2 \begin{Bmatrix} \cos^2\theta_C \\ \sin^2\theta_C \end{Bmatrix} \frac{(m - m')^5}{60\pi^3} (m'/m)^{3/2} \cdot \quad (13)$$

$$\cdot \left[g_V^2(0) + 3g_A^2(0) + \frac{(m-m')^2}{2mm'}(0.96g_V^2(0)+1.18g_A^2(0)) \right], \quad (13)$$

where m and m' are the masses of the barions B and B', respectively.

In these decays, the absolute value of the ratio g_A/g_V can in primciple de determined from the lepton(ℓ)-neutrino(ν)angular correlation. In the rest frame of reference of the decaying barion, in fact, the lepton angular distribution is of the type

$$W(\cos\theta_{\ell\nu}) = 1/2\,(1 + \alpha_{\ell\nu}\cos\theta_{\ell\nu}), \quad (14)$$

where $\theta_{\ell\nu}$ is the angle between the lepton and the neutrino,

$$\alpha_{\ell\nu} = \frac{g_V^2 - g_A^2}{g_V^2 + 3g_A^2} - 2\delta \quad (15)$$

and

$$\delta = (m-m')/(m+m'). \quad (16)$$

In Eqs. (14)-(16), $\alpha_{\ell\nu}$ is a correlation coefficient, while (-2δ) provides the recoil correction (at first order in the difference of the barion masses).

The magnitude and sign of g_A/g_V, in its turn, can be obtained (a) from the polarization distribution of the barion in the final state; (b) from the charged lepton spectrum for unpolarized initial leptons, or (c) from the asymmetry of the decay products for polarized hyperons.

Within the framework of the Cabibbo theory, however, all the barion decay rates can be expressed in terms of the three parameters θ_C, F and D. F and D respectively are the axial vector matrix elements which correspond to antisymmetric (F) and symmetric (D) couplings of two octets to form a third octet. θ_C is the Cabibbo angle which describes the weak mixing between strange and non-strange quarks. As it is shown in Table I, the ratio g_A/g_V depends only on F and D. If the results of experiments on the hyperons semi-leptonic weak decays are analysed in terms of the quoted SU(3) parameters, independent information can be obtained on the axial-vector coupling constant.

(v) **The muon nuclear capture rates by protons or deuterons.** With the notations developed so far, the rate of nuclear muon capture by a proton in either hyperfine structure state F of the muon-proton system can be written

$$\lambda_F(p) = R_\mu(p) \left(\frac{G \cos\theta_C M_p^2}{2\pi}\right)^2 \alpha^3 \left(\frac{m_\mu}{M_p}\right) \left(\frac{m_\mu}{m_\mu + M_p}\right)^3 \frac{M_n}{m_\mu + M_p}$$

$$\cdot \left[1 + \sqrt{1 + \left(\frac{E_\nu}{M_n}\right)^2}\right] \left(\frac{E_\nu}{m_\mu}\right)^2 \frac{m_\mu}{h} \Gamma_F , \quad (17)$$

where F=0 corresponds to the singlet and F=1 to the triplet total spin state of the muon-proton system, respectively. The hyperfine-structure factors Γ_F can be expressed by

$$\begin{cases} \Gamma_{F=0} = (C_V - 3C_A + C_P)^2, \\ \Gamma_{F=1} = (C_V + C_A)^2 - \frac{2}{3} C_P (C_V + C_A) + C_P^2, \end{cases} \quad (18)$$

while C_V, C_A and C_P are defined as follows:

$$\begin{cases} C_V(q^2) = g_V(q^2)\left(1 + \frac{E_\nu}{2M_n}\right) - g_M(q^2) \frac{E_\nu^2}{2M_p M_n [1+\sqrt{1+(E_\nu/M_n)^2}]}, \\ C_A(q^2) = g_A(q^2) - \frac{E_\nu}{2M_n} \left\{g_V(q^2) + g_M(q^2) \frac{1}{2}\left(1 + \frac{M_n}{M_p}\right)\right\}, \\ C_P(q^2) = \frac{E_\nu}{M_n [1+\sqrt{1+(E_\nu/M_n)^2}]} \\ \qquad \cdot \left\{g_P(q^2) - g_A(q^2) - \left[g_V(q^2) + g_M(q^2) \frac{1}{2}\left(1 + \frac{M_n}{M_p}\right)\right]\right\}, \end{cases} \quad (19)$$

while

$$\begin{cases} E_\nu = \frac{(m_\mu + M_p)^2 - M_n^2}{2(m_\mu + M_p)} = 0.938 m_\mu , \\ q^2 = m_\mu^2 \left(\frac{2E_\nu}{m_\mu}\right) - 1 = 0.877 m_\mu^2 . \end{cases} \quad (20)$$

In eq.(17), finally, $\alpha = 1/137$, and $R_\mu(p)$ takes into account the influence of the finite dimensions of the proton on the muon wave function. If we set the numerical values into eq. (17), this reduces to

$$\lambda_F = 14.88 (G \cos\theta_C)^2 \Gamma_F , \quad (21)$$

where $G \cos \theta_c$ is understood in units of 10^{-49} erg cm^3. The rate $\lambda_F(d)$ of the capture process of negative muons by deuterons

$$\mu^- + d \rightarrow n + n + \nu_\mu \qquad (22)$$

according to the hyperfine-structure state (namely the $F = \frac{3}{2}$, quartet, and $F = \frac{1}{2}$, doublet state) of the muon-deuteron system, can be written as

$$\lambda_F(d) = R_\mu(d) \frac{(G \cos \theta_c M_p^2)^2}{\pi(2\pi)^3} \alpha^3 \left(\frac{m_\mu}{M_p}\right)^2 \left(\frac{M_n}{M_p}\right)^2 \frac{m_\mu}{h} \left[\frac{1}{1+m_\mu/M}\right]^3 . \qquad (23)$$

$$\cdot \int \Gamma_F(d) dE_n = 44.16 R_\mu(d) (G \cos \theta_c)^2 \int \Gamma_F(d) dE_n ,$$

where M_d is the deuteron mass, E_n is the energy of one of the neutrons released in process (22). The hyperfine-structure factors Γ_F can in this case be expressed by

$$\begin{cases} \Gamma_{F=\frac{1}{2}}(d) = \{[(C_V' -2C_A')^2 + \frac{2}{3} C_P' (2C_V' - 4C_A' + C_P')] I_t + \\ \qquad + \frac{1}{3}(3C_A' - C_P')^2 I_s\} + \Gamma_{\frac{1}{2},vel} , \\ \Gamma_{F=\frac{3}{2}}(d) = \{[(C_V' + C_A')^2 - \frac{2}{3} C_P' (C_V' + C_A' - C_P')] I_t + \frac{1}{3} C_P'^2 I_s\} + \Gamma_{\frac{3}{2},vel} . \end{cases} \qquad (24)$$

The factor $R_\mu(d)$ accounts for the finite dimensions of the deuteron and for their effect on the muon wave function: it is very close to unity and is currently assumed to be equal to 1. $\Gamma_{\frac{1}{2},vel}$ and $\Gamma_{\frac{3}{2},vel}$ take into account the velocity terms of the muon-nucleon coupling. I_t and I_s are integrals over the neutrino momentum, defined by

$$I_{t,s} = \int |M_{t,s}|^2 p_\nu^2 dp_\nu , \qquad (25)$$

where the matrix elements $M_{t,s}$ is the nuclear matrix element

$$M_{t,s} = <f_{t,s}(r,k) | \exp[-ip_\nu r/2] | \Psi_d(r)> . \qquad (26)$$

In eq.(26), $\Psi_d(r)$ is the deuteron wave function and $f_{t,s}(r,k)$ is the wave function of the relative motion of the two neutrons (the indices t and s labelling their triplet and singlet spin state, respectively) with a relative momentum k.

The weak-interaction coupling constants, dropping second-class currents, enter the theoretical expression for the capture rate through the terms

$$\begin{cases} C'_V = g_V(q^2)\left[1 + \dfrac{E_\nu}{2M}\right], \\ C'_A = g_A(q^2) - [g_V(q^2) + g_M(q^2)]\dfrac{E_\nu}{2M}, \\ C'_P = \{[g_P(q^2) - g_A(q^2)] - [g_V(q^2)+g_M(q^2)]\}\dfrac{E_\nu}{2M}. \end{cases} \quad (27)$$

If one keeps into account Eqs.(6) to (8), as well as the assumption $g_T(q^2) = 0$, it is seen from Eqs.(17) to (27) that the results on the rates of reactions (1) and (22) can in principle be unfolded to get information both on g_P and on g_A. Nevertheless, given the reduced sensitivity of the available experimental rates to g_P[6], they will be considered in what follows as sources of information for g_A. The results which will be given are understood for the value of g_P provided by the Partially Conserved Axial Current hypothesis (PCAC)[7], which yields[4]

$$g_P(0.88 m_\mu^2) = -(8.1 \pm 0.1). \quad (28)$$

From the theoretical point of view, the value of $g_A(0)$ is provided by the Goldberger-Treiman relation[8]. As recently discussed by Dominguez[9], this gives

$$g_A(0) = (1+d)(f_\pi f_{\pi np}(0)/M_N) \quad (29)$$

where $f_\pi = (93.2 \pm 0.1)$ MeV is the pion decay constant, $f_{\pi np}(0)$ is the pion-nucleon coupling constant at zero momentum transfer, and $d = (2.5 \pm 0.5)\%$ is a corrective term provided by quantum chromodynamics calculations, which includes the effects of the approximate validity of chiral symmetry in strong interactions. M_N is the nucleon mass. Keeping in mind that

$$f_{\pi np}(q^2)/f_{\pi np}(0) = 1/(1 - q^2/\Lambda^2) \quad (30)$$

where[10] $\Lambda = (1150 \pm 350)$ MeV, and remembering that

$$\dfrac{f^2_{\pi np}(q^2 = m_\pi^2)}{4\pi} = (14.28 \pm 0.18), \quad (31)$$

one gets the value

$$(g_A(0))_{G.T.} = 1.28 \pm 0.02 \quad (32)$$

which we shall take as a reference theoretical prediction for the value of the axial-vector coupling constant.

3. RESULTS FROM EXPERIMENT

If now one looks at the results on g_A which the various experimental approaches have made available, they may be summarized as follows (see Table II):

(i) As far as regards the measurements of the neutron decay rate[11-16], these yielded results for g_A which disagree among them widely beyond the estimated errors. Underestimates of the systematic errors in some of the quoted experiments may well be present. Forthcoming measurements[17] of the free neutron decay rate by improved methods will be therefore extremely welcome.

(ii) The results on g_A obtained by correlation experiments[18], instead, are in good agreement among them, although they are not consistent with those obtained from the neutron decay rate measurements.

(iii) With regard to the results on g_A obtained from experiments on the weak semileptonic decays of hyperons, one may note that the SU(3) parameters can be fitted in several ways to the experimental results. The neutron decay rate, for instance, can be imposed as a constraint (which means to assume the value of the ratio g_A/g_V). For the present purposes, it is essential that the three parameters θ_C, F and D are fitted to the experimental results without any constraint. If this is done using all the data obtained in the CERN-SPS experiments[19], one gets

$$F = 0.453 \pm 0.011 \tag{33}$$

$$D = 0.729 \pm 0.017 \tag{34}$$

$$\sin\theta_C = 0.235 \pm 0.004 \quad . \tag{35}$$

This provides the value $g_A(0)/g_V(0) = F + D = 1.18 \pm 0.02$.

On can then see that there is no consistency among the results obtained on g_A starting from neutron or hyperon semileptonic decay experiments. If now one looks at the values one can extract from measurements of the muon capture rates by protons, the following points would be added:

(iv) The world-average result includes the values extracted from both gaseous and liquid hydrogen experiments[4]. The compatibility between the results from liquid and gaseous hydrogen experiments has significantly improved after the discovery of the ortho-para transition within the hydrogen mu-molecular ion $p\mu p^4$.

(v) The rate λ_{op} of the ortho-para transition was measured so far only in a pioneer experiment. If the value for this rate were the one subsequently foreseen by theoretical calculations, the value for $g_A(0)/g_V(0)$ one would estract from liquid hydrogen experiments would be 1.31 ± 0.04^4. Given the limited experimental accuracy available so far on λ_{op}, one would conclude that also the results of muon capture experiments in hydrogen provide for g_A values which are neither very precise nor well established.

(vi) An additional accent on this fairly undefined situation is set by the results on the muon capture rate in deuterium, which yield[20] the highest absolute values for g_A among those provided by experiment. In correspondence to these results, a discrepancy between experiment and theory on the rate of process (22) is present, at the level of two standard deviations. This discrepancy, which was first underlined in 1973, has recently beem confirmed[21] (*).

It is then clear that, besides the experimental field of weak semileptonic decays (with a special regard to the case of neutron decay), the processes of muon nuclear capture by protons and deuterons still demand to be explored globally from different and complementary standpoints. These ones will be discussed in the next lecture[1].

(*) A new result on the muon nuclear capture rate by deuterons, in the doublet spin state of the muon-deuteron system, has recently been communicated[22]. If this result is included, the ratio $g_A(0)/g_V(0)$ would change from 1.34 ± 0.04 to 1.32 ± 0.04, which does not affect the present conclusions.

TABLE I

FORM FACTORS OF BARIONS DECAYS IN TERMS OF
THE SU_3 PARAMETERS

Decay Mode $B \to B'$	$g_V(0)$	$g_A(0)$
$n \to p$	1	$F+D$
$\Sigma^{\mp} \to \Lambda$	0	$\sqrt{2/3}\, D$
$\Sigma^- \to \Sigma^0$	$\sqrt{2}$	$\sqrt{2}\, F$
$\Xi^- \to \Xi^0$	-1	$-F+D$
$\Lambda \to p$	$-\sqrt{3/2}$	$-1/\sqrt{6}(3F+D)$
$\Sigma^- \to n$	-1	$-F+D$
$\Sigma^0 \to p$	$-1/\sqrt{2}$	$1/\sqrt{2}(-F+D)$
$\Xi^- \to \Lambda$	$\sqrt{3/2}$	$1/\sqrt{6}(3F-D)$
$\Xi^0 \to \Sigma^+$	1	$F+D$
$\Xi^- \to \Sigma^0$	$1/\sqrt{2}$	$1/\sqrt{2}(F+D)$

TABLE II

Results on the ratio $g_A(0)/g_V(0)$ as obtained from different experiments.

Experiment	$-g_A(0)/g_V(0)$	Reference	Year
Neutron decay rate	1.17 ± 0.019	11	1959
	1.239 ± 0.011	12	1972
	1.279 ± 0.006	13	1978
	1.255 ± 0.055	14	1978
	1.230 ± 0.014	15	1980
	1.287 ± 0.077	16	1980
Angular correlations in neutron decay (typical value)	1.262 ± 0.005	18	1986
Hyperon decays	1.233 ± 0.016	19	1983
	1.18 ± 0.02 [a]		
Muon capture in hydrogen	1.24 ± 0.04 [b]	4	1984
	1.31 ± 0.04 [c]	4	1984
Muon capture in deuterium	1.34 ± 0.04	20	1986
Goldberger-Treiman relation	1.28 ± 0.02		

[a] Value obtained fitting the experimental results without imposing the neutron decay rate as a constraint.

[b][c] Value obtained assuming the experimental(theoretical) result on the ortho-para transition rate.

REFERENCES

1) A.Bertin and A.Vitale: Muon Capture in Hydrogen and Deuterium:Next Generation Experiments, lecture presented at this School.
2) E.Sudarshan and R.E. Marshak: Phys. Rev. 109,1860(1958).
3) R.P.Feynman and M.Gell-Mann: Phys.Rev.109,193(1958).
4) See e.g. A.Bertin and A.Vitale, Rivista del Nuovo Cimento 7,n.8(1984);and references therein.
5) N.J.Baker,A.M.Cnops,P.L.Connolly,S.A.Kahn,H.G.Kirk,M.J. Murtagh,R.B.Palmer, N.P.Samios, and M.Tanaka:Phys.Rev. D,23,2499(1981).
6) A.Vitale, A.Bertin and G.Carboni, Phys.Rev. D,11,2441(1975).
7) M.Gell-Maun and M.Lévy: Nuovo Cimento 16,705(1960).
8) M.L.Goldberger and S.B.Treiman: Phys.Rev.110,1178(1958).
9) C.A.Dominguez: Rivista del Nuovo Cimento 8,n.6(1985).
10) K.Holinde,Phys.Rep.68,3(1981).
11) A.N. Sosnovsky,P.E.Spivak,Y.A.Prokofiev,I.E.Kutikov, and Y.P. Dobrinin,Nucl. Phys.10,395(1959).
12) C.J.Christensen,A.Nielsen,A.Bahnsen,W.H.Brown, and B.M. Rustad:Phys.Rev. D5,1628(1972).
13) L.N. Bondarenko,V.V.Kurguzov,Yu.A.Prokof'ev,E.V.Rogov, and P.E.Spivak:Pis'ma Zh.Eksp.Teor.Fiz.28,328(1978);English translation: JETP Letters 28,303(1978).
14) See P.Liaud,K.Schreckenbach,J.Chauvin,P.de Saintignon and A.Bussiére:J.de Physigne 45,C3-37(1984); and references therein.
15) J.Byrne,J.Morse,K.F.Smith,F.Shaikh,K.Green and G.L.Greene:Phys. Letters 92B, 274(1980).
16) Yu.Yu.Kosvintsev,Yu.A.Kushnir,V.I.Morozov, and G.I.Terekhov,Pis'ma Zh.Eksp. Teor.Fiz.31,257(1980); English translation:JETP Lett.31,236(1980).(See ref.14).
17) A.Bussiére, private communication.
18) P.Bopp, D.Dubbers,L.Hornig,E.Klemt,J.Last,H.Schütze,S.J.Freedman,and O.Schär pf:Phys.Rev.Letters, to be published.
19) M.Bourquin,R.M.Brown,Y.Chatelus,J.C.Chollet,A.Degré,D.Froidevanx,A.R.Fyfe, J.M.Gaillard,C.N.P.Gee,W.M.Gibson,R.J.Gray,P.Igo-Kemeney,P.W.Jeffreys,B.Merkel,R.Morand,R.J.Ott,H.Plothow,J.P.Repellin,B.J.Saunders,G.Sauvage,B.Schiby, H.W.Siebert,V.J.Smith,K.T.Streit,R.Strub and J.J.Thresher,Z.Phys.C 21, 27 (1983)
20) A.Bertin,M.Capponi,S.De Castro,I.Massa,M.Piccinini,M.Poli,N.Semprini-Cesari, A.Vitale, and A.Zoccoli: Proc.of the Workshop on Fundamental Muon Physics: Atoms,Nuclei and Particles,LA-10714-C,184(1986).
21) G.Bardin,J.Duclos,J.Martino,A.Bertin,M.Capponi,M.Piccinini and A.Vitale: Nucl.Phys.A,453,591(1986).
22. M. Cargnelli,W.H.Breunlich,H.Fuhrmann,P.Kammel,J.Marton,J.Werner,J.Zmeskal,and C.Petitjean(unpublished).Value presented by E.Zavattini, in Proc. of the Workshop on Fundamental Muon Physics:Atoms,Nuclei and Particles, LA-10714 -C,182(1986).

MUON CAPTURE IN HYDROGEN AND DEUTERIUM: NEXT GENERATION EXPERIMENTS

A. BERTIN and A. VITALE

Dipartimento di Fisica dell'Università di Bologna, and
Istituto Nazionale di Fisica Nucleare, Sezione di Bologna, Italy

1. GENERAL REMARKS

The experiments on the nuclear muon capture by protons[1-7] and deuterons[8-11] (see Tables I and II) were carried out with targets at different temperature and density conditions. The measurements were performed both by detecting the neutrons[1-4,6,7-10] released in the reactions

$$\mu^- + p \rightarrow n + \nu_\mu \qquad (1)$$

$$\mu^- + d \rightarrow n + n + \nu_\mu \qquad (2)$$

and by the lifetime technique[5,11]. To understand clearly what was done and what remains to be done in this field, the following points should be retained:

i) The V-A theory of weak interactions[12] foresees for both reaction rate values which are strongly dependent on the spin state of the muon-nucleon system. In particular, in the higher hyperfine structure states (triplet for (μp) and quartet for (μd)) the rates of capture are greatly depressed with respect to those corresponding to the lower ones (singlet for (μp) and doublet for (μd))[13].

ii) Even in the most favourable cases, processes (1) and (2) occur with a probability of about 10^{-3} with respect to the dominant muon decay (which takes place at a rate of about $4 \times 10^5 \text{ s}^{-1}$)[14]. Therefore, the observation of these reactions in hydrogen and deuterium is difficult: refined techniques are to be developed in order to avoid that the measured effect be concealed by overwhelming backgrounds.

iii) The spin state of the system where the muon capture takes place depends on the life story of the muon after it has been stopped within the hydrogen or deuterium target. Such a story is made complex by the atomic and molecular processes where the muon is involved before decaying or undergoing muon capture[13,15]. These phenomena eventually constitute an independent chapter of low-energy muon physics. They are also connected to the problem of muon catalysis in the nuclear synthesis between hydrogen isotopes, which still represents a very appealing field for research on new energy sources[16].

iv) For the purpose of singling out the weak interaction coupling constants, measuring the rates of processes (1) and (2) in different spin states of the muon-nucleon systems is one of the main objectives of muon capture in hydrogen and deuterium. In fact, this would provide relations depending on significantly different combinations of the coupling constants themselves[13]. Despite the remarkable variety of the physical systems where reactions (1) and (2) were observed, however, most of the corresponding capture events were occurring in the lower (singlet or doublet, respectively) hyperfine structure states. This situation needs to be overcome: observing the capture processes (1) and (2) in the triplet and quartet spin state, respectively, remains an outstanding goal to be aimed at, after three decades of experimental research in this field.

In the following, we shall shortly review the main possibilities which are to be considered to add significant experimental information along this direction and to clarify the presently available information to the greatest extent.

2. EXPERIMENTS: PAST AND FUTURE

Some significant experimental lines which apparently are worth being undertaken are the following:

a) Muon capture in gaseous hydrogen. The measurements performed so far[6,7] were carried out by detecting the neutrons released in process (1). Therefore, the results obtained are dependent on the calculation of the efficiency of the neutron-detecting apparatus. This fact alone limits the accuracy of the experimental results at a maximum level of about 5%. The experimental conditions at which reaction (1) occurred in both the existing measurements are interpreted as corresponding to those of μp muonic atoms in the singlet spin state. This attribution is based on the experimental results obtained on the cross section for the elastic scattering process[15,17]

$$\mu p + p \to \mu p + p \quad . \qquad (3)$$

These results are in fair agreement with the values foreseen for the coherent scattering (3) of μp systems in the singlet spin state.

Two directions of progress can be foreseen along the line of gaseous hydrogen measurements. First, the measurement of the rate of reaction (1) should be performed by the lifetime technique. This would be achieved[5] by measuring the lifetime of negative muons stopped in gaseous hydrogen at a very high level of accuracy (10^{-5}), and by comparing it to the lifetime of positive muons. In this way, the systematical difficulties connected to the detection of neutrons would be avoided. Obviously, a pulsed muon beam of suitable characteristics (high intensity, low energy and low momentum dispersion) should be available. A beam which might be suitable for this purpose is presently being built at the Rutherford Appleton Laboratory (RAL)[18]. As a second point, we should like to underline that measurements in the low-pressure range would be of the greatest interest for the purpose of observing muon capture events coming from different admixtures of triplet and singlet states[13]. A systematic study of process (3) in the low-density direction would also represent a very significant-one should say also: necessary - test of the present interpretation of the available results on muon capture in gaseous hydrogen[19].

b) Muon capture in gaseous deuterium. Until now, only one result was published[9] on the rate of nuclear muon capture in gaseous deuterium. This was obtained by detecting one of the neutrons released in process (2). Such method requires to use gaseous deuterated hydrogen as a target, to avoid the neutrons coming from the fusion of the $d\mu d$ molecules which are formed in pure deuterium[13]. Also this result is dependent on the efficiency of the neutron detectors used, and on the neutron energy spectrum assumed to calculate the efficiency itself. Indirect measurements indicate that the experimental conditions were corresponding to μud systems mostly in the doublet spin state. Historically, this experiment was the first to show out a significant discrepancy with the theoretical predictions on the rate of muon capture by deuterons.

Also in gaseous deuterium, new measurements of the muon capture rate by the lifetime technique would be quite welcome. In particular, this method would yield the following advantages: (i) the systematic uncertainties due to the efficiency calculation for the neutron counters would be avoided; (ii) a pure deuterium gas target could be used. Here one would underline that the energy gap between the quartet and doublet states in the μud muonic atom (0.049 eV) is only slightly larger than the thermal energy (0.038 eV) at room temperature.

Therefore, by heating the gaseous deuterium target at a temperature of the order of 100 °C, one might achieve physical conditions favourable to the conservation of the initial statistical mixture of quartet and doublet states of the μud atoms formed[20]. For the reasons discussed previously, a measurement in these experimental conditions would indeed provide a very significant result.

c) **Muon capture in liquid hydrogen**. In liquid targets, the capture reaction (1) occurs predominantly within pμp molecular ions, which are formed in the ortho-state (L=1). The latter is separated by 148 eV from the lower-energy para-state (L=0). The relative spin orientation in the ortho-state is such that process (1) occurs in the singlet state for 3/4, and in the triplet state for 1/4 of the cases. The opposite is true for the para-state. On the basis of theoretical considerations, the transition rate from the ortho - to the para-state was assumed in the past to be negligible[13]. In 1981, however, experimental evidence was obtained for the existence of a long-term ortho-para transition occurring at a rate $\lambda_{op}=(4.1\pm1.4)\times10^4$ s^{-1} [21]. The result was obtained by looking at the time distribution (dn/dt) of the delayed capture neutrons released from the pμp atoms formed in a ultrapure liquid hydrogen target. The result, which is significant for the interpretation of all the muon capture experiments previously performed in liquid hydrogen[21,22], is affected by a fairly large experimental error.

A more accurate determination of λ_{op} could be achieved by observing the time distribution dn/dt with large statistics and over a time interval more extended than the one which was allowed in the pioneer experiment[21]. A muon beam like the one which is being developed at RAL[18] would be suitable for this purpose. A new measurement of the ortho-para transition rate is then considered as the first step to be achieved in the field of muon capture in liquid hydrogen.

d) **Muon capture in liquid deuterium**. The first measurement of the rate of reaction (2) in a liquid target was carried out at Columbia[8] using a deuterated hydrogen target and detecting one of the neutrons released in the capture reaction. Recently, the measurement has been carried out in ultrapure liquid deuterium by the lifetime method[11]. This measurement, performed by a Saclay-Bologna collaboration, is accurate, independent on the efficiency of neutron counting and on molecular effects, and can unambiguously be referred to the doublet spin state of the μud muonic atom. From the point of view of interpretation, it can be considered as the most accurate result obtained in the field of muon nuclear capture by protons and deuterons.

Future research in this field, on the experimental standpoint, includes detecting in coincidence the two neutrons released in process (2), and measuring their energy and angular correlation. In this way, the capture reaction would be observed in such a way to obtain information both on the different coupling constants involved in the interaction, and on some nuclear parameters of the highest interest[13]. Indeed, a complete set of results would thereby be available, exploiting the fact that the deuteron represents the transition target between the elementary reaction (1) and the more complex field of muon capture by nuclei.

TABLE I - Summary of the experimental results on the rate of muon nuclear capture by protons.

Authors	Hydrogen target	Technique[a]	Capture rate (s^{-1})	Reference
Hildebrand, et al.	liquid	bubble chamber	428± 85	1
Bertolini, et al.	liquid	bubble chamber	450± 50	2
Bleser, et al.	liquid	counters	515± 85	3
Rothberg, at al.	liquid	counters	464± 42	4
Bardin, et al.	liquid	counters, lifetime	460± 20	5
Alberigi Quaranta, et al.	gas, 8 atm.	counters	651± 57	6
Bystritskii, et al.	gas, 41 atm.	counters	686± 88	7

[a] In all experiments except the one by Bardin et al. the capture events were detected by looking at the released neutrons.

TABLE II

Experimental results on the rate of nuclear muon capture by deuterons in the doublet spin state of the muon-deuteron system (λ_c^d)

Experiment	Detected particles	Capture rate (s^{-1})	Reference
Columbia	neutrons	365±96	8
Bologna-CERN	neutrons	553±80 (^{a})	9
Vienna-SIN	neutrons	409±40	10
Saclay-Bologna	electrons	470±29	11

(^{a}) Corrected value, referring to λ_d^c. The direct experimental result was λ_{exp} = (445±60)s^{-1}.

REFERENCES

1) R.Hildebrand: Phys.Rev.Lett., 8,34(1962); R.Hildebrand and J.H. Doede: in Proceedings of the International Conference on High Energy Physics, Geneva (CERN, Geneva, 1962),p.418; J.H.Doede and R.Hildebrand, quoted by C.Rubbia: in Proceedings of the International Conference on Fundamental Aspects of Weak Interactions (Brookhaven, 1963),p.277(unpublished).

2) E.Bertolini,A.Citron,G.Gialanella,S.Focardi,A.Mukhin,C.Rubbia and F.Saporetti: in Proceedings of the International Conference on High-Energy Physics, Geneva (CERN,Geneva,1962),p.421;C.Rubbia: in Proceedings of the International Conference on Fundamental Aspects of Weak Interactions (Brookhaven,1963), p. 278.

3). E.J.Bleser,L.M.Lederman,J.L.Rosen,J.E.Rothberg and E.Zavattini:Phys.Rev.Lett. 8,288(1962).

4) J.E.Rothberg,E.W.Anderson,E.J.Bleser,L.M.Lederman,S.L.Meyer,J.L.Rosen and I.T.Wang:Phys.Rev.,132,2664(1963).

5) G.Bardin,J.Duclos,A.Magnon,J.Martino,A.Richter,E.Zavattini,A.Bertin,M.Piccinini,A.Vitale and D.Measday:Nucl.Phys.A,352,365(1981).

6) A.Alberigi Quaranta,A.Bertin,G.Matone,F.Palmonari,G.Torelli,P.Dalpiaz,A.Placci and E.Zavattini: Phys.Rev.177,Pt.I,2118(1969).

7) V.M.Bystritskii,V.P.Dzhelepov,P.F.Ermolov,K.O.Oganesyan,M.N.Omel'Yanenko, S.Yu.Porokhovoi,V.S.Roganov,A.Rudenko and V.V. Fil'chenkov: Z.Eskp.Teor.Fiz., 66,43(1974)(English translation: Sov.Phys. JETP, 39,19 (1974).

8) I.T.Wang,E.W.Anderson,E.J.Bleser,L.M.Lederman,S.L.Meyer, J.L.Rosen and J.E. Rothberg:Phys.Rev.,139,B1528(1965).

9) A.Placci,E.Zavattini,A.Bertin and A.Vitale:Phys.Rev.Lett.,25,475(1970); A.Bertin,A.Vitale,A.PLacci and E.Zavattini:Phys.Rev. D, 8,3774(1973).

10) M.Cargnelli,W.H.Breunlich,H.Fuhrmann,P.Kammel,J.Marton,J.Werner,J.Zmeskal, and C.Petitjean(unpublished): value presented by E.Zavattini, in Proc.of the Workshop on Fundamental Muon Physics: Atoms,Nuclei and Particles, LA-10714-C,182 (1986).

11) G.Bardin,J.Duclos,J.Martino,A.Bertin,M.Capponi,M.Piccinini and A.Vitale, Nucl.Phys.A,453,591(1986).

12) E.Sudarshan and R.E.Marshak: Phys.Rev.109,1860(1958).

13) See for instance: A.Bertin and A.Vitale, Riv.Nuovo Cimento 7,n.8(1984);and references therein.

14) See for instance: A.Bertin and A.Vitale,Riv.Nuovo Cimento 7,n.7(1984); and references therein.

15) See for instance: A.Bertin,A.Vitale and A.Placci,Riv.Nuovo Cimento 5,n.3, 423(1975); and references therein.

16) See e.g. S.Jones: Nature 321,127(1986); and references therein.

17) A.Bertin,M. Capponi,I.Massa,M.Piccinini,G.Vannini,M.Poli and A.Vitale:Nuovo Cimento A,72,225(1982).

18) G.H.Eaton, A.Carne,D.H.Reading, and E.G.Sandels: Nucl.Instr.and Meth. 214, 151(1983).

19) A.Bertin,M.Capponi,S.De Castro,I.Massa, M.Piccinini,M.Poli,N.Semprini-Cesari,A.Vitale and A.Zoccoli: Proc. of the Workshop on Fundamental Muon Physics: Atoms, Nuclei and Particles, LA-10714-C, 191(1986).

20) M.Piccinini, in Fundamental Interactions in Low-Energy Systems, edited by P.Dalpiaz,G.Fiorentini and G.Torelli, Plenum, 71(1985).

21) G.Bardin,J.Duclos,A.Magnon,J.Martino,A.Richter,E.Zavattini,A.Bertin,M.Piccinini and A.Vitale: Phys. Lett. B,104,320(1981).

22) See e.g. A.Bertin and A.Vitale, Present knowledge of the Axial-Vector Weak Interaction Coupling Constant, lecture presented at this School; and references therein.

PHYSICS WITH JET TARGETS AT THE SPS $\bar{p}p$ COLLIDER

UA-6 Collaboration
L. DICK, W. KUBISCHTA
CERN, Geneva, Switzerland

1. INTRODUCTION.

The simplest method to use the beam of a high energy accelerator is to send it on an internal target made of materials like copper, tungsten or others, chosen mainly for their mechanical or heat conduction properties. With the improvement of beam extraction techniques, allowing a wider choice in target material and improved secondary beam designs, internal targets became less attractive and were if at all only used for test and measurement purposes.

Recently, however, internal targets are making a come-back in the form of gas jet targets. By "gas jet targets" we understand a well localized region of sufficiently high gas density crossed by the accelerated or stored beam and designed such as to keep the perturbation to the accelerator vacuum to the strict minimum.

The features that make jet targets attractive for fixed-target experiments on a particles storage ring are :
 i) targets are small and well-defined
 ii) low density, which permits acurate detection of low- energy particles and electromagnetic final states;
 iii) parasitic operation
 iv) heavier gases can be used e.g. N_2, O_2, A, Xe.
 v) very low background can be achieved in storage or collider mode of operation.

Additional advantages of polarized jet targets are :

 i) clean target (pure hydrogen or deuterium)
 ii) high polarization ~ 95%
 iii) low systematic errors (typical spin reversal frequency ~ 1 kHz);
 iv) small instrument asymmetry (weak magnetic field on the target, ~ 20 G to maintain the polarization);
 v) flexibility : polarization can be easily oriented in any direction with respect to the beam.

2. LUMINOSITY.

The luminosity of a target beam crossed by an accelerated or stored beam is expressed by

$$L = N\, n\, d\, G\, f_{rev}\ cm^{-2}\ s^{-1}$$

where
N = number of circulating particles
n = number of nucleons per cm^3
d = target dimension in beam direction in cm
G = geometrical factor (beam/target overlap) ~ 1
f_{rev} = revolution frequency of the circulating beam in s^{-1}.

It is this last factor, usually a few $10^4\ s^{-1}$ which makes the use of gas jets as internal targets attractive.

The optimal luminosity is a function of the experiment to be performed. In the case of UA-6, for instance, the stored beam consists of 6 bunches of 4 ns length each. Since in this case the event rate has to be limited to not more than one event per bunch, the maximum luminosity L_{max} has to obey the relation

$$L_{max} < N_b\, f_{ref}/\sigma_{tot}$$

N_b being the number of bunches. With σ_{tot} = 4 x 10^{-26} cm² for 315 GeV/c p or p̄, and f_{rev} = 4 x $10^4\ s^{-1}$ we get

$$L_{max} < 6\ \text{x}\ 10^{30}\ cm^{-2}\ s^{-1}.$$

A typical number of particles per bunch being 2.5 x 10^{11} protons, the maximum target density in this case is

$$n_{max} = 10^{14}\ cm^{-2}.$$

3. BASIC DESIGNS.

Gas targets were used in nuclear physics experiments in the form of gas cells with thin windows[1], differentially pumped systems[2], and supersonic beams[3,4]. The high gas load caused by these targets required very small apertures and a multistage differential pumping system in the beam line. A first gas target to be used on a high energy accelerator was the target at Fermilab[5] producing rather localized gas densities of about 10^{17} atoms/cm³. It was later replaced by a supersonic nozzle beam with similar performances[6]. Both were pulsed and produced pressure bumps of up to 10^{-3} mbar in the vicinity of the target. Targets compatible with the vacuum of a strage ring were finally built using the technique of cluster beams[7,8], with densities up to several

10^{14} atoms/cm³ and background pressures of about 10^{-7} mbar. Targets of this type have been used at several accelerators and storage rings (SATURNE[9], CERN SPS[10], CERN ISR[11]) and others are being built or proposed (CERN LEAR[12], CESAR[13]). Up to now, only hydrogen has been used as target gas, but the method is applicable to any other gas too.

A rather different approach is required for polarized atomic beam targets. Historically, polarized atomic beams were first developed as first stages of polarized ion sources[14]. Later it was realized, that due to the multiple traversal factor, these beams could well be used as internal targets in circular accelerators or storage rings[15]. The development of techniques to cool atomic hydrogen to very low temperature[16] may lead to substantial improvements in atomic beam densities, useful as well for polarized ion sources as for polarized jet targets.

4. THE CLUSTER BEAM TARGET.

The principle of the cluster beam production is shown in Fig. 1 (from ref. 7). Gas expands through a nozzle (diameter 0.01 to 0.2mm) into vacuum. The temperature of the nozzle is chosen such that, in the p-T diagram, the vapour pressure is crossed. Further expansion leads to a subcooled state, and finally the gas condenses partially, forming small droplets of 10^5 to 10^7 molecules. These clusters have rather uniform velocity and are not scattered by background gas. A few cm from the nozzle, a skimmer removes the uncondensed gas and defines the cluster beam profile. One or two more differential pumping stages are required to reduce the pressure in the accelerator to an acceptable level. After crossing the accelerator beam, the cluster beam is captured in a beam dump.

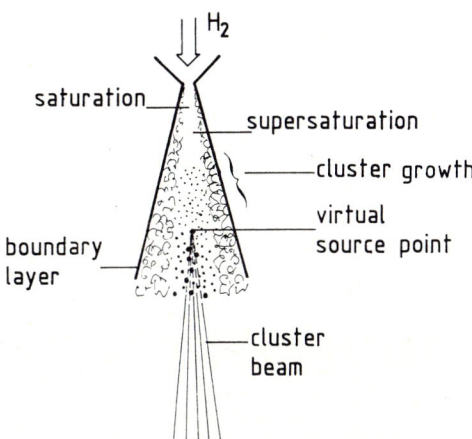

1. Principle of cluster beam source.

The performance of the system is shown in Fig. 2. The flux density was measured with a compression gauge. The cluster velocity was not measured, it is probably between 600 and 700 m/s. The target thickness was determined independently by low-t elastic scattering. The slight curvature of the target thickness curve can be explained by the increase of the cluster velocity with increasing nozzle gas flow[8].

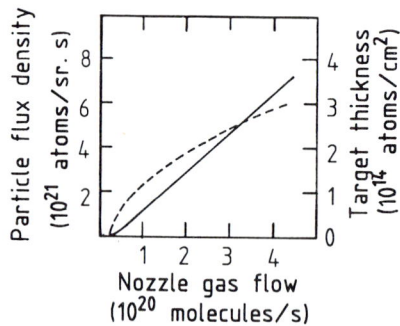

2. Particle flux density and target thickness (dotted curve) as a function of nozzle gas flow.

Figure 3 shows a cross section of the UA-6 cluster target. The nozzle (throat diameter 0.1 mm, length 30 mm, trumpet shaped) is cooled to about 28 K by a closed cycle Helium refrigerator. The skimmer has an oval opening of 1.6 x 0.7 mm, resulting in a cluster beam profile at the crossing point of 8 x 3 mm, adapted to the dimension of the proton beam.

Nozzle gas throughput is rather high, therefore an elaborate pumping system is required. Table 1 shows the main caracteristics of the UA-6 target vacuum system.

Pumping stage	type of pumps	pumping speed (nominal, 1/s)	typical pressure (mbar)
1	turbomolecular	14000	1.10^{-2}
2	turbomolecular	2000	1.10^{-4}
3	turbomolecular	2000	1.10^{-6}
4 (SPS)	ionization	1600	6.10^{-7}
5 (dump)	refrigerator cryopump	-	-

Table 1 : Main components of the UA-6 target vacuum system.

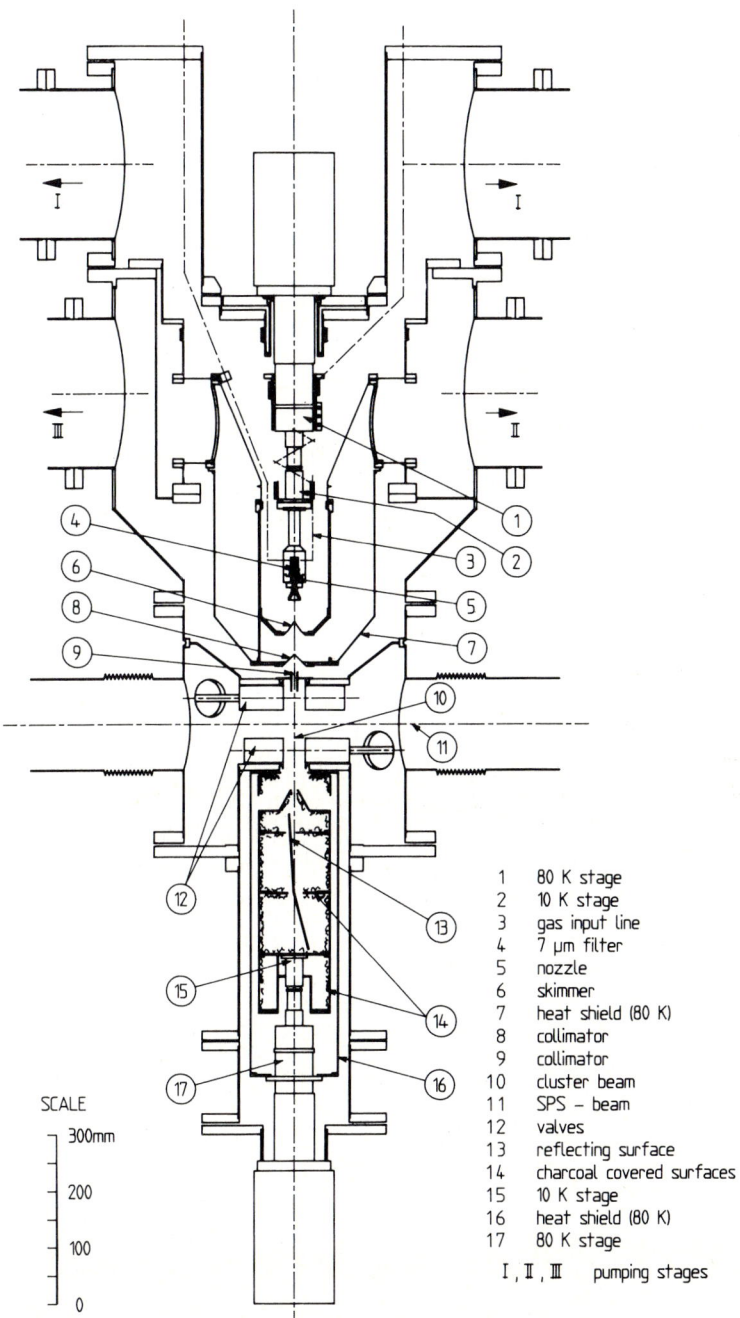

3. Cross section of the CERN UA-6 cluster beam target.

5. THE POLARIZED ATOMIC BEAM TARGET.

Polarized atomic beams are produced by a method similar to the Stern-Gerlach experiment. The principle can most easily be understood with the aid of the energy level diagram of the hydrogen atom ground state (Fig. 4). In an inhomogenous field, atoms in states 1 and 2 move in the direction of decreasing field, while atoms in states 3 and 4 go in the direction of increasing field.

The force on an atom with magnetic moment μ_{eff} in an inhomogenous field B is

$$F = -\text{grad } W = \mu_{eff} \text{ grad}|B|$$

with W being the potential energy of the atom in the field B.

Usually, quadrupoles or sextupoles are used as focusing elements. Trajectories in a constant aperture quadrupole have parabolic shape, while the trajectories in a sextupole are sine (states 1 and 2) or hyperbolic sine shaped (states 3 and 4).

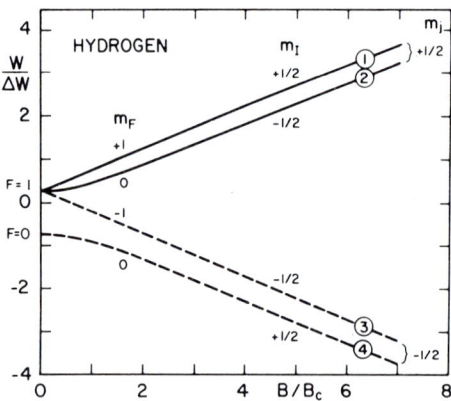

4. Energy level diagram of the hydrogen atom ground state.

A schematic view of a polarized atomic beam target is shown in Fig. 5[18].

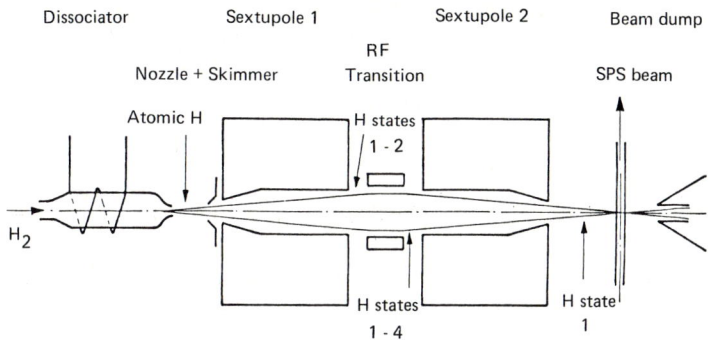

5. Schematic view of a polarized atomic hydrogen beam target.

Atoms are produced in an electrodeless high frequency discharge. The atoms are shaped into a beam by a nozzle followed by a skimmer. A first magnet selects states 1 and 2, states 3 and 4 are eliminated and are pumped away. The nuclear polarization of the beam of states 1 and 2, at high fields, is equal to 0, while it approaches 50% at low fields. In order to increase the nuclear polarization, radio frequency transitions are being used, in our case a transition 2 to 4. The second magnet again focuses atoms in state 1 and eliminates those in state 4. The beam after this magnet is almost pure state 1, and has a polarization of about 95% at any magnetic field level. After crossing the high energy beam, the atoms are captured in a dump, as in the case of the molecular beam target.

A polarized hydrogen atomic beam at room temperature with a mean velocity ~ 2700m/sec is expected to achieve a desity ~ 5.10^{11} atoms/cm³. By cooling the atomic beam at the nozzle level (~ 24° K) we obtain a mean velocity of the order of 1000 m/sec. Due to a supersonic effect observed at this temperature, Mach number ~ 3, the width of the Maxwellian distribution is considerably reduced[19]. This kind of cooled atomic beam will improve the hydrogen density in two ways :
    1) The lower mean velocity gives the possibility to increase the acceptance of the following elements.
    2) Due to the narrower velocity distribution we will obtain a sharper image at the target focussing point (chromatic abberation).

With these two combined effects we expect to achieve a density of up to ~ 5.10^{12} atoms/cm³. Further improvements of the low pressure gas dynamics at the injection level with a more efficient dissociator nozzle cooling and a better pumping speed may give densities up to 10^{13} atoms/cm³.

In the future, the final goal is to achieve a density of 10^{14} atoms/cm³ which is more or less the maximum density usefull for a high energy $\bar{p}p$ collider with the one interaction per bunch condition. A theoretical limit of 10^{14} atoms/cm³ for a polarized atomic beam is also given by the atomic mean free path in the beam[20].

6. UA-6 PHYSICS PROGRAM AT THE CERN $\bar{p}p$ COLLIDER.

The UA-6 experiment is a collaboration of CERN, the Universities of Lausanne, Michigan and the Rockefeller University, which has developed and installed a jet target facility at the SPS $\bar{p}p$ collider. The set-up of the experiment is given in Fig.6 and is located in the SPS tunnel at the medium straight section (MSS)514 in between UA-1 and UA-2 experiments. The target profile obtained by moving the hydrogen jet across the collider beam gives the lateral dimension of the target. The interaction rate is measured by a triple counter telescope looking at the target region (width 3 mm, length 8 mm (jet longitudinal size) and 1 mm hight (vertical circulating beam size)). It looks like a point target compared to the conventional liquid hydrogen target. For any interaction point inside the jet target we have a constant geometrical acceptance and this gives the possibility to build a rather compact experimental set up (total length about 10 metres).

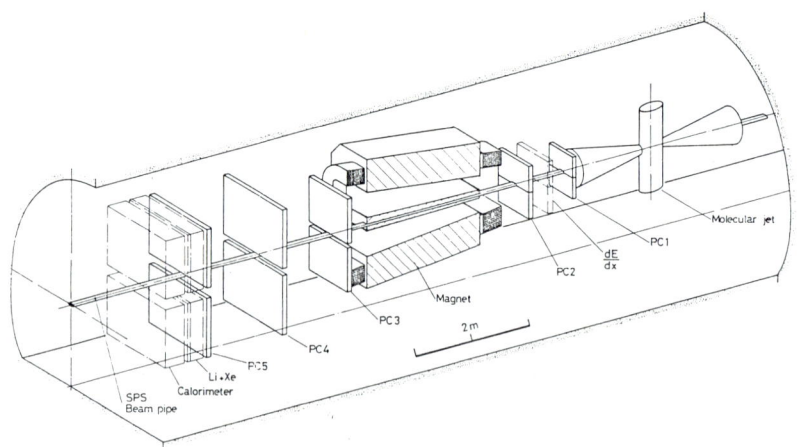

6. Schematic view of the UA-6 experimental layout.

The experimental layout is shown schematically in Fig. 6. The jet target is symmetric with respect to p and $\bar{p}$ direction. The components of the two arm spectrometer can be mounted on either side of the jet target in order to study both pp and $\bar{p}p$ interactions. Each arm covers a polar angle of 20 to 100 mrad and 75° in azimuthal angle. The geometrical acceptance is 1.8 sterad. The different spectrometer components are the following :

i) A two metre dipole magnet of 2.2 Tm bending power with the pole pieces pointed to the target in order to make the direction of the magnetic field perpendicular to the particle trajectories emerging from the target.

ii) Two sets of MWPC : two chambers before the magnet and three chambers after the magnet.

iii) An ionization chamber (dE/dx) to distinguish single gammas and electron pairs (Dalitz pairs).

iv) A transition radiation detector (TRD)[21] with 800 lithium radiator foils associated with a proportional chamber (Xe-He-CO_2).

v) A lead/proportional tube gas sampling calorimeter of 24 radiation lengths.

vi) A set of solid state counters located at 90° in a moving recoil arm.

The instantaneous luminosity is monitored by silicon detectors at 90° measuring the low-t elastic scattering.

With a jet target density of 4.10^{14} atoms/cm³ we obtain an instantaneous luminosity of 8.10^{29} cm²/sec for a 4.10^{10} circulating $\bar{p}$ beam.

One of the main part of the detector is the lead/proportional tubes sampling calorimeter[22]. Typical π^0 and η mass distributions are shown in Fig. 7, obtained by all two-cluster combinations with total $p_t > 2.5$. The quality (low background) of this spectrum is mainly due to the point like target[23].

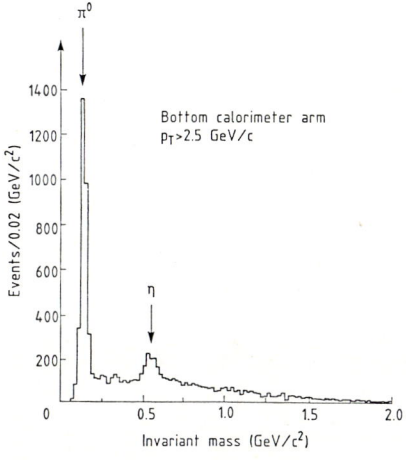

7. Two-cluster mass distribution.

A clear signal of $\bar{p}p$ direct photon production was measured and we have also identified e^+e^- pairs in the decay of J/ψ.

The physics programme of UA-6 with the hydrogen cluster jet is mainly to compare $p\bar{p}$ and pp interactions with a high luminosity and good efficiency for $\bar{p}$.

i) Large-p_T π^0, η nd single γ production (hard scattering). The pp and $\bar{p}p$ comparison would clarify the contribution from the different processes.

ii) The e^+e^- pairs continuum and J/ψ production. In the $\bar{p}p$ annihilation, q and $\bar{q}$ are valence quarks and it is expected to be dominant in the e^+e^- pair production.

iii) Polarization of Λ^0 and $\bar{\Lambda}^0$.

In pp collisions, a large polarization effect has been found and no centre-of-mass energy dependence has been observed from 6 GeV to 63 GeV. In a review article K.Heller has pointed out the main features of this effect[24]. This process does not require any polarisation in the initial state since the Λ^0's decay plane in respect to the production plane reveals the polarisation.

The final interest of UA-6 is to study spin effects in these reactions taking advantage of the ACOL programme, which will increase the $\bar{p}$ intensity stored in the SPS $\bar{p}p$ collider by a factor of 10. For this period we expect to set up an improved version of the hydrogen polarized jet we built in 1980[18] with a density of $\sim 5.10^{12}$ atoms/cm². The corresponding luminosity will be $\sim 2.10^{29}$ for $\bar{p}$ and more than 10^{30} for p with 95% polarisation.

The main intriguing effect is that quark and gluon spin does not wash out even in the multiparticle states and seems to be preserved in the apparent chaos by the well defined collective structure. This is well demonstrated by the large polarization effect in the hyperons productions independent of $\sqrt{s}$. Having the direction of the spin under control in the initial state, the measurement of the spin correlation between the initial and final state will be of great importance for the study of the large spin memories in hadronic collisions. With the polarization of hyperons or mesons in the final state we can measure the strength of the spin coupling in the production mechanism.

In this framework it is conceivable that spin effect is related to the properties of the colour field which is responsible for the confinement of quarks and gluons in hadronic matter. This kind of effect will be very constraining for the theory. Quantitative predictions are up to now, rather poor for those reactions and our need is for more further theoretical work.

REFERENCES

1. W.J. Wallace et al., Nucl. Instr. Methods **68**, (1969) 337.

2. C.M. Jones et al., Nucl. Inst. Methods **68**, (1969) 77.

3. J. Ulbricht, G. Clausnitzer, G. Graw, Nucl. Inst. Methods **102**, (1972) 96.

4. W. Tietsch, Thesis, University of Frankfurt, 1976.

5. V. Bartenev et al., Adv. Cryog. Eng. **18**, (1973) 460.

6. P. Mantsch, F. Turkot, Fermilab Report TM-582 (1975).

7. E.W. Becker, K. Bier, W. Henkes, Zeitschr. f. Physik **146** (1956) 333.

8. E.W. Becker, R. Klingelhofer, P. Lohse, Z. Naturforschung **17a** (1962) 432.
   J. Gspann, K. Korting, J. Chemical Physics **59** (1973) 4726.

9. R. Burgel et al., Nucl. Inst. Methods **204** (1982) 53.

10. CERN-Lausanne-Lund-Michigan-Rockefeller Collaboration CERN/SPS/78-23 (14.02.1978).

11. C. Baglin et al., Proc. 12th Int. Conf. on High Energy Accelerators, Fermilab, Batavia, 1983 (Fermilab, Batavia, 1984) p. 251.

12. P. Lefevre, D. Mohl, G. Plass, Proc. 11th Int. Conf. on High Energy Accelerators, Geneva, 1980 (Birkhauser, Basel 1980).

13. C. Ekstrom, B. Holmquist, H. Sterner, IUCF Workshop on Nuclear Physics with Stored Cooled Beams, Bloomington, Indiana, USA, Oct. 15-17, 1984.

14. W. Haeberli, Ann. Rev. Nucl. Sci. **17**, (1967) 373.
    W. Haeberli, Proc. Int. Workshop on Polarized Sources and Targets, Montana, 1986, Helvetica Physica acta **59** (1986) 513.

15. CERN-Lund-Michigan-Rockefeller Collaboration CERN SPSC/77-77 26 August 1977.

16. I.F. Silvera, Proc. Int. Conf. Low Temp. Physics, Physica 109 and 110B (19820 1499. T.J. Greytak, D. Kleppner, Les Houches, Session XXXVIII, 1982; G. Grynberg and R. Stora, eds. Elsevier Science Publishers, 1984.

17. W. Obert, Proc. 6th Cryogenic Engineering Conf., K. Mendelssohn, ed., Grenoble 1976 (IPC Sci. Techn. Press, England) p.219.

18. L. Dick, J.-B. Jeanneret, W. Kubischta, J. Antille, Proc. Int. Symp. on High Energy Physics with Polarized Beams and Targets, 1980, eds. C. Joseph and J. Soffer (Experiencia 38, 1981) 212.

19. J.-C. Berney, L. Dick, W. Kubischta, in Proc. Int. Workshop on Polarized Sources and Targets, held at SIN-Montana, CH-1986 HPA 59, (1986) p. 578-581.

20. T.O. Niinikoski, A. Hershcovitch, in Proc. Int. Workshop on Polarized Sources and Targets, held at SIN-Montana, CH-1986 HPA 59, (1986) p. 596.

21. A. Vacchi, in Proc. of the Wire Chamber Conf., Vienna 1986.

22. G.R. Snow, in Proc. of the Gas Sampling Calorimeter Workshop, Fermilab 1985.

23. CERN-Lausanne-Michigan-Rockefeller UA-6 Collaboration, paper presented at the Int. Conference on High Energy Physics, Berkeley 16-23 July 1986. To be published.

24. K. Heller, Proc. 6th Int. Symp. on High Energy Spin Physics, Marseille, 1985, ed. J. Soffer (J. Phys. France) 46 (1985) C2-121.

III

NUCLEAR PHYSICS WITH HADRONS AT INTERMEDIATE ENERGY

PERSPECTIVES OF THE NUCLEAR PHYSICS: THE ROLE OF THE HADRONIC PROBES

Tullio BRESSANI

Istituto di Fisica Superiore dell'Università di Torino, Torino, Italy
Istituto Nazionale di Fisica Nucleare, Sezione di Torino, Italy

1. INTRODUCTION

In this lecture I will follow the lines of previous talks that I have given recently [1]. The Intermediate Energy Nuclear Physics with Hadrons, born as an independent sector of the Physics about twenty years ago, is still today a sort of interface between the Particle and the Nuclear Physics. It is a discipline which grows considerably and has very brilliant perspectives; on the other hand it is not so easy to find a clear and simple definition for it.

Traditionally with "Intermediate Energy" we mean energies up to some GeV (but the limit has the tendency of increasing in the years!), with "Hadrons" obviously nucleons, antinucleons, pions and kaons, but also light ions (up to ^{4}He). Also the word "Nuclear", which seems to be the most straightforward one, is today sometimes used to indicate elementary interactions at low energies. This is partially due to the modern description of the interaction between "elementary" particles in terms of the interaction between the constituent substructures, the quarks. Following this point of vue, the interaction between two "elementary" protons and that between two ^{4}He nuclei would not be described in a very different way, the main distinction beeing the number of the constituent and interacting quarks. The difference would be then at the same level of the old, but still yet used, distinction between "few nucleon" and "many nucleon" systems of the traditional Nuclear Physics. Maybe a most concise and elegant definition of the discipline like "Physics of the Confinement" [2] will be accepted in the future.

In this lecture indeed I will limit my considerations only to the true "nuclear" (or many-quark) systems. A modern classification of the properties, interactions and models of the Nuclear Physics is that one which takes into account the degrees of freedom necessary in the description, and more precisely:

a) traditional Nuclear Physics: the elementary constituents are only the nucleons
b) mesonic degrees of freedom: nucleons and mesons have to be considered as elementary constituents
c) color degrees of freedom: quarks and gluons are the elementary consti-

tuents.

Clearly each of the descriptions b) and c) contains completely the phenomena observed in the preceding one but explains new aspects. I will devote only a few words on the points a) and b), whereas I will pay more attention to c), which may be considered as the frontier of the modern Nuclear Physics.

2. TRADITIONAL NUCLEAR PHYSICS

In the last few years, following the setting-up and the operation of big experimental facilities at the "meson factories" (SIN, TRIUMF, LAMPF) and the parallel progress in the theoretical interpretation, a huge amount of clean data was produced and carefully analyzed. They furnished informations as reliable and precise than those up to now obtained with the "clean" electromagnetic probes on:

1) nuclear matter distibutions and transition densities;
2) spectroscopy and structure of excited nuclear states;
3) internal momentum distributions of nucleon and clusters.

Seth has given excellent reviews of the field [3]. I will present only one example of the quality and uniqueness of the data that may be obtained (concerning the spectroscopy of excited nuclear states). It illustrates the potentiality of the pion double charge exchange reaction is selecting and identifying double analog states ($\Delta T = T(analog) - T(ground\ state) = 2$).

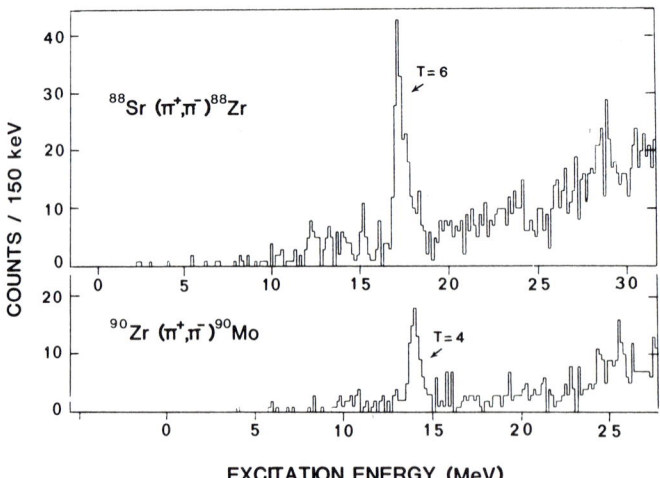

FIGURE 1

Missing mass spectra for (π^+, π^-) reactions on Sr^{88} (top) and ^{90}Zr (bottom) at $\theta = 5^0$ and $T(\pi^+) = 292$ MeV, taken from Ref. 4.

Fig. 1 shows missing mass spectra for (π^+, π^-) reactions on ^{88}Sr and ^{90}Zr, very recently measured [4]. One is immediately struck with the dramatic selectivity of the reaction in populating the double analog states. For the ^{88}Zr case, essentially no states are populated till 17 MeV excitation energy, and then the double analog state uniquely stands out, deep in the continuum. The data are so clean that one immediate question is whether a nucleus could show well defined excitations even with $\Delta T > 2$, at higher energies.

3. MESONIC DEGREES OF FREEDOM

The description of a nucleus as a system of interacting nucleons and pions is now rather well accepted, and effects due to the presence of Δ's in nuclei are visible in practically all the reactions at Intermediate Energy. At first sight one could believe that the pion-induced reactions are the best suited to this purpose. Indeed the necessity of introducing several non-trivial approximations and corrections may obscure the interpretation of the data. A good discussion of these aspects is due to Weise [5]. It may then happen that other hadronic probes could give a cleaner answer.

I will discuss, as example, a very rare but perhaps very selective reaction: the "doubly coherent" pion production by light ions (often called also "pionic fusion"):

$$A_B + A'_C \rightarrow (A+A' \mp 1)_D + \pi^\pm \tag{1}$$

In this process <u>all</u> the kinetic energy carried by the nucleons in the entrance channel is transferred to a pion in the exit channel. We have then a well defined energy transfer between two separate degrees of freedom of the nucleus, with exclusion of all the thermalization processes. Following the Erlangen group [6,7] the pion field could act as a "reservoir" of the energy which is transferred between the two degrees of freedom. Fig. 2 helps to explain the mechanism. We may assume the existence of doorway collective Δ-hole excitations (Giant Isobaric Resonances), at some hundreds of MeV, which decay by the emission of a real pion to the ground or low-lying excited states of the nucleus D (right part of the figure). The mechanism is similar to the "photonic fusion", represented in the left part of the figure, where the well-known collective particle-hole excitations (Giant Resonances) act as doorway states.

A recent experiment [8] seems to confirm the above mechanism. The reactions:

$$^3He + ^6Li \rightarrow ^9C + \pi^- \tag{2}$$

and

$$^{3}\text{He} + {}^{7}\text{Li} \rightarrow {}^{10}\text{C} + \pi^{-} \qquad (3)$$

were studied at an incident energy of 910 MeV (303 MeV/nucleon).

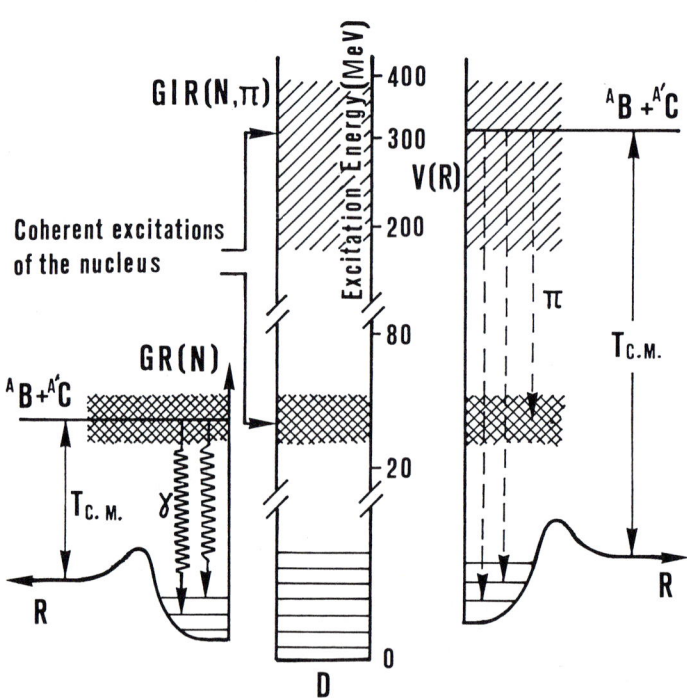

FIGURE 2

Scheme of a proposed reaction mechanism for the doubly coherent production of pions by ions. GR(N)= giant resonance (only nucleons), GIR(N,π)= giant isobaric resonance (nucleons and pions). The other details are explained in the text.

Fig. 3 shows the measured pion spectra, corresponding to cross-sections of the order of the pb/sr, the lowest up to now measured for ion induced nuclear reactions, to my knowledge. One can notice a weak production of final bound states, and a preferential production of discrete unbound states, very likely collective particle-hole states.

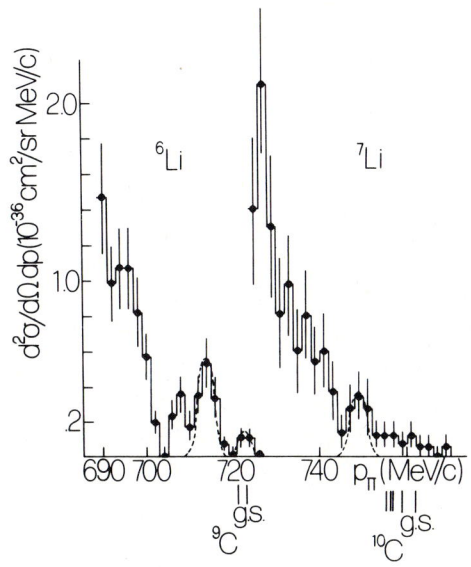

FIGURE 3

Measured pion spectra from ^{6}Li and ^{7}Li targets bombarded with ^{3}He ions of 910 MeV, taken from Ref. 8. The bottom lines represent pion momenta corresponding to the formation of ^{9}C and ^{10}C in known levels. The dashed lines represent Gaussian fits corresponding to the momentum resolution to observed pion peaks.

The structure of these states is the most similar one to that of the parent collective Δ-hole states (only a pion is emitted, without otherwise changing the nature of the excitation), and their preferential production seems a confirmation of the described reaction mechanism. Further experiments are indeed necessary to elucidate all the aspects of the problem.

4. COLOR DEGREES OF FREEDOM

The third stage of the Nuclear Physics, for which the nuclei have to be considered as systems of many quarks and gluons, with possible collective and coherent effects not otherwise observable, is the extension of the historical development of the Nuclear Physics, following which the most recent approaches to the description of the elementary interactions between particles are applied to the complex nuclear systems. The description of nucleons and mesons in terms of quarks, and the success of the QCD for the explanation of the high energy phenomena led to the question whether such an approach could be extended even at low energies and to the nuclear systems, despite the formidable difficulties due to the increasing of the coupling costant $\alpha_s = g_s/4\pi$

at low energies and then to the applicability of perturbative methods.

It is not yet clear whether the quark description of nuclei is an elegant, even if extremely complicated, approach unifying the particle and nuclear physics, or a necessary method able to explain phenomena and properties not otherwise predicted. The well-known EMC effect [9] is up to now the best indication for the existence of color degrees of freedom in nuclei and the main objective for the future years will be that of elucidating the degrees of freedom truly necessary for the explanation of the nuclear properties and the nature of the quark confinement in nuclear matter.

To achieve this program, a big experimental effort is needed, with modern facilities supplying excellent beams of electrons, heavy ions and hadrons, able to explore the nuclear response in a complementary way. The electrons are obviously the best probe of the electromagnetic response. Indeed, they have access directly only to the charged quarks and not to the neutral gluons, which are assumed to carry about a half of the momentum content of the nuclear matter. Relativistic heavy ions have access to both quarks and gluons and may led to extreme conditions [10] in which the many-body quark-gluon system exhibits new properties, but it is not yet clear which is the unambiguous signature for the new phase of the quark-gluon plasma. The powerful advantage of the hadrons is that of the selectivity. They have access, like the heavy ions, to both quarks and gluons and the experimenter, by choosing between the various N, $\bar{N}$, π, K beams may vary in a controlled way the quark content of the probe and then the specific response of the nucleus. I will now discuss with some details a few typical experiments.

4.1. Insertion experiments

The general approach of these measurement is that of substituting one (possibly two) of the u- and d-quarks of a nucleus with a "marked" s-quark. The quark structure of a nucleus is then studied at $Q^2 \simeq 0$, and the method is similar to the "radioactive marked nucleus" technique of the molecular physics. In other words, this class of experiments belongs to the Hypernuclear Physics. I will give a complete review of this field in another lecture [11].

4.2. Scattering experiments

The elastic and inelastic scattering of hadrons from nuclei has been for a long time a source of frustration for the engaged physicists, when they compared their conclusions with the clean results of the electron scattering. Now the situation is changed, mainly due to a complete theoretical approach recently developed, the Dirac formalism [12], which has brought the hadron scattering at the same level of the electron scattering. This class of experiments is the most simple for explaining the unique properties of hadrons in studying the modern features (quark structure) of nuclei. Simply by changing in a controlled way the quark content of the probe (a nucleon, an antinucleon, a kaon, a pion) one can explore, in the same kinematical conditions, the response of the nucleus, and then study the underlying quark structure. Among all, K^+ scattering is of first importance. The reason can be understood if one looks at the graphs illustrating the difference between the elementary quark interactions of a K^+ with a neutron, and those of all the other hadrons usable as projectiles (see Fig. 4, where a K^- - p process is illustrated as counterpart).

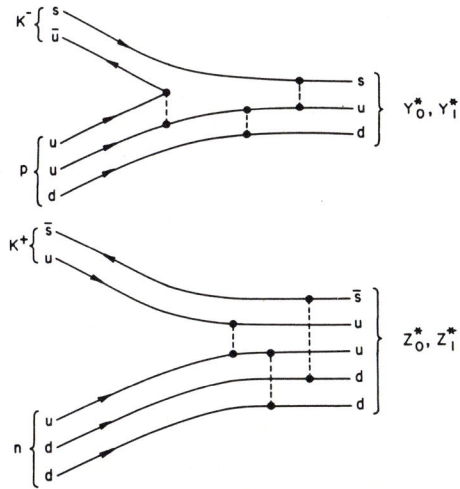

FIGURE 4

Quark diagrams illustrating resonance formation in K^-p and K^+n interactions, taken from Ref. 14.

The K^+ is the only hadron containing an $\bar{s}$-quark, that cannot annihilate with a real s-quark. Only exotic configurations ($4q$-$\bar{q}$), the Z^* particles, can contribute to the interaction vertex; even if they exist at all, their configuration probability is very low. The opposite is true for other hadrons, like the K^-, where many baryon resonances can contribute in the interaction vertex.

The peculiar properties of the K^+ among the hadronic probes was recognized from a long time simply by observing that its mean free path into the nuclear matter was an order of magnitude larger than for the others. This feature, toghether with the simplicity of the elementary K^+ - N interaction at low energies, suggested that it could be considered as a sort of "heavy electron" or "weakly interacting hadron" in order to obtain quantitative informations on nucleon (mainly neutron) distributions in nuclei. The remark [13] that even the elementary K^+ - N interaction is well described in terms of the constituent quark interactions opens the possibility that K^+ scattering could give directly informations comparable with high-energy electro-weak probes (EMC effect) at very lower Q^2. A part some exploratory experiments, no systematic data for K^+ - Nucleus scattering exist, due to the lack of suitable beams.

Costa [15] will devote a complete lecture to the hadron-nucleus scattering experiments.

4.3. Deposition experiments

As mentioned before, the leading objective of the relativistic heavy ion facilities is that of achieving, in a nucleus-nucleus collision, such extreme conditions of density and temperature that a new phase of the hadronic matter, the quark-gluon plasma, is reached. Whithin a smaller volume (a portion of the nucleus), the same density and temperature conditions could be reached by means of $\bar{p}$ annihilation in the interior of a heavy nucleus. Antiproton beams could then be used in order to deposit a known amount of energy (at least 2 GeV) in a very localized volume and then follow the evolution of the locally heated nuclear matter. Even precursor phenomena of the quark-gluon plasma could appear with such a method.

The experimental information on $\bar{p}$ annihilations nuclei is very scarce and the first results are produced now at the LEAR facility at CERN. Piragino [16] will review the status of the $\bar{p}$-Nucleus studies in another lecture.

Following the very interesting results of the first-generation experiment, an ambitious second-generation experiment has been proposed, as a part of the large experimental program of the OBELIX facility [17], to be installed at LEAR after ACOL (the new antiproton accumulator at CERN). I will describe with some details this apparatus, in order to give an idea of the experimental problems that have to be solved in present and future studies of Nuclear Physics with Hadrons. OBELIX is a ~4π magnetic spectrometer, able to measure with good efficiency and energy resolution all the particles (charged and neutral) emitted following antinucleon annihilations on Nuclei.

Fig. 5 shows a scheme of the apparatus. The Open Axial Field Magnet (OAFM) provides a large unobstructed volume (L=1.5 m; 2.5 m distance from beam axis to bottom yoke) with a central field of 0.5 T. The field lines are obviously contained in a plane passing through the symmetry axis of the magnet (beam line). Conical holes inside the two poles with 15° angle to the beam leave, upstream and downstream of the poles, an aperture with ∅=48 cm for insertion of the internal detectors.

Moving out from the centre of the magnet, one finds the following components:
- a gaseous target (but also solid targets could be used);
- a spiral projection chamber [18] (SPC) (imaging vertex detector with three-

dimensional readout for charged tracks);

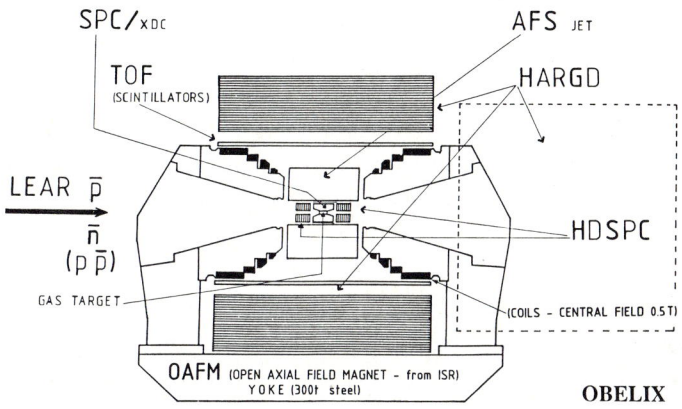

FIGURE 5
Schematic view of the OBELIX detector, taken from Ref. 17.

- a layer of thin (0.5 cm) scintillators giving the start signal for the TOF measurement on the charged particles;
- the AFS jet drift chamber (pictorial drift chamber from the CERN experiment R 807, with large size (∅~1.6 m), high resolution (σ~200µm), three dimensional readout, and dE/dx measurements with up to 42 points per track;
- a layer of 90 thick (3 cm) scintillators giving the stop signal for the TOF measurements and allowing also an independent measurement of dE/dx and of the multiplicity;
- a high angular resolution gamma detector (HARGD) (four moduli made of layers of 3x4 m^2 converter foils enclosed by planes of drift tubes parallel to the beam axis, and of limited streamer tubes in the transverse direction);
- two high-density spiral projection chambers (HDSPCs) are positioned upstream and downstream of the vertex detector inside the AFS jet chambers; they complete the solid angle for gamma and prong detection in the end-caps.

The proposed configuration permits charged particles to be detected over ~4π and to have high-resolution momentum, dE/dx and TOF over >2π (with the SPC + AFS jet chambers and the scintillators) owing to the large size of the volume instrumented in the radial direction. Gamma rays detection is made over nearly 4π with high efficiency and granularity, excellent angular resolution, and three-dimensional reconstruction of the showers. Such a powerful instrument, coupled to the unique properties of the p̄ beam at LEAR, will allow a real breakthrough in the new field of the $\bar{p}$-Nucleus Physics.

5. CONCLUSION

I hope that I have given an idea of the interest and richness of the Intermediate Energy Nuclear Physics with Hadrons and that the other lectures will reinforce my statement. Unfortunately, apart some aspects of the $\bar{p}$ physics, the aforementioned experiments and programs are now proceeding very slowly, and sometimes they are downright stopped, due to the lack of the main tools, the beams. For this reason, many european physicists, active in the field and that contributed strongly to the experimental and theoretical progress, spontaneously gathered during some workshop and meetings, and originated the Study Group of the E.H.F. The acronym stands for "European Hadron Facility", a project still under discussion and elaboration [19] whose aim is that of building in Europe a vanguard machine for the needs of the Intermediate Energy Physics up and beyond the year 2000. My personal opinion is that the many-quark (or Nuclear) aspects of the hadronic matter, with possible collective or coherent effects, could be a field of research unique of the E.H.F. and that alone could justify the construction. Unfortunately the theoretical frame is not so clear, due to objective, enormous difficulties in the calculations. A close collaboration between theoreticians and experimentalists is necessary in order to overcome this obstacle.

ACKNOWLEDGEMENT

I would like to thank Miss L. Ceretta for the careful preparation of the typed version of the manuscript.

REFERENCES

1) T. Bressani, in Perspectives on Theoretical Nuclear Physics, Proceedings of the Workshop held in Cortona (Italy), September 16-18, 1985, eds. L. Bracci et al. (ETS Editrice, Pisa, 1986) pp. 75-83
    T. Bressani, Perspectives of the Nuclear Physics at Intermediate Energy with Hadronic Probes, summary talk given at the LXX Meeting of the Italian Physical Society, Genova (Italy), October 4-9, 1984, unpublished
    T. Bressani, Nuclear Physics with Hadrons, invited talk given at the I.N.F.N. National Conference, Castelgandolfo, July 10-12, 1985, unpublished.
2) G. Preparata, this volume.
3) K. K. Seth, in Proc. of the Tenth International Conference on Particles and Nuclei, Heidelberg, July 30 - August 3, 1984, eds. B. Povh and G.

Zu Putlitz, Nucl. Phys. A434 (1985) pp. 287-328

K. K. Seth, in Proc. of the Workshop on Nuclear and Particle Physics at Intermediate Energies with Hadrons, eds. T. Bressani and G. Pauli, Conf. Proceedings Vol. 3 (Italian Physical Society, Bologna, 1986) pp. 1-41.

4) K. K. Seth et al., Phys. Lett. 155B (1985) 339.

5) W. Weise, in Proc. of the Ninth International Conference on High Energy Physics and Nuclear Structure, Versailles, July 6-10, 1981, eds. P. Catillon, P. Radvanyi and M. Porneuf, Nucl. Phys. A374 (1982), pp. 505-519.

6) M. G. Huber and M. Dillig, in Proceedings on High Resolution Heavy Ion Physics, eds. M. Martinot and C. Volant (unpublished).

7) K. Klingenbeck, M. Dillig and M. G. Huber, Phys. Rev. Lett. 47 (1981) 1654.

8) T. Bressani et al., Phys. Rev. C30 (1984) 1745.

9) E. Predazzi, this volume.

10) R. A. Ricci, this volume.

11) T. Bressani, this volume.

12) See e.g. L. Roy, in Proc. of the Third LAMPF II Workshop, Rep. LA 9933C (1983) p. 419.

13) H. J. Pirner and B. Povh, Phys. Lett. 114B (1982) 308.

14) C. B. Dover and G. E. Walker, Phys. Rep. 89 (1982) 1.

15) S. Costa, this volume.

16) G. Piragino, this volume.

17) R. Armenteros et al., in Physics with Antiprotons at LEAR in the ACOL Era, eds. U. Gastaldi, R. Klapish, J. M. Richard and J. Tran Thanh Van (Ed. Frontières, Gif-sur-Yvette, 1985) pp. 369-380

R. Armenteros et al., proposal CERN/PSCC/86-4/PSCC/P95 (1986).

18) U. Gastaldi, Nucl. Instr. Meth. 157 (1978) 441

U. Gastaldi, Nucl. Instr. Meth. 188 (1981) 459.

19) F. Bradamante, this volume

RELEVANT ASPECTS AND PERSPECTIVES OF LOW-ENERGY HEAVY-ION
REACTIONS

Renato Angelo RICCI

University of Padova and Laboratori Nazionali di Legnaro-
INFN, Padova

1. INTRODUCTION

Among the most relevant aspects of the actual nuclear physics investigation, the interplay between the structure and dynamical behaviour of atomic nuclei, as revealed by different types of reactions, has an important impact.

The phenomenological richness in testing such properties is characterized by the various probes one can use in different kind of interactions with atomic nuclei. Among these interactions induced either by electromagnetic (electrons, photons, muons) or hadronic (nucleons, antinucleons, pions, ions...) projectiles, the ones induced by heavy ions play an essential role.

The major conceptual aspect of this kind of interaction is that it is not to be related to a sort of primary (fundamental) process (in the elementary sense) but, instead, to the result of combined structural and dynamical effects for interacting particles which are bound and/or confined.

In the nucleus-nucleus collisions, which could be considered as the phenomenological signature of a sort of complex "super-nuclear" interaction, the dominant characteristic is that the interacting systems are complex and charged and consequently the phenomenological area covered by the various possible projectile-target combination is very large.

Moreover, depending on the bombarding energy, one has a large variety of excitation and decay energies at disposal, which can give rise to different kind of phenomena dealing with new physical aspects of the nuclear matter.

An usual way to represent the possible nuclear area to be covered is that represented in Fig. 1, where the expected *new nuclear* species far away from the stability are indicated[1].

On the other hand the energy-mass diagram of Fig. 2 (Scott-diagram) shows different boundaries connecting fundamental physical

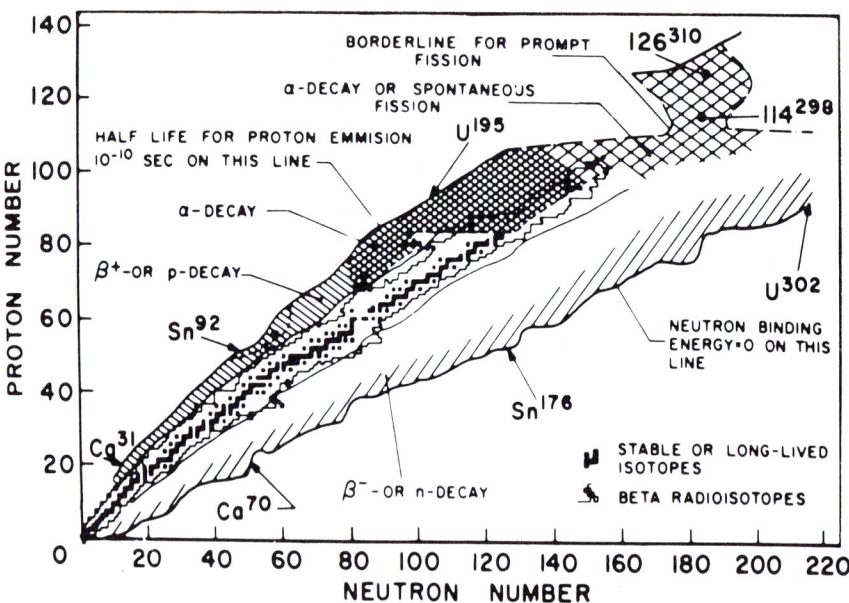

Fig. 1. Map with the existing nuclei (stability valley and new possible species).

domains. The three center-of-mass energies of 20, 140 and 930 MeV, mark the supersonic, mesonic and relativistic threshold, respectively: they correspond to different domains of the nuclear density ρ as compared to the normal density ρ_0 and to the possible occurrence of new nuclear-matter phases such as pion condensate and quark-gluon plasma. On the mass-side ($A^{1/3}$) the transition from microscopic (single, A∼few unities) effects to collective macroscopic behaviours ($A^{1/3} >> 1$) is indicated so as the occurrence at $Z \cong \frac{1}{2} \cdot 170$ of the projectile-target combination $Z_p + Z_t = 170$ (instead of $137 = 1/\alpha$, due to nuclear-size effects) where supercritical field phenomena are expected to occur[2].

The complexity of heavy-ion reactions, together with the possibility of testing nuclear reaction mechanisms using a large variety of combinations, leads, on the other hand, to more difficult interpretations.

The first significant question is how to select the most probable root, in the reaction space, connecting given initial and final states (phenomenological selection). Under certain conditions (energy, interaction, initial and final configuration) many different channels can be open (different types of excitation modes, nuclear shape transitions, nucleon or cluster transfer ...). As pointed out by Fesbach[3] this is a crucial problem in nucleus-nucleus collisions.

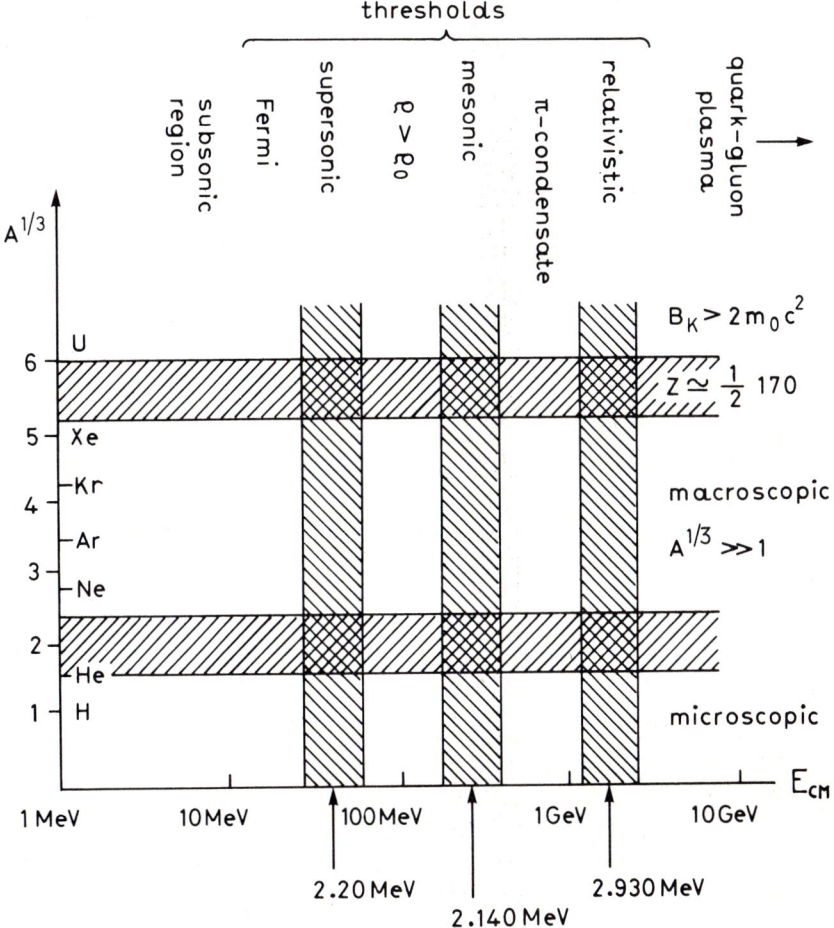

Fig. 2. Scott Diagram

The second important question is which kind of fundamental understanding one can derive from such interaction (cognitive selection). Once again following Fesbach and Weidenmüller the major contribution is the better understanding of the mechanism of "energy and particle transfer and dissipation into small strongly interacting systems"[3].

This means that the nucleus-nucleus interacting process can be considered as a part of a more general physical problem i.e. the statistical description of systems formed by a relatively small number of strongly interacting particles.

In this context, the hadronic character of heavy-ions and of

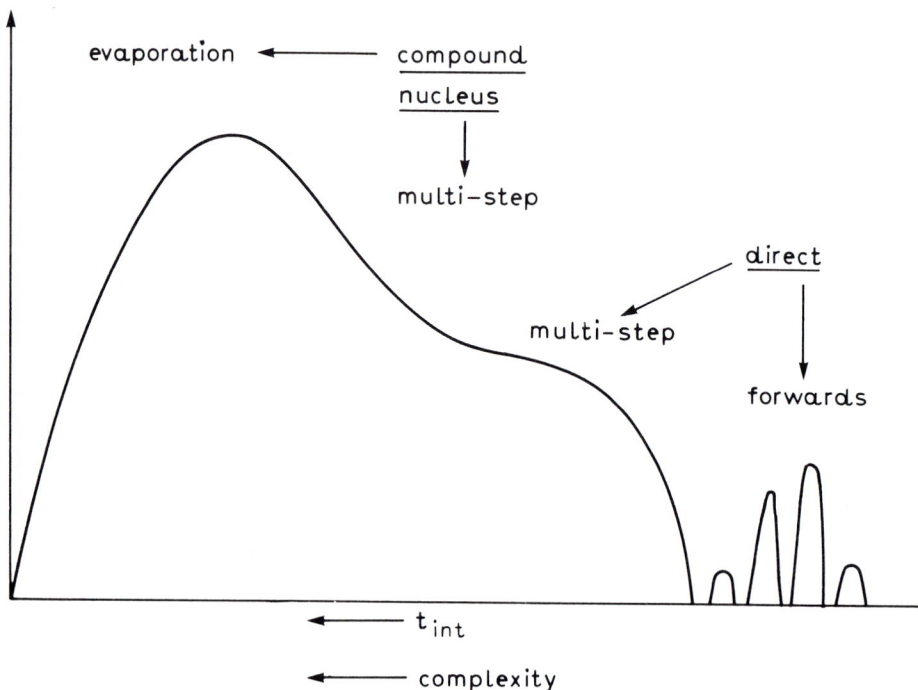

Fig. 3. The Fesbach diagram. The evolution, with increasing interaction time, from direct to compound (evaporation) process, through multi-step intermediate stages, is indicated.

the nucleon constituents as partner of the complex interactions is a peculiar feature to find interesting relations with the simpler hadron-nucleus interaction, which constitutes the main topical subject of this School.

2. DISSIPATIVE PHENOMENA IN HEAVY-ION REACTIONS

The complex nature of the interacting partners make the nucleus--nucleus collisions, at relatively low and moderate energies, strongly absorbing processes.The large absorbtion can be related to the existence of different stages of the evolution of the composite system quite similar to multi-step processes evidenced in nucleon-nucleus interactions. In such a case the Fesbach diagram of Fig. 3 holds,where different reaction mechanisms between the two extreme processes such as the simplest direct and compound nucleus modes, are indicated as a function of the interaction time[3].

A part from the fact that in heavy-ion reactions many more degrees of freedom are needed, a similar situation can occur. Strong

absorbtion means a large overlap of the two system which no longer
remain in their original states, especially for central collisions.
This gives rise to strong energy dissipation and mass transfer
which, on the other hand, will be distributed in different degrees
of freedom following the interaction time.

A classical picture of classifying heavy-ion reactions is shown
in Fig. 4, as a function of the impact distance of the two colliding nuclei. This picture leads to a simplified distinction between

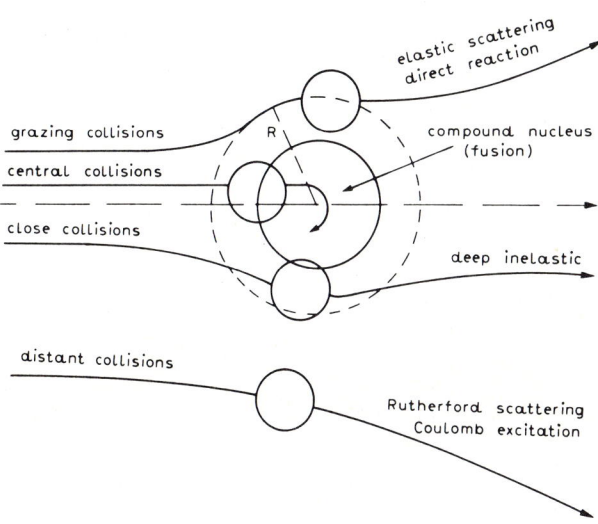

Fig. 4. Classical picture of different types of nucleus-nucleus
collisions.

peripheral (grazing and direct) collisions and close and central
(deep inelastic and fusion) collisions, a part from the pure (distant) Rutherford scattering.

An immediate approach in the description of the different types
of interaction mechanisms consists in a classification on the
ground of the angular momentum involved[4]. The simplest division one
can make is between *strongly-damped* and quasi-elastic reactions

where the last are the most pheripheral. As the former mechanism is concerned, two classes of processes are easily observed: the central collisions involving the lowest partial waves, which leads to *compound nucleus* (CN) formation which can be followed by evaporation or fission, and the more pheripheral collisions with higher partial waves, in which not all the degrees of freedom are relaxed and there is a sort of dinucleus formation; (these processes are the so-called *deep-inelastic collisions* (DIC) since the strong energy dissipation does not destroy the individuality of the two colliding partners (see Figures 5 and 6).

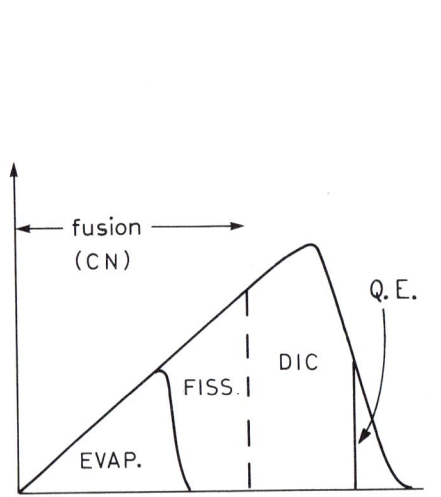
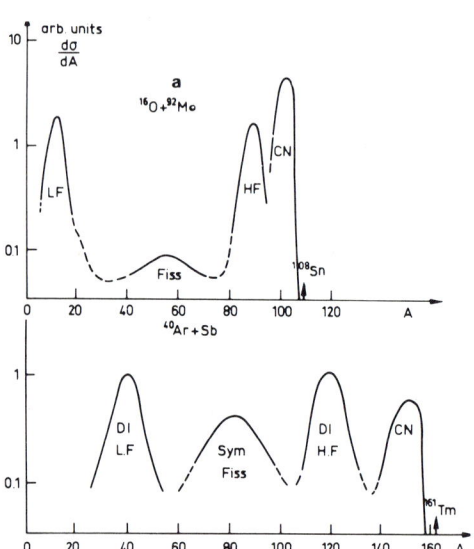

Fig. 5. Schematic representation of the heavy-ion reaction cross section as a function of the angular momentum ℓ.

Fig. 6. Schematic representation of the fragment-mass distribution (exit-channel) for a) DIC and b) CN and fusion followed by fission. (from ref. 8)

However this picture turns out to be over-simplified. In fact, especially in the last years, many experimental results indicated that the CN formation ($\sim$complete fusion) and DIC are not the unique damped processes, but there is a *continuum* of situations between these two extremes. In other words, it seems established the existence of *intermediate* (or transitional) mechanisms which may present features of *both* the above processes.

The first examples of these new phenomena were discovered in the heavy-mass region and explained in the framework of dynamical models as the "quasi-fission" and "fast-fission" ones[5]. These approaches define a class of phenomena originating from a partially equilibrated system which occurs in a time scale intermediate between DIC (10^{-21}s) and CN formation ($10^{-16} \div 10^{-18}$s).

More recent experimental investigation in the mass region of A$\sim$100 at bombarding energies $\sim$1.5 times the Coulomb barrier ($\sim$5 MeV/A) as performed at the 16 MV XTU Tandem of the Laboratori Nazionali di Legnaro and at the 18 MV Tandem of the Strasbourg Laboratory, pointed out the presence of extended transitional processes even in this region. Such intermediate mechanisms are evidenced by the following peculiar features:[6]

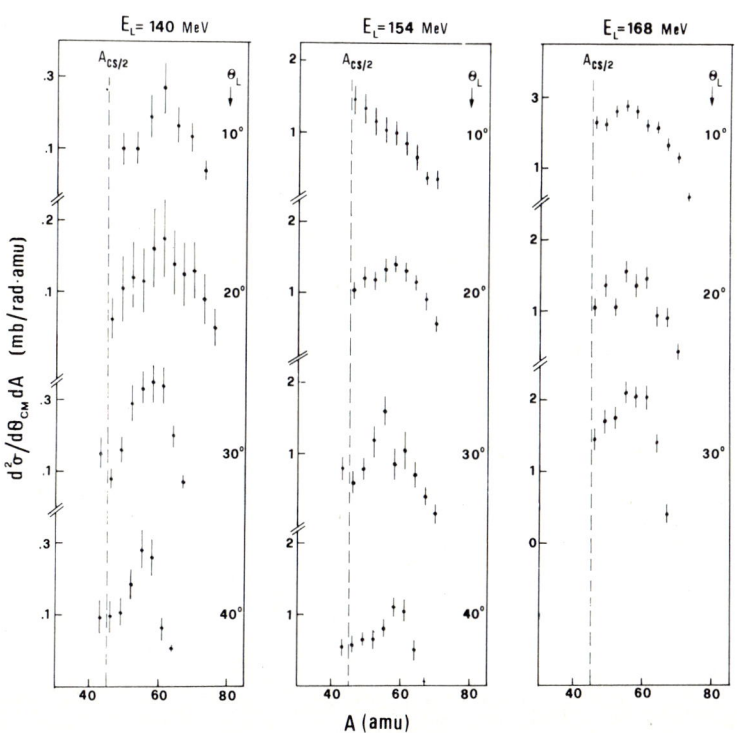

Fig. 7. Mass distribution of fission-like fragments from the ^{32}S+^{59}Co reaction.

a) evolution of the fragment-(fission-like)-mass distributions with the observation angle i.e. relaxation of the mass-asymmetry in a time scale comparable with the rotation period of the dinuclear system;

b) complete energy dissipation (fully damped kinetik energies) independent on bombarding energy and observation angle in agreement with fission processes.

Fig's. 7 and 8 show such a behaviour for the mass asymmetry and energy relaxation respectively, for the case of the $^{32}S+^{59}Co$ reaction at different bombarding energies.

This evidence of intermediate dissipative processes in low-energy heavy-ion reactions can be interpreted in terms of a diffusive relaxation process leading the *dinuclear system* from the entrance channel to more or less equilibrated configurations depending of its lifetime (see Fig. 9).

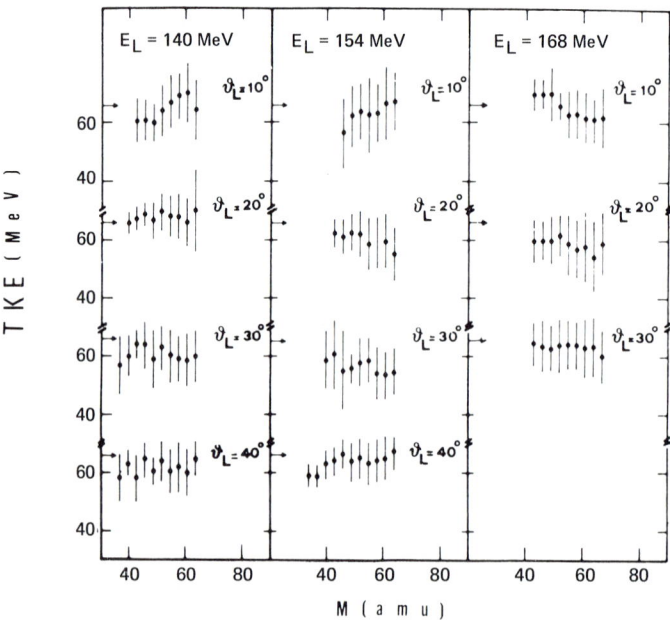

Fig. 8. Kinetic energy distributions of fission-like fragments from the $^{32}S+^{59}Co$ reactions.

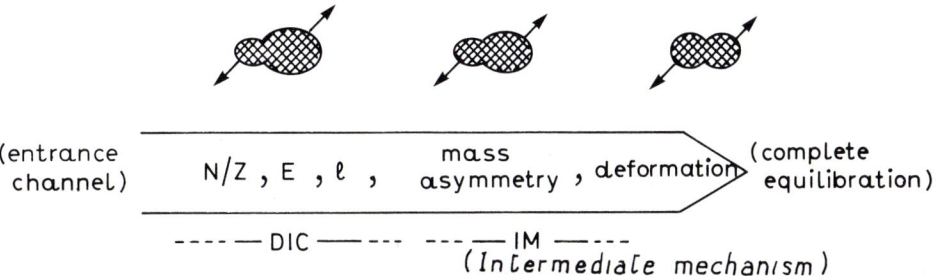

Fig. 9. Time evolution of dissipative phenomena

The transition from fusion to deep inelastic process with increasing energy is due to the fact that the compound nucleus can no longer accept such a large energy and angular momentum and their dissipation, together with the nucleon transfer, will occur in very short timescales in an increasingly complex reaction. The mechanism for this fast dissipation and the following equilibration is not yet well understood even if peculiar features have to be related to collective flows of nuclear matter from a participant to the other.

An important contribution to this rapid energy dissipation seems to arise from the excitation of giant resonance modes. This is one of the most relvant signature of the interplay between dynamical and spectroscopic features of nucleus-nucleus collisions. In fact we are dealing here with typical *doorway states* with resonant character and the experimentally observed *fine structure*, as in the $^{12}C+^{27}Al$ reaction, suggests the dissolution of the giant resonance strength into the underlying continuum states[7].

3. EVOLUTION OF THE ENERGY-DISSIPATION WITH INCREASING ENERGY

At low energies ($\sim$10 MeV/A) the collisions between individual nucleons are neglected and the mean field behaviour dominates as shown by the almost full relaxation achieved.

If the energy increases the influence of the nucleon-nucleon collision terms begins to be important giving rise to the following features:[8]

a) For peripheral collisions the deep inelastic interaction, which is still a slow binary process, evolves to a quasi-elastic or *slightly inelastic fast perturbation* leading to break-up and/or fragmentation (three-body process).
b) For central collisions, the fusion-cross section decreases, though the presence of *quasi-fusion* processes cannot be excluded, after fast-light-particle emission, which will limit the linear momentum transfer.

In the first case, occurring at impact parameters intermediate between grazing and central collisions, the energy region where the DIC really disappear will depend on the mass asymmetry of the colliding system. A reasonable statement is that this limit is achieved when the relative velocity exceeds 0.2 c, where c is the velocity of light (i.e. for the center of mass energy $E_{CM}=22$ MeV/A

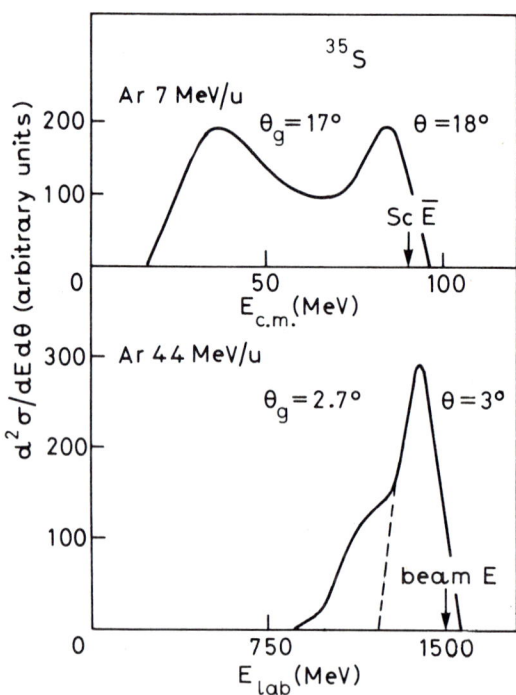

Fig. 10. Energy spectra of ^{35}S fragments close to the grazing angle in ^{40}Ar+^{58}Ni at 7 MeV/A and 44 MeV/A, respectively.(ref.9)

for Ca and 40 MeV/A for Pb).

In that case the collision is too fast for the energy dissipation to occur in a large extent.

Fig. 10 shows the energy spectra of the projectile-like fragment ^{35}S following the $^{40}Ar+^{58}Ni$ at two different energies (7 MeV/A and 44 MeV/A respectively)[9]. At the higher energy the bump corresponding to the strong dissipative events to the D.I.C. has disappeared.

The main process which could occurr at these intermediate impact parameters and energies, is the creation of a *hot transient nuclear ball* (∼500 MeV excitation energy), made of the projectile plus a piece of the target, exploding into several light fragments (*multifragmentation*) (see Fig. 11).

In the second case (central collisions) one reaches, at an excitation energy E* of about 4÷5 MeV per nucleon, the limit where the fusion process takes place and the statistical or fission decay follows.

The observation of linear momentum transfer allows to deduce the excitation energy and the corresponding limit for complete and incomplete fusion. A critical limit ranging from 6 to 4 MeV per nucleon at a composite mass of 60 and 260 respectively, is indeed found[9]. The corresponding temperatures, using thermodynamical con-

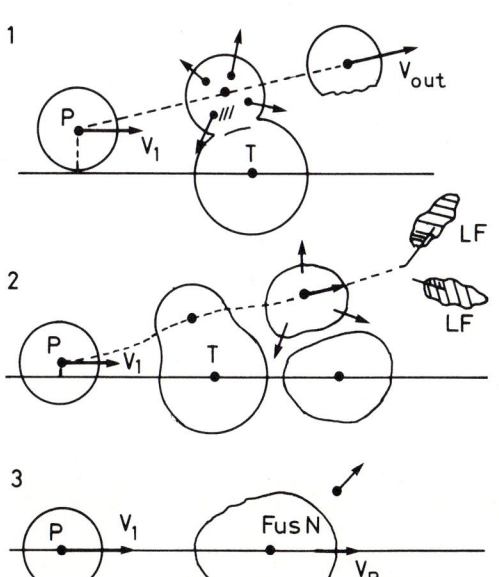

Fig. 11. Grazing abrasion, DIC with multifragmentation and quasi-fusion for heavy-ion reactions at Fermi energies.

cepts, $T=\sqrt{kE^*/A}$, reach the limit where the *nucleus can boil* and the nucleons are emitted into a *gas phase*.

The large enrgy deposit could be dissipated, in that case, via a multifragmentation process. Fig. 11 shows a schematic representation of the 3 classes of collisions at energies around the Fermi energy (30÷40 MeV/A) i.e. the grazing *abrasion* of the projectile (1); the deep inelastic *multifragmentation* (2); and the *quasi-fusion*.

On the other hand, the multifragmentation process opens already a new root for the nucleus-nucleus collisions, which, together with pre-equilibrium fragmentation and sequential decay, will develop into "new" reaction mechanisms.

A spectacular view of the multifragmentation phenomena is represented in Fig. 12, as recorded in a photographic emulsion[10]. Here

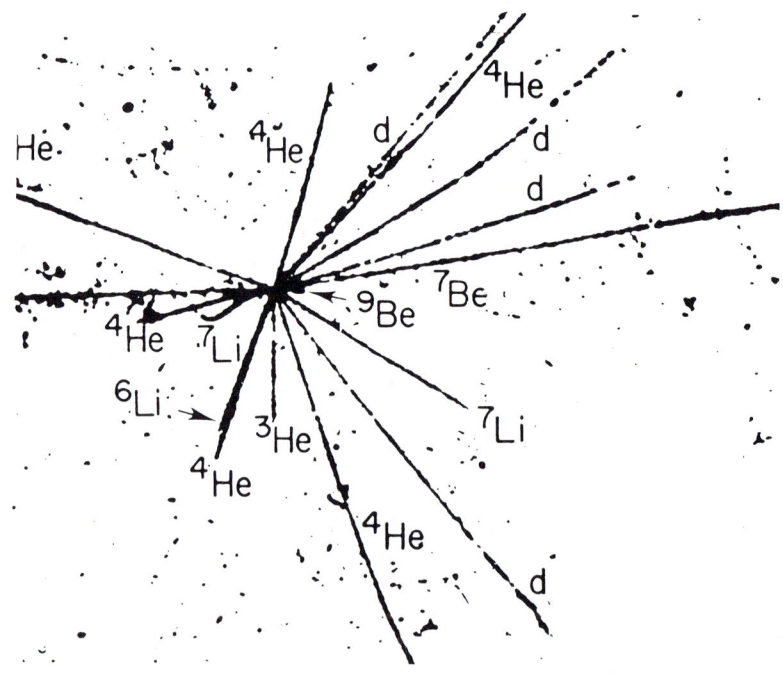

Fig. 12. Multifragmentation from a $^{12}C+^{108}Ag$ nuclear collision at 70 MeV/A.

we are already at a bombarding of 70 MeV per nucleon and a ^{12}C projectile has struck a silver nucleus of the emusion acting as a target. 16 fragments of nuclear matter have been recognized as emerging from the collision point, i.e. nuclei of deuterium, helium, lithium and berillium with masses from 2 to 9.

With this "nuclear boiling" experiment one achieves the upper limit of the "liquid phase" of nuclear matter. However, the subsequent multifragmentation can be interpreted as due to a simple evaporation phenomenon, where the peculiar nucleonic structure of the nuclear matter is still conserved.

REFERENCES

1) D.A. Bromley, Proceedings of the Int. Conf. on Nuclear Physics, eds. P.Blasi and R.A.Ricci (Compositori, Bologna 1983) p.9.

2) see for instance W.Greiner et al. in Proc. of the Conf. on Nuclear Structure with Heavy-Ions, eds. R.A.Ricci and C.Villi (Editrice Compositori, Bologna 1986) pp. 405-465.

3) H. Fesbach in Heavy-Ion Physics and Nuclear Physics, eds. L.Salvadori et al., (North-Holland, Amsterdam, 1983) p. 423c.

4) See for instance W.Nörenberg, J. Phys. 37 (1976) C5-141.

5) See W.J. Swiatecki, Phys. Scripta 24 (1981) 113; and C.Gregoire, C. Ngô and E.Tomasi, Nucl. Phys. A387 (1982) 137c.

6) See for instance: P.Boccaccio and R.A.Ricci in Fundamental Problems in Heavy-Ion Collisions, eds. N.Cindro, W. Greiner and R.Caplar (World Scientific, 1984) pp. 223-232.

7) See for instance: D.A.Bromley, Il Nuovo Cimento 81A (1984) pp. 362-401 and references quoted herein.

8) See M.Lefort: Proc. of the Conf. on Nuclear Structure with Heavy-Ions, eds. R.A.Ricci and C.Villi (Editrice Compositori, Bologna, 1986) pp. 467-488.

9) J.Galin et al.: Phys. Rev. 182 (1969) 1267 and Nucl. Phys. A159 (1970) 461; V.V.Volkov et al.: Nucl. Phys. A126 (1969) 1.

10) From B.Jakobsson et al., Phys. Lett. 102B (1981) 121.

COLOR DEGREES OF FREEDOM IN NUCLEAR PHYSICS

F. CANNATA

Dipartimento di Fisica dell'Universitá and INFN, I 40126 Bologna, Italy.

We illustrate the limitations of a purely nucleonic description of few body systems and nuclear matter. The relevance of quark and gluon degrees of freedom is then discussed with special attention to propagation of narrow exotic glueballs and hybrid mesons in nuclear matter.

1. INTRODUCTION

Traditionally nuclear physics is formulated in terms of pointlike nucleons interacting via two body potentials.

Microscopically the understanding of the medium range nucleon nucleon attraction by a meson exchange (σ–meson) has not been successful and one has been forced to consider also the excitation of the Δ isobar[1] e.g. among others the graphs of figures 1 and 2 :

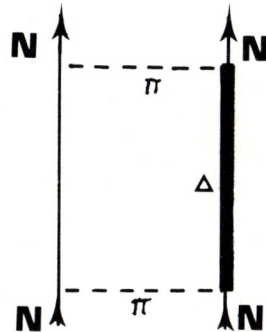

FIGURES 1 and 2

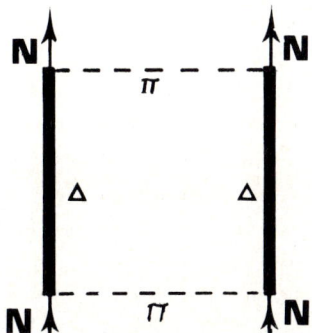

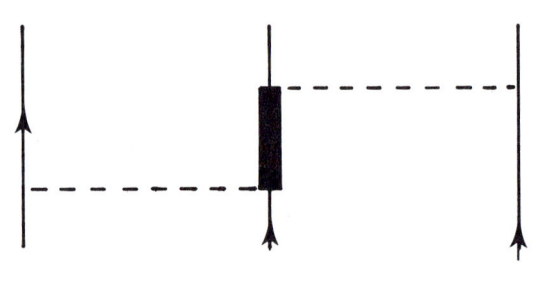

FIGURE 3

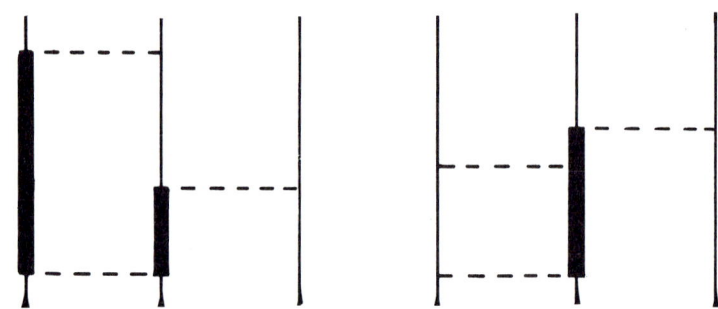

FIGURES 4 and 5

From the point of view of field theory an obvious consequence is that the same "type" of diagrams for three nucleons generate a three body force, figure 3.

Of course there are also more complex diagrams like in figures 4 and 5.

We stress that it is a very general fact that if we need isobars to explain the intermediate range attraction we are forced to obtain a three body force. This just follows from the fact that the nucleon can be excited to an isobar when interacting with another nucleon and the fact that the force is given by a meson exchange : clearly the isobar can decay by exchanging a meson with a third nucleon.

Few body physics offers a testing ground for the practical use of such ideas. One can keep isobar degrees of freedom frozen and calculate the binding energy of the triton with the best empirical two body force and a theoretical three body force [2]. In such a way one obtains satisfactory results for the binding energy. On the other side what one should really make is a coupled channel calculation including nucleons and Δ's. In such a scheme one has to recalculate everything, including the two body force since nucleons and Δ's are always coupled[3]. Constraints on such a calculation are given by π production and absorption and elastic π deuteron scattering. This provides immediately a link between low energy and intermediate energy nuclear physics. The calculation along these lines of the triton binding energy gives results which are at variance with the first procedure and with the experimental value for approximately half an MeV.

A possible conclusion is that probably we do not yet understand fully the intermediate range Nucleon–Nucleon attraction and the associated three body force. These results are

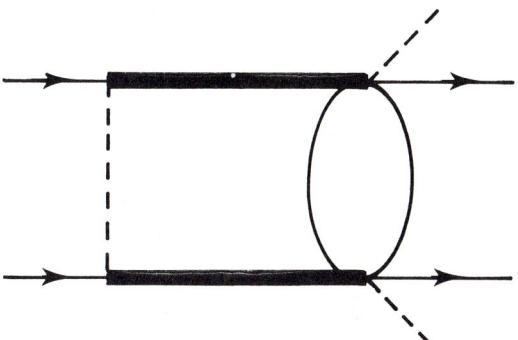

FIGURE 6

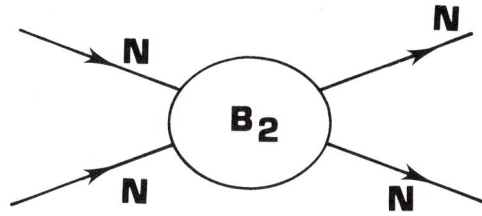

FIGURE 7

not academic [4] in view of the fact that three body forces are relevant in dense nuclear matter and determine the breathing modes and the nuclear compressibility. Astrophysical applications concern neutron star matter.

Here we would like to stress that in intermediate energy nuclear physics one can study Δ's degrees of freedom explicitly. For example with a few GeV machine with an intense proton beam of good monochromaticity one could study very accurately the reaction (figure 6)

$$NN \to \Delta\Delta \to N\pi N\pi$$

An extensive measurement of this reaction should provide useful information to constrain the model dependence of the three body force. If however one should not be able to explain correctly the triton binding energy only with nucleons and Δ's this may lead to the very important consequence that new degrees of freedom like quark and color degrees of freedom have to be dealt with explicitly. This may be related for example to the existence of dibaryons (figure 7).

The most exciting interpretation of the dibaryon is that of a color hidden state (color octet, color octet coupled to color singlet)[5].

Before entering a more general discussion concerning specifically the relevance of color degrees of freedom for nuclear physics we would like to elaborate on effects related to compositeness. Somewhere else we have discussed the effects concerning π and K nucleus interactions[6]. An important aspect of compositeness is that there is a Pauli exclusion effect operating between the quarks constituents of the projectile and the quarks constituents of the nucleus. Here we would like to stress that compositeness does not always produce a hindrance but it can lead in some circumstances to very specific predictions. Take for example a very simple microscopic model to describe $\bar{p}$ annihilation i.e. start from a $q\bar{q}$ annihilation vertex ^3Po (triplet p wave coupled to zero in order to have the vacuum quantum numbers). This simple model predicts rather peculiar effects (e.g. for $\bar{p}$ annihilation in s wave) for annihilation of the three antiquarks by the quarks of the nucleus. Think e.g. of a He^4 with all twelve quarks[7] in s shell or six quarks in s shell and six quarks in p shell. Clearly the annihilation would depend on the p to s shell population in the quark shell model.

2. QUARK AND GLUON DEGREES OF FREEDOM

The basic starting point fo speculations concerning the relevance of color degrees of freedom in nuclear physics is the vacuum structure of a quantum gauge field theory with non abelian character[8].

The physical vacuum is generally assumed to be a complex medium in which functions of gluon and quark fields have a non zero expectation value[9]. According to common wisdom the difference between the perturbative vacuum and the physical vacuum should lead to an explanation of the confinement of color and asymptotic freedom. Preparata has argued that there may be an infinite energy difference between physical vacuum and perturbative vacuum leading as a consequence to the non applicability of the concept of asymptotic freedom.

One general way to study complicated many body systems is to excite them. In the same spirit one can study vacuum excitations studying gluonia which are mesonic states formed by gluons without an appreciable quark–antiquark content. One can make sure that the $q\bar{q}$ content is vanishing if the quantum numbers are exotic : such states should be narrow[10].

The propagation of such mesons (if their width is not too large ≤ 50 MeV) in nuclear matter will be a very stimulating subject to study since the absorption of such mesons can be due to a very peculiar mechanism. Take e.g. a two gluon state (G) gluonioum (figure 8).

It can decay in nuclear matter into two gluons subsequently absorbed by two nucleons which transform into nucleon octets recombining finally to standard nucleons by gluon exchange. This absorption mechanism can be "resonant" if dibaryons exist and are to be interpreted as color hidden states. Alternatively one can also have absorption on short range color hidden components of a nucleon pair[11] (figure 9).

It is clear from these qualitative considerations that nuclear physics (i.e. at least two nucleons) can give relevant informations which could not be obtained with production of such mesons on hydrogen.

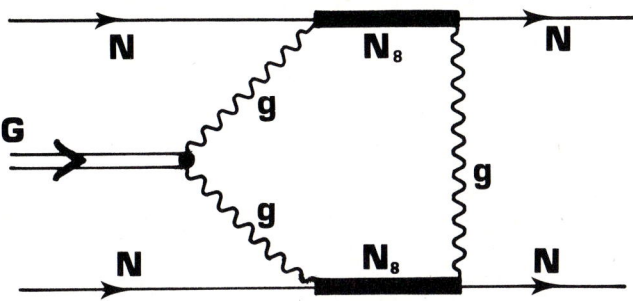

FIGURE 8

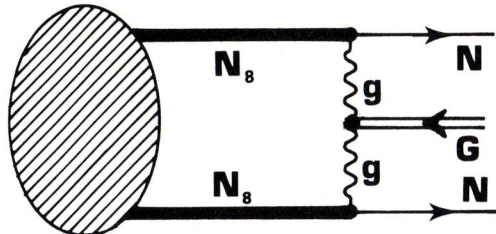

FIGURE 9

From what we have argued the gluonia could have rather large absorption in nuclear matter even in absence of gluonium nucleon resonances.

A rather interesting possibility is also given by hybrid mesons made by quarks, antiquarks and gluons which could also have anomalous properties when interacting with nuclear matter.

Coming to more conventional states the $2q\,2\bar{q}$ mesons are rather peculiar because not much is known about their structure, size etc.

E.g. the S^* (975) meson which is very close to $K\overline{K}(\sim 990$ MeV) threshold could be interpreted as a weakly bound $K\overline{K}$ state[12–13]. Due to its loose binding it could have a rather large extension. Here again the interaction of such mesons with nuclear matter could provide information about its size much in the same way as in the old days deuteron nucleus scattering was found to depend on the size of the deuteron[14].

The structure of $2q\,2\bar{q}$ mesons may be relevant for studying the QCD vacuum since the agent of their binding is not obvious. According to non relativistic quark models the binding may be due to quantum mechanical exchange phenomena (figure 10)

much in the same way as the molecular binding of H_2 [15].

We would like to stress that such an explanation for the binding of the S^* is far from

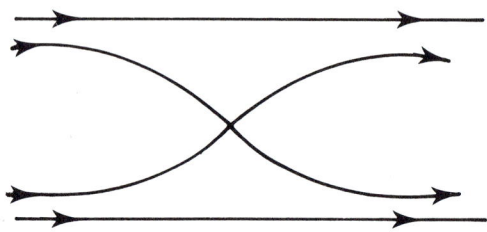

FIGURE 10

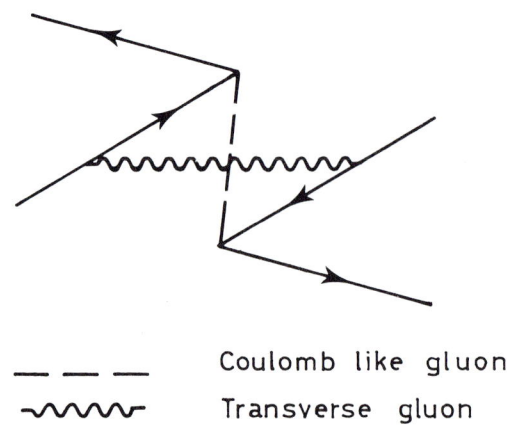

FIGURE 11

trivial since there are on the other side recent theoretical suggestions that quark exchange processes are strongly suppressed due to special properties of the physical vacuum[16].

The binding energy of such a meson when propagating in nuclear matter could be largely affected since many body effects might play an importante role, e.g. if the S^* is a very extended object it could interact with two nucleons at the same time. Calculations of multiple scattering of S^* in nuclear matter would be very important to test these ideas. At the end of the discussion we would like to mention that among the standard pseudoscalar mesons the η' seems to be particularly interesting[6] since it may have an admixture (to the standard $q\bar{q}$ state) of a configuration like in figure 11.

If it is this admixture that shifts the mass of the η' in respect to the mass of the η [17] one could expect large modifications of the mass of the η' in nuclear matter due to possible non linear coupling of the gluon of the η' with the gluons of the nucleus[13]. The combination of transverse and longitudinal gluons is very interesting since it is also the key ingredient in the standard explanation of asymptotic freedom in QCD[18].

As a last topic we would like to point out a somewhat more controversial case which from the few body point of view is however quite appealing.

In strong interactions there are not many cases of three body bound states with the

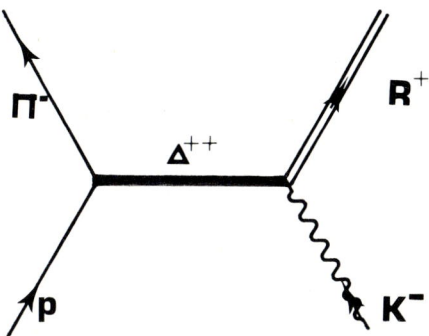

FIGURE 12

exception of the $A = 3$ system.

A possible phenomenological candidate could be a $\Sigma\varphi\varphi$ bound state if one interprets a such an experimental observation of a narrow hyperon state of mass 3,17 GeV (called R) with a width of less than 20 MeV[19-20]. From what is known experimentally one can argue that in contrast to normal hyperons with one strange quark this could be a multiparticle state with many strange quarks (φ is an $s\bar{s}$ state). In such a case multistrange decays are not suppressed.

As an example le us take the reaction

$$K^- p \to \pi^- + R^+$$

A possible scheme of the reaction mechanism involving only hadronic degrees of freedom is given in figure 12.

From a microscopic point of view one can describe the vertex[21]

$$\Delta^{++} + K^- \to R^+$$

as in figure 13.

the R^+ is a $u^2 s^3 \bar{s}^2$ or possibly also a $\Sigma\varphi\varphi$ bound state. Experimentally it is tentatively concluded that the "normal" decay modes cannot be significantly larger than the multistrange modes[19]. These characteristics make the $R(3170)$ rather interesting.

Although the existence of the specific R narrow resonance is still controversial the search for multistrange quark states is extremely important because of the information it can provide on new states of nuclear matter (strange matter).

3. CONCLUSIONS

We hope to have illustrated the interesting connections between few body nuclear physics[22] and specifically few cluster physics applied to hadron spectroscopy in QCD (multiquarks states).

In nuclear matter the nucleons are sufficiently close that important consequences for the two phase model (physical vacuum and perturbative vacuum) and in particular for the

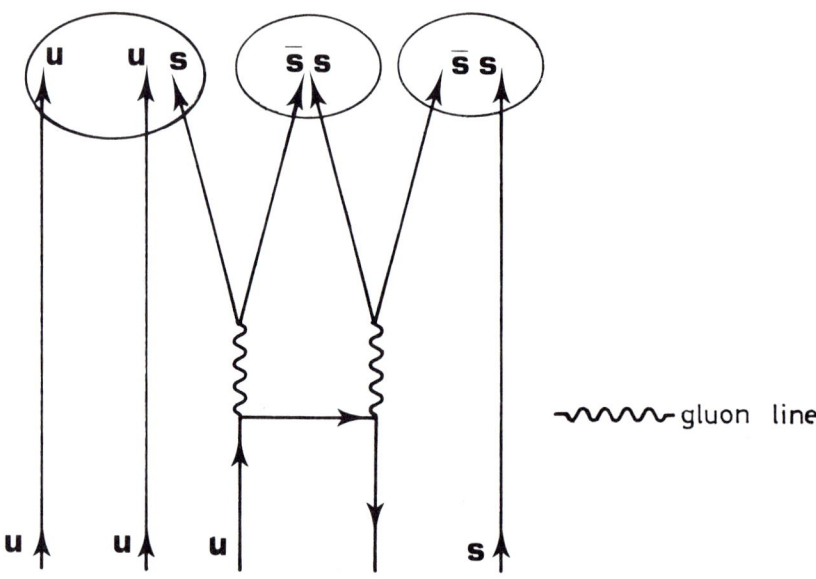

FIGURE 13

sharpness of the transition physical vacuum → perturbative vacuum, may be derived just from the fact that the nucleon independent particle model appears to work rather well[23]. When three nucleons are rather close to each other (i.e. the physical situation where the three body force is relevant) it appears that one can have a promising testing ground for quark effects[24]. This would be particularly relevant if the $N\Delta$ coupled channel treatment would not give the correct result for the triton binding energy and the charge and momentum distribution.

The propagation of narrow glueballs and hybrid mesons in nuclear matter may enlighten specific QCD properties like color confinement.

ACKNOWLEDGMENT

I would like to thank F. Lenz and G. Preparata and W. Sandhas for useful discussions.

REFERENCES

1) G.E. Brown and A.D. Jackson, The nucleon–nucleon inteaction (North–Holland, Amsterdam 1976).
2) T. Sasakawa and S. Ishikawa, Few body systems 1, (1986) 3.
3) P. Sauer, invited talk at the International Symposium on the three–body foce in the three-nucleon system, the George Washington Univ. Washington D.C. April 24-26 (1986).
4) V.R. Pandharipande, invited talk at the International Symposium on the three–body force in the three–nucleon system, the George Washington Univ. Washington D.C. April 24-26 (1986).
5) M.P. Locher, M.E. Sainio, A. Švarc, SIN PR -85-12 (October 1985) Submitted to Advances in Nuclear Physics.
6) F. Cannata, in Nuclear and Particle Physics at Intermediate Energies with Hadrons Eds. T. Bressani and G. Pauli, Conference Proceed. vol 3, Soc. It. Fis. 1986.
7) H.R. Petry, H. Hofestädt, S. Merk, K. Bleuler, H. Bohr and K.S. Narain, Phys. Lett. $\underline{159B}$ (1985) 363.
   L.A. Kondratynk and M. Zh. Shmatikov, Z. Phys. A321 (1985) 301.
8) J.M. Cornwall, Comm. Nucl. Part. Phys. $\underline{15}$ (1986) 223.
9) V.A. Novikov, M.A. Shifman, A.I. Vainshtein and V.J. Zakharov, Nucl. Phys. $\underline{B181}$ (1981) 301
   T.H. Hanson, Phys. Lett. $\underline{166B}$ (1986) 343.
10) P.M. Fishbane, S. Meshkov, Comm. Nucl. Part. Phys. $\underline{13}$, 325 (1984).
11) V.A. Matveev, P. Sorba, Lett. Nuovo Cimento $\underline{20}$ (1977) 435.
12) T. Barnes : Phys. Lett. $\underline{165B}$, (1985) 434.
13) F. Lenz, in "The future of medium and high energy physics in Switzerland" Les Rasses, Switzerland May 17,18 (1985) Proc. SIN CH5234 Villigen Switzerland.
14) R.J. Glauber, Phys. Rev. $\underline{99}$ (1955) 1515
    A.I. Akhiezer, A.G. Sitenko, Phys. Rev. $\underline{106}$ (1957) 1236.
15) F. Lenz, J.T. Londergan, E.J. Moniz, R. Rosenfelder, M. Stingl, K. Yazaki, SIN 85-09 Quark confinement and hadronic interactions to be published in Ann. of Phys. (N.Y.).
16) G. Preparata, lectures at this School and preprint LNF - 86/2 (1986).
17) J.F. Donoghue, H. Gomm, Phys. Rev. $\underline{D28}$ (1983) 2800.
18) N.K. Nielsen, Am. J. Phys. $\underline{49}$ (1981) 1171.
19) J. Amirzadeh et al., Phys. Lett. $\underline{89B}$ (1979) 125.
20) D. Aston et al., Inclusive production of multistrange Hyperons from 11GeV/C K^-p interaction SLAC-PUB-3702 June 1985 (T/E) submitted to Phys. Rev. D.
21) M. Hirata, pivate communication.
22) F. Cannata, Proceed. of the X. European Symposium on the dynamics of Few-body systems 3-7 June 1985 Balatonfüred Hungary (invited talks).
23) B.L. Birbrair, A.B. Gridnev, M.B. Zhalov, E.M. Levin, V.E. Starodubsky, Phys. Lett. B $\underline{166B}$ (1986) 119.
24) A.I. Veselov, I.L. Grach, Yu. S. Kalashnikova and I.M. Narodetskii, Sov. Journ. Nucl. Phys. $\underline{42}$ (1985) 347.

HYPERNUCLEAR PHYSICS

Tullio BRESSANI

Istituto di Fisica Superiore dell'Università di Torino, Torino, Italy
Istituto Nazionale di Fisica Nucleare, Sezione di Torino, Italy

1. INTRODUCTION

The Λ particle has a mass M_Λ of 1115 MeV (i.e. about 20% larger than the nucleon mass); it is neutral, with spin-parity $J^\pi=\frac{1}{2}^+$, isospin I=0 and strangeness S=-1. Since the strangeness is conserved in strong interactions, and the Λ is the lightest between the hyperons, it is stable against strong decay in the nuclear matter. The lifetime of the Λ is 2.6×10^{-10} s, and the dominant decay channel is $\Lambda \to N+\pi$, through the strangeness non-conserving weak interaction. If we consider that the strength of the Λ-N interaction is of the same order of magnitude than that of the N-N interaction, we may expect the existence of nuclear systems named hypernuclei, in wich one or more nucleons are replaced by one or more Λ particles. The same argument may be extended also to the other hyperons (Σ, Ξ,...), notwithstanding some additional difficulty that will be discussed in the following.

The symbol generally used to indicate an hypernucleus is that of the nucleus preceded by a lower case index labelling the hyperon(s) substituting one (more) nucleons. Then $^{12}_\Lambda C$, called hypercarbon Λ, indicates a nuclear system composed by 6 protons, 5 neutrons and a Λ; $^{12}_\Sigma Be$, called hyperberyllium Σ^-, a nuclear system composed by 5 protons, 6 neutrons and a Σ^-; $^{6}_{\Lambda\Lambda}He$, called hyperhelium $\Lambda\Lambda$ or double hyperhelium Λ, a nuclear system composed by 2 protons, 2 neutrons and 2 Λ's.

The hypernuclei were discovered in 1953 by M. Danysz and J. Pniewski [1], in the analysis of the events recorded by a stack of photographic emulsions exposed to the cosmic radiation at about 26 Km from the earth's surface with a balloon. They observed an event whose most reasonable explanation was the existence of a nuclear fragment containing a Λ particle, i.e. a Λ hypernucleus.

The observation and the interpretation was confirmed after a short time by other similar events obtained under the same experimental conditions.

Following the advent of K^- beams (1956), the experimental study of the hypernuclear physics started also at the particle accelerators. However, up to the '70 s, only visualizing techniques were used, namely photographic emulsions or bubble chambers (filled with ^{4}He or heavy liquids). The identification of the hypernuclei and their classification (mass and atomic numbers) were done by means of the kinematical analysis of the disintegration star (method of the decay of the hypernucleus). Two events were further identified as due to the disintegration of ΛΛ hypernuclei [2,3]. The main result from this series of experiments was the measurement of the binding energy of the Λ in the hypernuclei and the identification of the most frequent modes of decay.

From the '70 s the study of the hypernuclear physics was performed by means of electronic techniques (magnetic spectrometers), following the progress in the design of new intense K^- beams, and benefiting from the developement of a new particle detector, the multiwire proportional counter [4]. The physical informations were obtained from the kinematical analysis of the formation reaction (method of the formation of the hypernucleus), allowing then a great qualitative step in the hypernuclear physics. The main results that have been (and still continue to be) obtained from the electronic experiments are the discovery and the identification of Λ-hypernuclear excited states, and the related physical interpretation in terms of microscopic parameters describing the Λ- Nucleus potential. Furthermore narrow Σ- hypernuclear states were identified; their existence was not foreseen on the grounds of the simpler predictions that could be made about the Σ-N interactions in the nuclear matter. There is at present a great effort (and progress) on the theoretical interpretation of the results on Λ- and Σ- hypernuclear states, whereas there is a slowing-down on the flow of new experimental results due to the lack of suitable beams. Recent review articles on hypernuclear physics are due to Povh [5], Gal [6], Walcher [7] and Dover and Walker [8].

From the physical point of view the importance of the hypernuclei stems in the fact that they are nuclear systems composed of 3 baryons instead than 2, like the nuclei. Their mean life ($\sim 10^{-10}$s), compared to the typical time of the nuclear reactions and interactions ($10^{-22} \div 10^{-23}$s), ensures that they are stationary states of systems in which we may study the quantum-mechanical states of a single baryon (the hyperon) interacting strongly with the remaining (A-1) baryons without the complications arising from the Pauli exclusion principle. By comparing the properties of a hypernucleus $^A_\Lambda Z$ with the corresponding ones of a nucleus $^Z A$ one may study in a selective way the different deformations of the core of the (A-1) nucleons induced by a Λ and respectively a neutron. By the study of the Σ- hypernuclei it is possible to obtain quantitative tests of the quark models that may be used for the description of the nuclear matter. Furthermore, from future studies on Ξ and ΛΛ hypernuclei it will be possible to obtain informations of the interactions of elementary particles (like Ξ-N and $\Lambda\Lambda$), not otherwise measurable.

2. GROUND STATE OF THE Λ-HYPERNUCLEI

The ground state of a hypernucleus $^A_\Lambda Z$ is described as a nuclear core $(A-1)Z$ in the ground state with the Λ in a $s_{\frac{1}{2}}$ single particle shell model state. Up to now 24 Λ- hypernuclei in the ground state were unambiguously identified [9], mainly with the disintegration method (nuclear emulsions). The binding energy B_Λ of the Λ for a Λ- hypernucleus in the ground state is defined as:

$$B_\Lambda \text{ (g.s.)} = M_{core} + M_\Lambda - M_{Hy} \qquad (1)$$

where M_{core} is the mass of the nucleus $(A-1)Z$ in the ground state, as given by the tables of nuclear masses, M_Λ the mass of the Λ particle and M_{Hy} the measured mass of the hypernucleus $^A_\Lambda Z$.

Fig. 1 shows the hypernuclei up to now unambiguously recognized in a representation [10] knows as "Λ- hypernuclei table" similar to the Segrè representation for the nuclei.

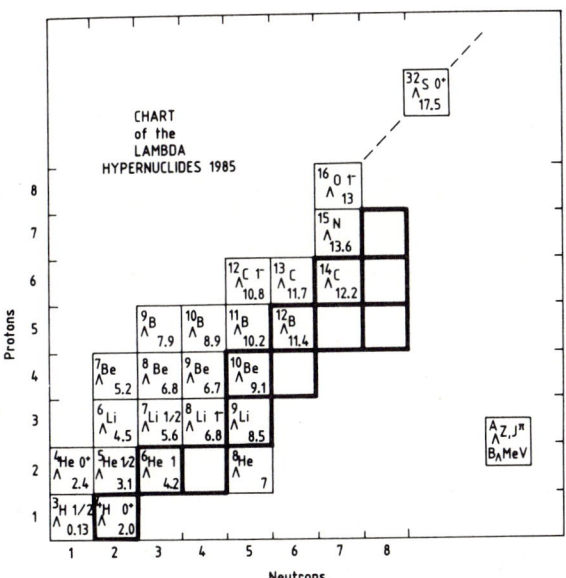

FIGURE 1

The Λ-hypernuclei chart, taken from Ref. 11. The heavy squares indicate the hypernuclei that could be formed by two-body (K^-, π°) reactions.

We may notice that this table is up to now limited to the light hypernuclei (A<16), mainly for experimental reasons. In fact the method up to now most precise and reliable is that based on the kinematical analysis of all the nuclear fragments produced in the disintegration of the hypernucleus. In order to apply the above method it is necessary that all the fragments and the produced mesons are charged and then clearly recognizable in the nuclear emulsion. It is straightforward to understand that these conditions are as much satisfied as lighter is the hypernucleus under study. Fig. 2 shows the variation of B_Λ(g.s.) as a function of the mass number of the hypernucleus.

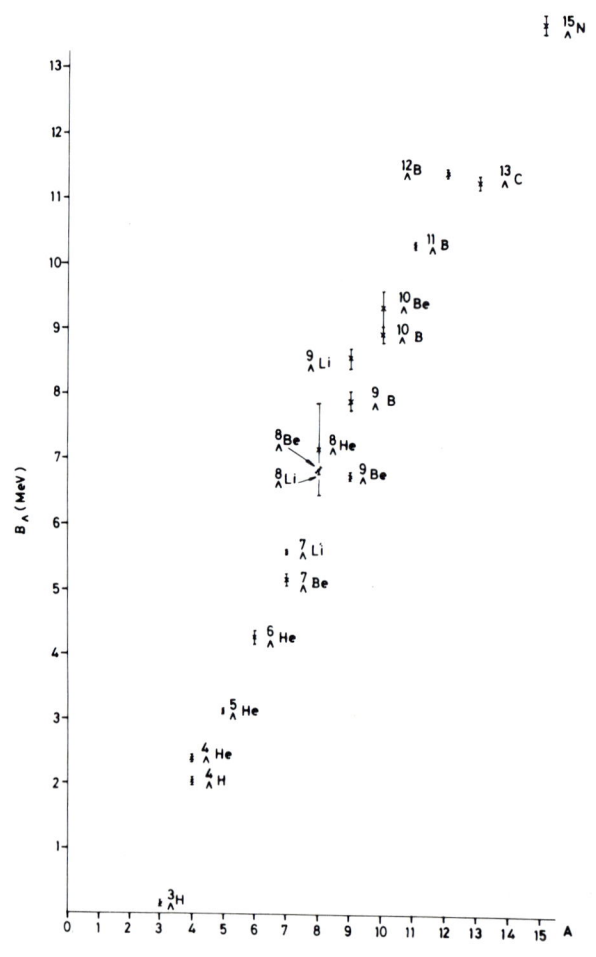

FIGURE 2
Variation of the B_Λ (g.s.) values with the hypernuclear mass number A, taken from Ref. 9.

For heavy hypernuclei an upper limit for B_Λ(g.s.) of 22.7 ± 0.4 MeV was estabilished [12]. The values of B_Λ may be reproduced in a rather satisfactory way by a simple theoretical model following which the Λ is confined in a potential well of radius R approximately equal to the radius of the nuclear core, and of depth D_Λ independent from the hypernuclear mass, in close analogy with the simple models of the nuclear physics. Some discrepancies (namely the $^5_\Lambda$He case) will be discussed in the following in the framework of the quark models for the hypernuclei. We obtain then for heavy hypernuclei a value of D_Λ of ~27 MeV, to be compared with the 50 MeV for the heavy nuclei. This difference is consistent with that, at the elementary interactions level, of the N-N and Λ-N strengths. The "long" range (≤ 2 fm) component of the N-N interaction is in fact dominated by vector bosons (π and ρ) exchange, whereas that of the Λ-N interaction is dominated by K or 2π exchange. The "short" range ($\leq .8$ fm) components are at the contrary comparable. These features of the elementary interactions determine also other macroscopic properties of the hypernuclei, like the spin-orbit potential, that will be discussed in the following.

3. EXCITED STATES OF THE Λ-HYPERNUCLEI

Since a hypernucleus is composed by 3 baryons, the number of configurations of them that might produce an excited state is much larger than for a nucleus, composed only by 2 baryons. We may try to classify the excited states of a Λ-hypernucleus in the following way, by distinguishing the relative configurations of the (A-1) nucleons and those of the Λ:

a) the (A-1) nucleons are in the ground state corresponding to the ground state of the $^{(A-1)}Z$ nucleus and the Λ in one of the single-particle shell model excited states;

b) the (A-1) nucleons are in an excited state (of single particle or collective) and the Λ in the ground state;

c) both the (A-1) nucleons and the Λ are in an excited state.

Fig. 3 represents schematically the above configurations, for the case of the $^{12}_\Lambda$C. For simplicity, only different neutron configurations are shown as representatives of the whole of the nucleons.

This classification, simple to describe, is on the other hand extremely hard to study experimentally, not only for instrumental difficulties but also for interpretative ones, due to the possible existence of configuration mixings. For the study of the spectrum of the excited states of a hypernucleus $^A_\Lambda Z$ one must take into account the spectrum of the nuclei $^A Z$ and $^{(A-1)}Z$ (the core), in order to extract the first and simpler informations. Up to now it has been possible to study the structure of some hypernuclear excited states in the s and p shells with counter techniques.

With the de-excitation γ-ray spectroscopy technique, following the capture of K^- at rest, some low-lying excited hypernuclear states were identified and studied, with a life-time for electromagnetic decay smaller or of the same order of the Λ particle life-time.

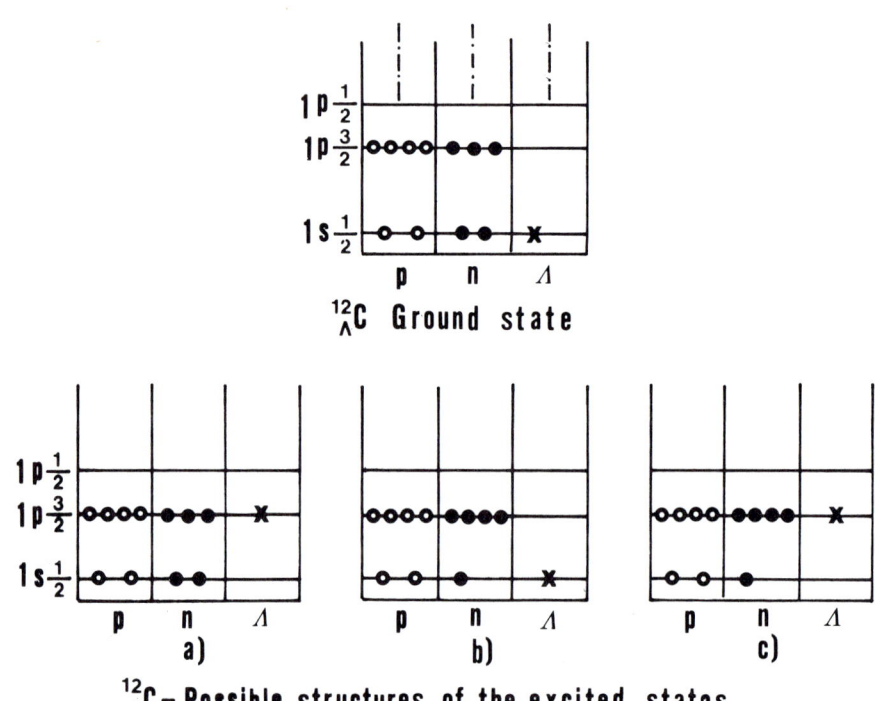

FIGURE 3
Schematic representation of the possible configurations for a Λ-hypernuclear excited state. a) The Λ is in an excited state; b) the nuclear core is in an excited state; c) both the Λ and the nuclear core are in excited states.

These states belong to the categories a) and b). Of a particular interest must be considered the observation and the measurement of the excitation energy [13] for the first excited states in the simple systems $^4_\Lambda H$ and $^4_\Lambda He$, both described as arising from a different coupling (to $J^\pi = 1^+$) of the spin of the Λ in the s-state with the 3 nucleons. Very recently, de-excitation γ-rays were measured in coincidence [14] with the formation reaction (K^-, π^-), opening really a new field for the study of low-lying hypernuclear excited states.

In the formation experiments, the existence of hypernuclear excited states belonging to the categories a) and c) was ascertained. Among all, the so-called "substitutional states" were studied with particular care. As indicated by their definition, they are excited states described as the results of the substitution of a neutron in the nucleus $^A Z$ with a Λ, without otherwise changing the structure of the spatial and spin wave function. The substitutional states were observed for nearly all the light hypernuclei. Fig. 4, taken from a recent experiment [15], shows the quality of the results so far obtained.

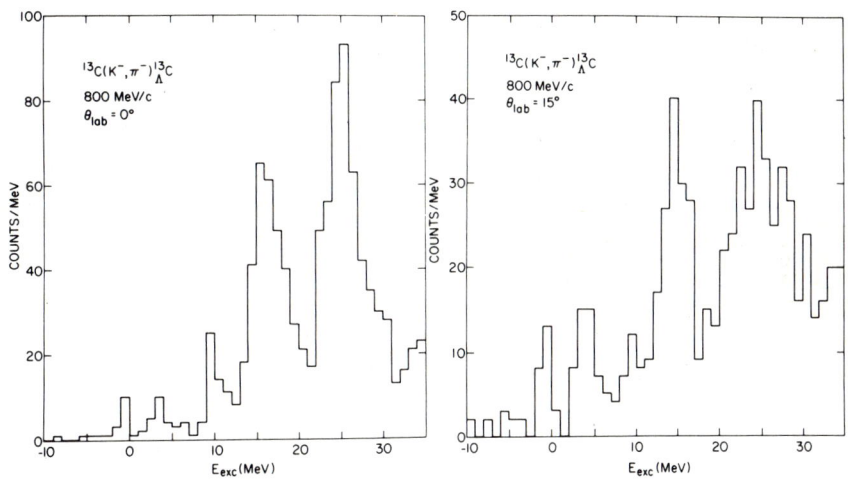

FIGURE 4

An example of the excitation spectrum of the $^{13}_\Lambda C$, from Ref. 15, obtained in a formation experiment with different kinematical conditions (scattering angle).

The study of the substitutional states led to a very important conclusion: the spin-orbit potential of a Λ in a hypernucleus is very small [16], whereas for a nucleon it is of the order of some MeV. It is another manifestation, in the frame of a standard shell model, of the difference at a microscopic level of the long range components of the Λ-N and N-N elementary interactions.

4. PRODUCTION REACTIONS FOR THE Λ-HYPERNUCLEI

For the experiments on the decay of the hypernuclei the knowledge of the detailed formation reaction is of minor importance, whereas it is the most important in the formation experiments. The simplest two-body reactions were then taken into account, and the most studied and used up to now was the "strangeness exchange" reaction:

$$K^- + {}^A Z \rightarrow {}^A_\Lambda Z + \pi^- \qquad (2)$$

which allows also the utilization of a specific kinematical feature [17] of the "elementary" (or "free") reaction

$$K^- + n \rightarrow \Lambda + \pi^- \qquad (3)$$

If we consider the kinematics of the exothermic reaction (3) (Q value = 178 MeV), with the additional condition of detecting the π^- at 0°, we see that there is in the laboratory frame a "magic" value of the K^- momentum, 530 MeV/c, for which the Λ is produced at rest, and the π^- carries a momentum equal to the incident one. If now the (3) occurs on a neutron bound in a nucleus, it is easy to realize that the product of the reaction will be an hypernucleus, in which one of the neutrons of the target nucleus has been replaced by a Λ, without otherwise changing the spatial and spin wave function, i.e. a hypernucleus in a substitutional excited state.

From an analysis of the kinematics of (3) it is also evident that, for incident K^- momenta $\neq$ 530 MeV and/or π^- emission angles $\neq$ 0°, the Λ may be produced with a well known and controlled value of its momentum. Fig. 5 shows the forward kinematics for (3).

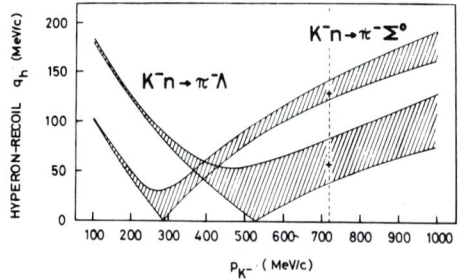

FIGURE 5

Kinematics in the forward direction for the Λ and respectively Σ° production, taken from Ref. 7. The curves which limit the hatched areas downwards are for a π^- emission angle of 0°, those which limit upwards for 6°.

It is then possible, by a simple change of the experimental conditions, to produce hypernuclei by the reaction (2) also in non-substitutional states, allowing a jump of shell between the struck neutron and the produced Λ. Production of p-shell hypernuclei even in the ground state was then observed by this method. It is worthwhile to remind that, in order to exploit the capabilities of (2) for studies of hypernuclear physics, one needs a rather sophisticated and powerful magnetic spectrometer, with good momentum resolution, large acceptance, able to work with high rates of particles. Then the size (and the cost!) of these spectrometers has grown considerably from 1972, when the reaction (2) was first observed experimentally [18], to now. Fig. 6 allows a comparison between the first [19] and the latest [7] of the magnetic spectrometers used for hypernuclear physics.

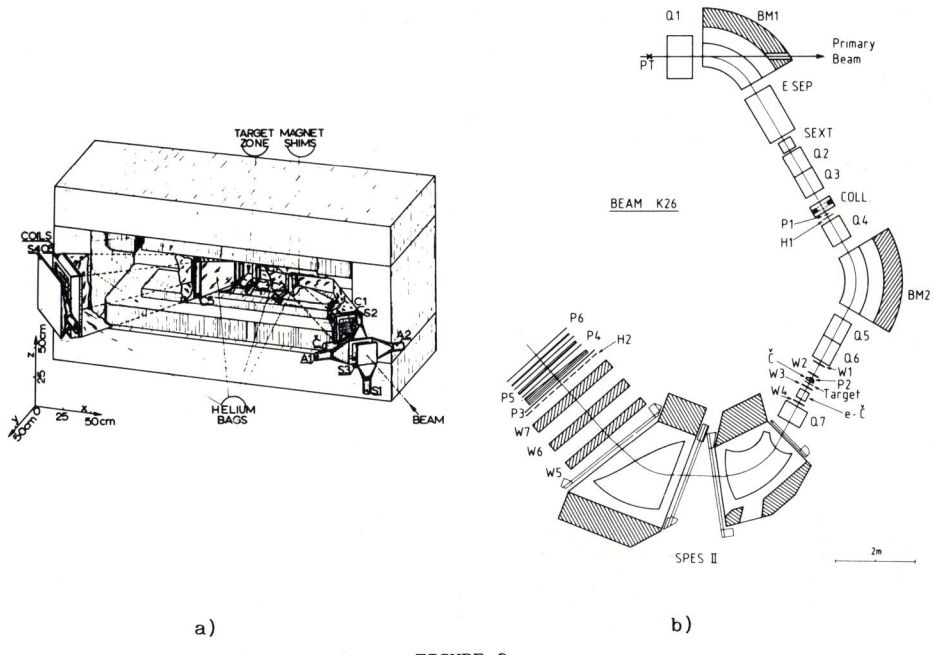

a)   b)

FIGURE 6

A schematic view of the first (a, from Ref. 19) and the latest (b, from Ref. 7) of the magnetic spectrometers used for studies of hypernuclear physics with formation experiments. Details of the experimental arrangements can be found in the original papers. Here we intend only to give an idea of the relative dimension.

The reaction (2) was used also with K^- at rest, and it appears that in this case the Λ is produced with a momentum of 250 MeV/c, i.e. the same order of magnitude of the Fermi momentum of the nucleons in the nucleus. Since we do not expect a particular kinematical selection of the hypernuclear states produced in this way, this technique of production was mostly used in the experiments [13] on the decay of the hypernuclear excited states (γ-ray-spectroscopy).

Recently the reaction:

$$\pi^+ + {}^A Z \rightarrow {}^A_\Lambda Z + K^+ \tag{4}$$

induced by π^+ of ~ 1 GeV was used to produce and study hypernuclei [30]. The free production reaction is in this case:

$$\pi^+ + n \rightarrow \Lambda + K^+ \tag{5}$$

and from the analysis of the kinematics it appears that the Λ is produced with a momentum larger than 250 MeV/c. Since the (5) is an endothermic reaction (Q_{value} = -530 MeV), it follows that there are not "magic" values like for (3). The hypernuclei are produced not in substitutional states but perhaps in high-spin excited states [21]. The reactions (2) and (4) are then complementary for the study of the hypernuclear excited states. The differential cross-sections for the reaction (5) are one order of magnitude lower than the corresponding ones for (3), but the π^+ beams that one might use are at least one order of magnitude larger than the K^- beams, and then the yields of hypernuclei in experiments using (2) or (4) may be considered rather equivalent.

Other reactions for the production of hypernuclei, up to now not so widely used, involve beams of relativistic ions [22] (^{16}O of ~ 2 GeV/nucleon) and of $\bar{p}$ at rest [23].

5. LIFE-TIME AND DECAY MODES OF THE Λ-HYPERNUCLEI

The lifetime of the Λ-hypernuclei is of particular interest since it is related to the weak component of the Λ-N interaction. The free Λ decays mainly through the mesonic decays:

$$\Lambda \begin{array}{l} \nearrow p + \pi^- \quad \text{branching ratio: 64\%} \\ \searrow n + \pi^0 \quad \text{branching ratio: 36\%} \end{array} \tag{6}$$

with a mean life τ_Λ = 2.6 x 10^{-10} s. The nucleon produced in the decay of the free Λ has a kinetic energy of 5 MeV (~ 100 MeV/c in the centre-of-mass frame). When the Λ is embedded in the nuclear matter (an hypernucleus), the main decay modes (6) are strongly depressed by the blocking effect due to the Pauli exclusion principle. The momentum of a nucleon in a nucleus is in fact of the order of 200 MeV/c and then, as more nucleons are present in the hypernucleus, as lower is the number of quantum-mechanical energy states accessible to the nucleon produced by the Λ decay. Furthermore inside the nuclear matter the Λ may weakly interact, without the necessity of conserving the strangeness, with the other nucleons following the reactions:

$$\Lambda \begin{array}{l} + p \to n + p \\ + n \to n + n \end{array} \tag{7}$$

and producing the so-called non-mesonic hypernuclear decays. It is then rather straightforward to expect an increase of the ratio of the non-mesonic decays (7) to the mesonic ones (6) with the mass number of the hypernucleus.

In spite of the great intrinsic interest of the lifetime and decay modes of the hypernuclei, the experimental information is still very scarce. The ratio between non-mesonic and mesonic decays was measured for some light hypernuclei with visualizing techniques, with low statistics and then large errors. The lifetime of some light hypernuclei was measured with different experimental techniques, both electronic and visualizing. The results up to now obtained are reported in Fig. 7, where one can notice the scarcity of the data and the large errors.

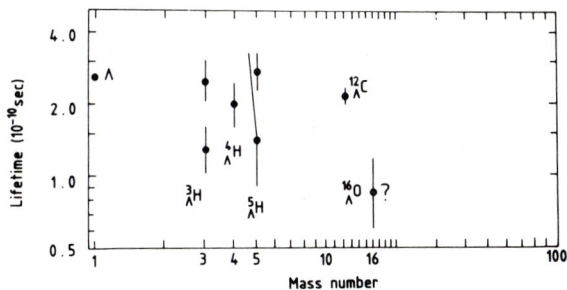

FIGURE 7

Measured lifetimes of the Λ particle and the Λ-hypernuclei, taken from Ref. 23.

It seems that τ_Λ in an hypernucleus is lower than the value for a free Λ, but it is also clear that no extrapolation for the heavy hypernuclei can be attempted at present.

The situation is rather disappointing since in the last few years τ_Λ in heavy hypernuclei has been considered to be directly linked to the possible existence of quark degrees of freedom in nuclei [24]. Following the most advanced (or most ambitious!) nuclear models which try to describe the properties and the interactions of the nuclei starting from the elementary interactions between quarks and gluons, one could observe a partial deconfinement of the quarks (a "percolation" from the bags in close contact) in the nuclear matter, with the consequent appearance of color degrees of freedom. Inside the nuclear matter one could then expect the existence of systems of 6 q, 9 q, bound by color forces. The value of τ_Λ in heavy hypernuclei, following the above models, would be directly related to the probability of occurrence of these multi-quarks configurations inside the nuclear matter.

6. Σ-HYPERNUCLEI

The discovery of the Σ-hypernuclei is very recent [25] (1980) and this fact may be considered rather surprising if one takes into account that the Σ hyperon, though appearing into three states of charge (Σ^+, Σ^-, $\Sigma°$), is for many aspects similar to the Λ, and was discovered and known from a long time. The reason was that Σ-hypernuclei were not expected to exist in well-defined energy states (experimentally detectable as narrow peaks in formation reactions) since, beeing the mass of the Σ about 80 MeV larger than the mass of the Λ, in the nuclear matter the conversion reaction Σ + N → Λ + N through strong interaction is allowed. The lifetime of a Σ in the nuclear matter was then expected to be of the same order of magnitude than the typical time of the strong interactions, $\sim 10^{-22}$ s and not 0.8×10^{-10} s as for the free Σ. From the Heisenberg principle $\Delta E . \Delta t \geq \hbar$ an energy spread of about 30 MeV was then expected for a hypothetical bound state corresponding to a Σ-hypernucleus.

On the contrary, and with great surprise, Σ-hypernuclei in well-defined energy states were discovered. The production reactions were:

$$K^- + {}^A Z \rightarrow {}^A_{\Sigma°} Z + \pi^- \tag{8}$$

$$K^- + {}^A Z \rightarrow {}^A_{\Sigma^+}(Z-1) + \pi^- \tag{9}$$

$$K^- + {}^A Z \rightarrow {}^A_{\Sigma^-}(Z-1) + \pi^+ \tag{10}$$

corresponding to the free reactions:

$$K^- + n \rightarrow \Sigma° + \pi^- \tag{11}$$

$$K^- + p \rightarrow \Sigma^+ + \pi^- \tag{12}$$

$$K^- + p \rightarrow \Sigma^- + \pi^+ \tag{13}$$

In analogy to what discussed in Sec. 4, even for the reactions (11), (12) and (13) it is possible to find, with the additional condition of observing the π at 0°, a "magic" value for the K^- momentum for which the Σ particle is produced at rest in the laboratory frame, and the π carries out all the incident momentum. Due to the mass difference between Σ^+, Σ^- and $\Sigma^°$ the magic values of the K^- momenta for the reactions (11), (12) and (13) are slightly different; as a typical value we may assume 280 MeV/c (see Fig. 5).

Continuing with the analogy, it is rather evident that in these kinematical conditions the Σ-hypernuclei will be produced mainly in substitutional states, for which a Σ hyperon would have replaced a nucleon without otherwise modifying the spatial and spin structure of the nuclear wave function. In these conditions about 10 Σ-hypernuclei were observed [26,27] in the p-shell. The width of the experimental peaks observed in the formation experiments was compatible with the experimental energy resolution of the apparatus; then only an upper limit for the natural width of the Σ-hypernuclear states of 3 MeV could be established. In spite of the considerable theoretical efforts [6,28], a clear-cut explanation for the occurence of narrow Σ-hypernuclear states has not yet been put forward. Fig. 8 shows an example of an experimental missing-mass spectrum in which the formation of $^{12}_\Sigma$-Be in different states is clearly visible. We notice that this spectrum was obtained with K^- at rest. As discussed in Sec. 4, and observing the kinematics of Fig. 5, we do not expect a preferential population of the substitutional states in Σ-hypernuclei with K^- at rest. It is suggested [29] that the experiments with K^- at rest are more advantageous for the study of the Σ-hypernuclei than for the Λ-hypernuclei.

The analysis of the substitutional states of the Σ-hypernuclei allowed the determination of the central and spin-orbit terms for the Σ-Nucleus potential. Table I summarizes the results for the different baryons (N, Δ, Λ, Σ) in a nucleus [30]. The most striking feature is the difference between the Λ and Σ spin-orbit potentials. It is explained in a very simple and elegant way by the additive quark model [31], which takes into account the detailed quark structure of the baryons. Following it, only the s- and d- quarks contribute to the spin-orbit potential, not the s-quark for several reasons, in particlar the large mass. If we look at the quark structure of the Λ and Σ hyperons (see Fig. 9), we observe that the u- and d-quarks are coupled to $\vec{s}=0$, $\vec{I}=0$ and $\vec{s}=1$, $\vec{I}=1$ respectively. It follows then immediately that the spin-orbit potential for the Λ is expected to vanish, whereas the factor 4/3 of the Σ compared to N is explained by a factor 2 for the spin structure of the baryon and a factor 2/3 for the average interaction with the u- and d-quarks of the nucleus (the s-quark do not contribute).

TABLE I
Experimental values for the central potential V_c for N, Δ, Λ, Σ in the nucleus. The esperimental values of the spin-orbit potential are normalized to the nucleon one and compared to the prediction of the additive quark model. Taken by Pohv [30].

System	N	Δ	Λ	Σ	Ξ
Central potential V_c (MeV)	50	30	30	15-20	
Spin-orbit potential (experiment)	1	1	0	1-5/3	
Spin-orbit potential (additive quark model)	1	1	0	4/3	-1/3

7. DOUBLE HYPERNUCLEI

Up to now two events ascribed to the existence of the $\Lambda\Lambda$ hypernuclei $_{\Lambda\Lambda}^{6}$He and $_{\Lambda\Lambda}^{10}$Be (or $_{\Lambda\Lambda}^{11}$Be) were found [2,3]. These observation are rather old (1963) and were obtained by analyzing the events registered by emulsion stacks exposed to K⁻ beams at high energy. A two-step production mechanism was proposed:

a) a Ξ^- (S=-2) is produced by the interaction of the incident K⁻

b) the Ξ^- is subsequently absorbed by a light nucleus in the emulsion and produces a double hypernucleus via the elementary reaction $\Xi^- + p \rightarrow \Lambda + \Lambda$. The double hypernucleus is identified by the two subsequent weak decays $\Lambda \rightarrow p + \pi^-$. The mass and atomic numbers of the double hypernucleus may be unambiguously determined if no neutral particles are emitted in the disintegration.

There is an apparent contradiction between the strong physics interest for the double hypernuclei, whose study could supply precious informations about the Λ-Λ elementary interaction, not otherwise possible, and the Λ statistics, and the complete lack of recent and more precise experimental results.

This is due to the fact that the photographic emulsion technique is today neglected, whereas the K⁻ beams into operation have intensities too low for an attempt of studying with electronic techniques two-body production reactions of double hypernuclei such as:

$$K^- + {}^Z_A \rightarrow {}^A_{\Lambda\Lambda}(Z-2) + K^+ \tag{14}$$

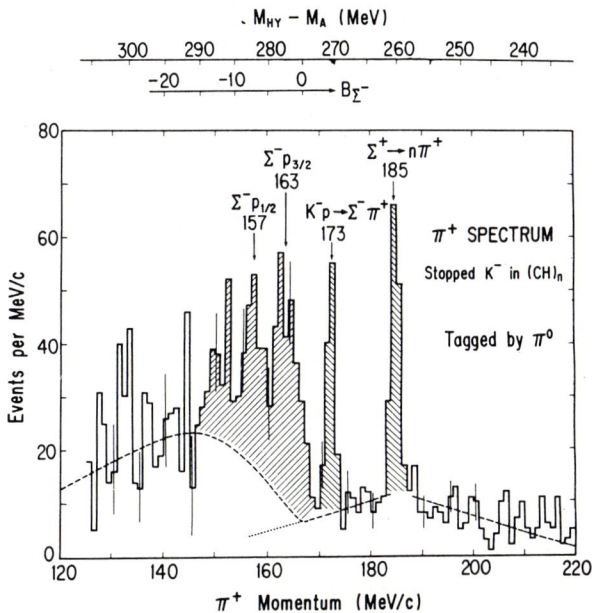

FIGURE 8

The π^+ spectrum from K^- stopping in a plastic $(CH_2)_n$ scintillator, taken from Ref. 29. The three peaks at 157, 163 and 173 MeV/c are attributed to the formation of ${}^{12}_\Sigma$Be in different states.

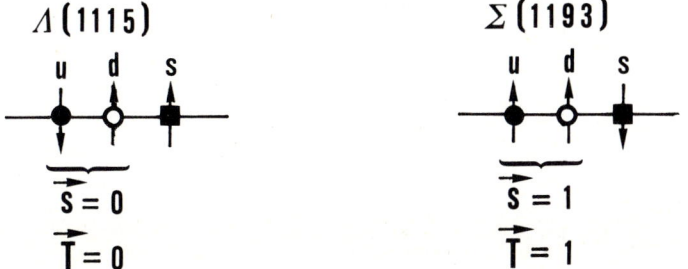

FIGURE 9

Quark structure of the Λ and Σ^0 hyperons.

The predicted cross sections are in fact at least 3 orders of magnitude lower than the Λ-hypernuclei cross sections [32].

The simplest ΛΛ hypernucleus could be considered the H particle, a bound (and stable) Λ-Λ system whose existence was predicted by QCD calculations in the framework of the MIT bag model [33]. It would be a di-baryon with quantum numbers $J^\pi = 0^+$, S=-2, electrically neutral, having a simple and fully symmetric configuration of 6 quarks (2 u, 2 d and 2 s) all in an s-state, in a flavor singlet. Among the few quark systems, it would be the analogous of the α-particle in the few nucleon systems. Experimental searches with non-dedicated apparatus did not succeed in ascertaining the existence of the H particle. A dedicated experiment is now planned at the AGS [34]. Very recently the existence of the H has been invoked [35] in order to explain the anomalous frequency of events coming from the pulsar Cygnus X-3.

8. CONCLUSIONS AND OUTLOOK

From the previous considerations it appears rather clearly that the hypernuclear physics is a field of research still in its infancy, with big potentialities for a better understanding of both the Nuclear and Particle Physics, and which may deserve big surprises. Its importance was recognized only in the last few years, following the developement of quark models for the description of the nuclei. We remind that it is the frontier of the modern Nuclear Physics, and follows its hystorical evolution, following which the properties of the nuclei were derived from the most advanced models used in the description of the interaction between the elementary constituents. The extension to the Nuclear Physics of the quark picture, so useful for the description of the elementary interactions, is extremely hard, and must answer to the fundamental questions:
a) which is the nature of the confinement of the quarks in the hadrons [36]?
b) there are color degrees of freedom in nuclei?

It is possible that a careful study of the hypernuclear physics may supply one of the most clear-cut answers to the above problem. Following the quark description, a hypernucleus differs from a nucleus only by the fact that an s-quark, with very specific properties and then easily recognizable, substitutes one of the many u- and d-quarks of an ordinary nucleus. The s-quark may be considered like a "marked" quark and, in analogy to what happens in molecular physics with the marked nucleus technique, one could hope that by a careful study of the properties of the s-quark one could gain a good understanding of the properties of the full assembly of u- and d-quarks (the nucleus).

Fig. 10 may help to understand in a simple way the above arguments.

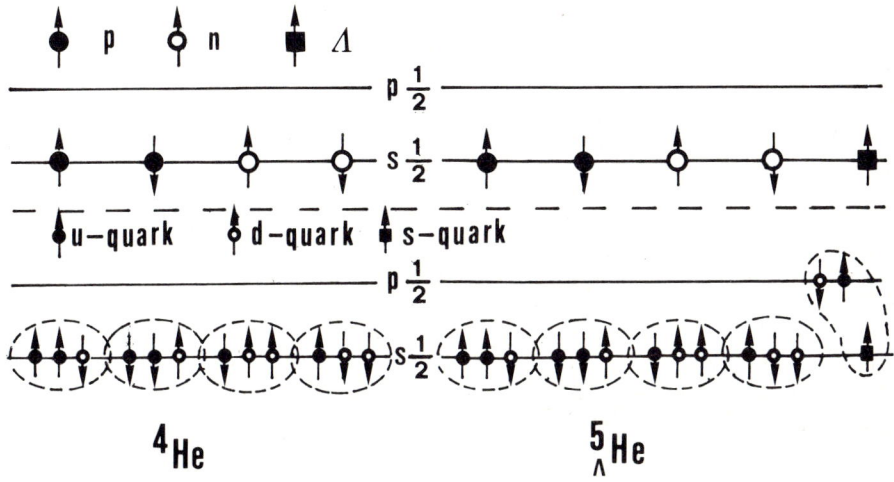

FIGURE 10

Baryon (upper row) and quark (lower row) configurations for the ^{4}He and $^5_\Lambda$He systems.

In the first row the baryon configurations for the simple nucleus ^{4}He and the simple hypernucleus $^5_\Lambda$He are reported. The baryons are assumed to be distinguishable, with the component quarks fully confined. In the second row the opposite hypothesis is assumed, i.e. that the quarks are fully deconfined so that they occupy quark shell orbitals (one must remember that in filling the quark orbitals the color has to be taken into account, carrying a factor 3 in the occupation number). For the ^{4}He, there is no substantial difference between the nucleon and quark description. For $^5_\Lambda$He, on the contrary, one can notice that the u- and d-quarks of the Λ cannot stay in the $s_{\frac{1}{2}}$ orbital, but in the $p_{3/2}$ one, whereas the s-quark may occupy the lowest energy state. In the baryon picture, all the 5 elementary components may occupy the $s_{\frac{1}{2}}$ orbital. Then, if there is a partial deconfinement of the quarks in nuclei, this effect could be visible more easily in hypernuclei. In fact it is known that B_Λ(g.s.) in $^5_\Lambda$He is too small, and it was suggested [37] that a partial deconfinement could be the answer to the problem.

Unfortunately the lack of suitable beams prevents at moment a substantial improvement of hypernuclear physics. Practically all the measurements so far mentioned come from first generation experiments. There is a general need of better statistics and better resolution and of a more wide systematics over the mass number. Other two-body reactions could be used for producing hypernuclei, benefiting also on the technological improvements of the experimental set-ups. A typical example is the $(K^-, \pi^°)$ reaction [11,38], that would produce the "mirror" hypernuclei with respect to the (K^-, π^-), represented by heavy squares in Fig. 1. High resolution γ-spectroscopy in coincidence with the formation reaction could allow the study, i.e., of isospin breaking effects in mirror hypernuclei, expecially sensitive to the quark substructure of the nucleus. Completely open are the fields of Ξ-hypernuclei and double hypernuclei. In conclusion, the hypernuclear physics is by itself a good justification for building a dedicated machine, like the proposed EHF [39].

ACKNOWLEDGEMENT

I would like to thank Miss L. Ceretta for the careful preparation of the typed version of the manuscript.

REFERENCES

1) M. Danysz and J. Pniewski, Phil. Mag. 44 (1953) 348
2) M. Danysz et al., Nucl. Phys. 49 (1963) 121
3) J. Prowse, Phys. Rev. Lett. 17 (1966) 782
4) G. Charpak, R. Bouclier, T. Bressani, J. Favier and Č. Zupančič, Nucl. Instr. Meth. 62 (1968) 262
5) B. Povh, Nuclear Physics with Strange Particles, in Progr. Part. Nucl. Phys., in print; preprint MPI H-1986-V3
6) A. Gal, in Proc. of the Tenth International Conference on Particles and Nuclei, Heidelberg, July 30 - August 3, 1984, eds. B. Povh and G. Zu Putlitz, Nucl. Phys. A434 (1985) 381
7) Th. Walcher, in Proc. of the Tenth International Conference on Particles and Nuclei, Heidelberg, July 30 - August 3, 1984, eds. B. Povh and G. Zu Putlitz, Nucl. Phys. A434 (1985) 343
8) C. B. Dover and G. E. Walker, Phys. Rep. 89 (1982) 1
9) M. Jurič et al., Nucl. Phys. B52 (1973) 1
10) B. Povh, Progr. Part. Nucl. Phys. 5 (1981) 245
11) D. Measday, in Proc. of the Workshop on Nuclear and Particle Physics at Intermediate Energies with Hadrons, eds. T. Bressani and G. Pauli, Conf. Proceedings Vol. 3 (Italian Physical Society, Bologna, 1986) 183
12) J. Lemonne et al., Phys. Lett. 18 (1965) 354

13) M. Bedjidian et al., Phys. Lett. 83B (1979) 252
14) M. May et al., Phys. Rev. Lett. 51 (1983) 2085
15) M. May et al., Phys. Rev. Lett. 47 (1981) 1106
16) W. Brückner et al., Phys. Lett. 79B (1978) 157
17) H. Feshbach and A. K. Kerman, in Preludes in Theoretical Physics, eds. A. De-Shalit, H. Feshbach and L. Van Hove, (North-Holland, Amsterdam 1966) 260
18) G. C. Bonazzola et al., Phys. Lett. 53B (1974) 297
19) G. C. Bonazzola et al., Nucl. Instr. Meth. 123 (1975) 269
20) M. Milner et al., quoted in Ref. 6
21) C. B. Dover, L. Ludeking and G. E. Walker, Phys. Rev. C22 (1980) 2073
22) K. J. Nield et al., Phys. Rev. C13 (1976) 1263
23) M. Berrada et al., in Physics with Antiprotons at LEAR in the ACOL Era, eds. U. Gastaldi, R. Klapish, J. M. Richard and J. Tran Thanh Van (Ed. Frontières, Gif-sur-Yvette), 1985, 627
24) C. Y. Cheung, D. P. Heddle and L. S. Kisslinger, Phys. Rev. C27 (1983) 335
25) R. Bertini et al., Phys. Lett. 90B (1980) 375
26) H. Piekarz et al., Phys. Lett. 110B (1982) 428
27) R. Bertini et al., Phys. Lett. 136B (1984) 29
28) J. Dabrowski, in Proc. of the Tenth International Conference on Particles and Nuclei, Heidelberg, July 30 - August 3, 1984, eds. B. Povh and G. Zu Putlitz, Nucl. Phys. A 434 (1985) 373
29) T. Yamazaki, in Nuclear Physics with Electromagnetic Probes, eds. A. Gerard and C. Samour, Nucl. Phys. A 446 (1985) 467
30) B. Povh, in Proc. of the Workshop on Nuclear and Particle Physics at Intermediate Energies with Hadrons, eds. T. Bressani and G. Pauli, Conf. Proceedings Vol. 3 (Italian Physical Society, Bologna, 1986), p. 43
31) H. J. Pirner, Phys. Lett. 85B (1979) 190
32) C. B. Dover, in Proc. of the 1979 Int. Conference on the Hypernuclear and Low Energy Kaon Physics, Nukleonika 25 (1980) 521
33) R. L. Jaffe, Phys. Rev. Lett. 38 (1977) 195
34) G. Franklin and P. D. Barnes, Bookhaven AGS proposal 813 (1985)
35) I. B. Khriplovic et al., preprint INP 85-117, Novosibirsk (1985)
36) G. Preparata, this volume
37) E. V. Hungerford and L. C. Biedenharn, Phys. Lett. 142 B (1984) 232
38) E. Chiavassa, this volume
39) F. Bradamante, this volume.

HADRON SCATTERING ON NUCLEI

Sergio COSTA

Istituto di Fisica Superiore della Università di Torino
Istituto Nazionale Fisica Nucleare - Sezione di Torino

The idea or, perhaps, the need of thinking of a European Hadron Facility (EHF) is a unavoidable consequence of the rapid evolution and success of intermediate energy nuclear physics (IENP) which is steadily gathering adepts among low-energy nuclear and particle physicist.

IENP, the site where these two field of research blend, grew typically in some meson factories, and now the times are ripening for a second generation of these facilities.

At Los Alamos people is preparing LAMPF II. EHF design seems, at present, not so unlike.

In this lecture I wish therefore to report some points, discussed at LAMPF workshops on hadron scattering.

Since Rutherford's times, scattering experiments have paved the road towards a better knowledge of the subatomic world. A variety of probes, exploiting every type of know interaction, have been used to gain informations on nuclear structure and dynamics, with a resolution increasing with the momenta transferred to the struck target.

The analysis of the data has been, of course, easier, in a certain sense, when the "rules of the game" were sufficiently well known, as it is the case of probes interacting electromagnetically.

Electron scattering has been indeed one of the main sources of reliable and accurate data on some nuclear properties, the most obvious example being the ground state charge density, $\rho(r)$. The electromagnetic interaction between the electron and the nucleus can be understood as due to the exchange of (space-like) virtual photons. In the simplest case a single photon is exchanged (first Born approximation) and the elastic scattering cross section is given by:

$$\left(\frac{d\sigma}{d\Omega}\right)_{el} = \left(\frac{d\sigma}{d\Omega}\right)_{Mott} |F(q^2)|^2$$

where the Mott cross section for point-like scatterers (hard scattering) turns out to be modified for the finite size of the nucleus via the "form-factor" $F(q^2)$:

$$F(q^2) = \int \rho(\vec{r})\, e^{i\vec{q}\cdot\vec{r}}\, d\vec{r}$$

The momentum transfer $q^2 = 4p_i p_f \sin^2\theta/2$, θ being the scattering angle and $p_{i,f}$ the initial and final momentum of the electron. Comparing the experimental and Mott cross sections one directly measures the form-factor, or, Fourier anti-trasforming, the charge density:

$$\rho(r) = \frac{1}{2\pi^2} \int_0^\infty F(q^2)\, \frac{\sin qr}{qr}\, q^2 dq$$

This procedure is independent of any nuclear model and only faces the problem of measuring $F(q^2)$ up to sufficiently high values of q. In Fig. 1-a the

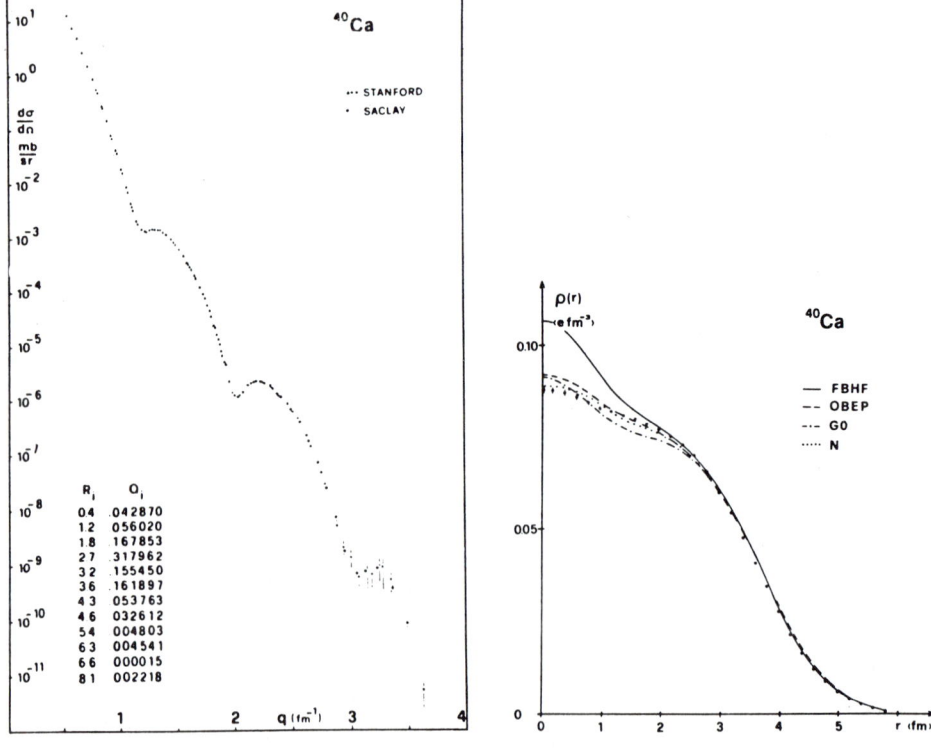

FIGURE 1

a) Elastic electron scattering cross section of ^{40}Ca measured at Stanford and Saclay.

b) Experimental charge density (dots) of ^{40}Ca compared with theoretical calculations. For details see ref.1

elastic cross section[1] is shown for ^{40}Ca up to $q \approx 3.7$ fm^{-1}. While with low-q experiments one would be able to extract only the gross features of the charge distribution, like the r.m.s. radius, here the measured cross section (which spans over 12 orders of magnitude) allows to obtain the very detailed charge density distribution shown in Fig. 1-b: $\rho(r)$ is determined with great accuracy even in the deep interior of the nucleus, where the test of theoretical calculations becomes more severe.

Inelastic electron scattering allows in a similar way to learn about properties of the nuclear dynamics like the transition charge (or current) densities which represent the charges (or currents) variations associated with a specific nuclear transition.

Unfortunately, electrons are almost blind to neutrons, and to obtain similar informations on nuclear matter one must use strongly interacting probes. For lack of "strong Maxwell equations", a "theory" of the hadron-nucleus scattering has to be constructed finding prescriptions for mixing together in a many-body calculation several phenomenological ingredients: the basic hadron-nucleon scattering amplitudes as well as effects associated with the nuclear medium, like nucleon densities, Fermi motion, absorption and so on.

Most calculation have been based either on multiple scattering methods or on derivation of optical model potentials. The usual starting point is the Lippmann-Schwinger operator equation and several approximations are introduced to simplify the calculations.

In the optical potential approach of Kerman-McManus-Thaler[2] (KMT) the two-body scattering amplitudes are involved in an iterative series to express the hadron-nucleus T-matrix. The optical potential is identified[3] in a multiple excitation series which contains all virtual hadron and nuclear excitations in intermediate states. At lowest order, these excitations are neglected (which roughly correspond to neglect two-body and higher correlation functions), and usually the impulse and factorization ($t \cdot \rho$) approximations are adopted.

In Fig.2 some LAMPF data on 500 MeV polarized protons elastically scattered off several targets are compared with KMT predictions[4].

These predictions are based on p-N amplitudes obtained from phase-shift analysis, proton densities from e$^-$-scattering, and neutron densities (3-parameter

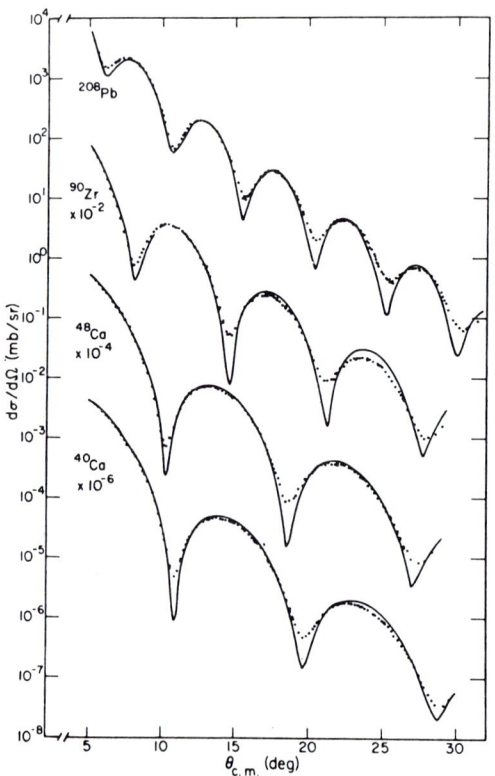

FIGURE 2

Elastic cross sections for 500 MeV proton scattering on ^{40,48}Ca, ^{90}Zr and ^{208}Pb. The solid curves are the results of KMT calculations[4].

Fermi distributions) adjusted to obtain best fits to the data. The fit is manifestly not perfect, but things seem even worse if one looks at the analyzing power[4].

These failures have been recently ascribed to the use of the non-relativistic Schrödinger equation in which relativity is only treated in the kinematics[5,6]. Including negative energy intermediate scattering states of the projectile by means of the Dirac equation in a relativistic impulse approximation (RIA) which employs the usual $t \cdot \rho$ form of the optical potential, improved descriptions of the experimental results have been achieved[6] (Fig.3).

A good fit of proton-nucleus scattering at 800 MeV has been obtained as well with a very careful, second order KMT analysis[7] from which point neutron density

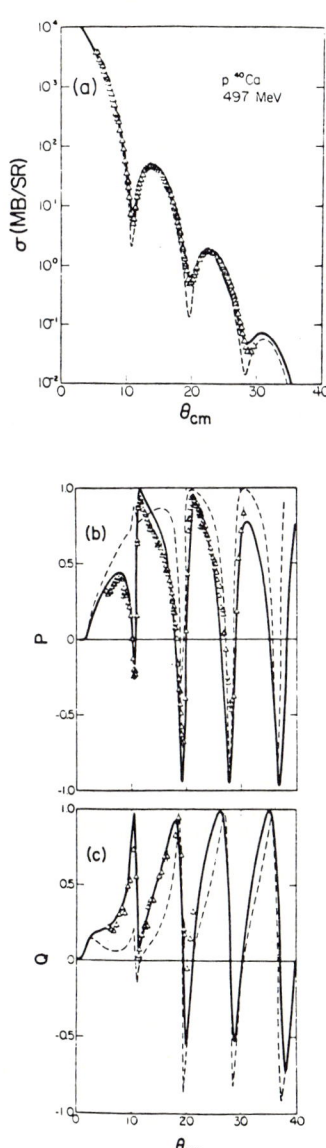

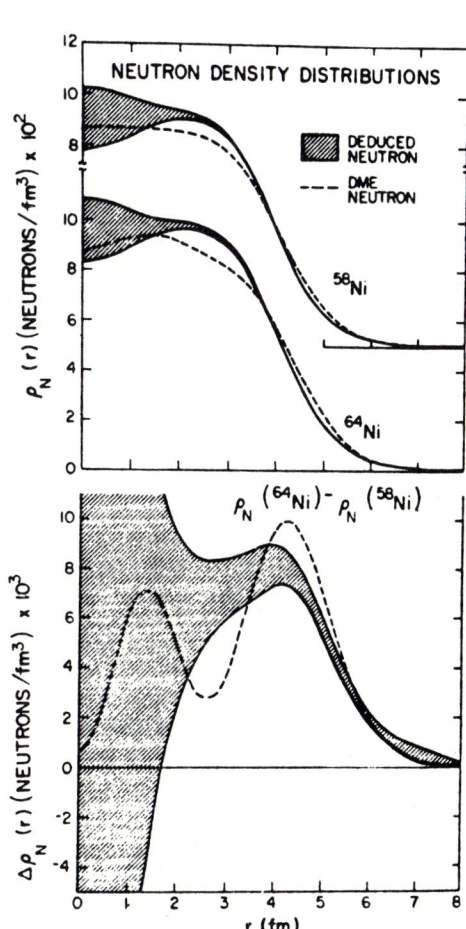

FIGURE 3
RIA (solid curves) and KMT (dashed curves) predictions[18] compared to data[4,19] for 500 MeV polarized protons on ^{40}Ca. a) elastic cross section; b) analyzing power; c) spin rotation parameter.

FIGURE 4
Point neutron density distributions for ^{58,64}Ne and their difference deduced (shaded bands) from second-order KMT analysis. The dashed curves are theoretical predictions by the Density Matrix Expansion.

distributions have been extracted. In the case of ^{58,64}Ni these are shown in Fig.4, but they seem still hardly comparable with the analogous proton densities obtained by e^--scattering.

It is not easy, indeed, to pick up informations on the nuclear interior using probes that interact mainly in the periphery of the nucleus.

A hadron with "semi-strong" interaction would be necessary.

Looking at Fig.5 one sees at once that, for that goal, p, π and K^- are more

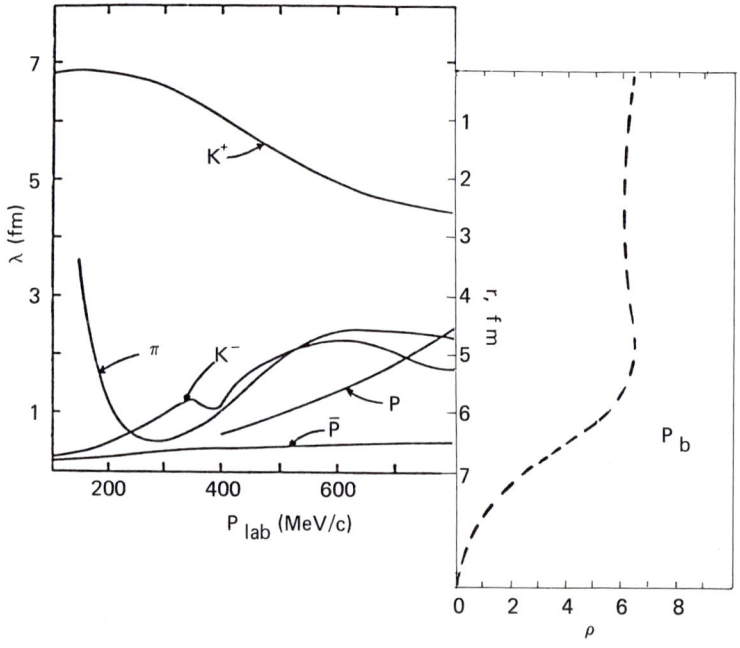

FIGURE 5

Mean free path λ of various hadrons in nuclei as a function of momentum[11]. The charge distribution of ^{208}Pb from e^--scattering[12] is shown to give an idea of the nuclear regions that may be explored with those probes.

or less in the same situation; there is however a possible candidate: the K^+

We will come back to kaons later. Now I wish first to report some comments concerning the interest of extending at higher energies (such as those available at EHF) the use of pion and proton beams to gain insight into simple modes of nuclear excitation.

Among the features of the π-N interaction that make the pion an interesting nuclear probe, one of the most outstanding is the isovector property: commuting from a π^+ to a π^- beam one interchanges in a scattering experiments the roles of the protons and of the neutrons.

In the elementary π-N system, π^+-p and π^--n form pure T = 3/2 states, while other combinations are mixtures of T = 1/2 and T = 3/2. At the Δ_{33} resonance the π-N cross section is dominated by the T = 3/2 channel, as can be seen from Fig.6 where the π-N cross section is shown in terms of isospin channels representation.

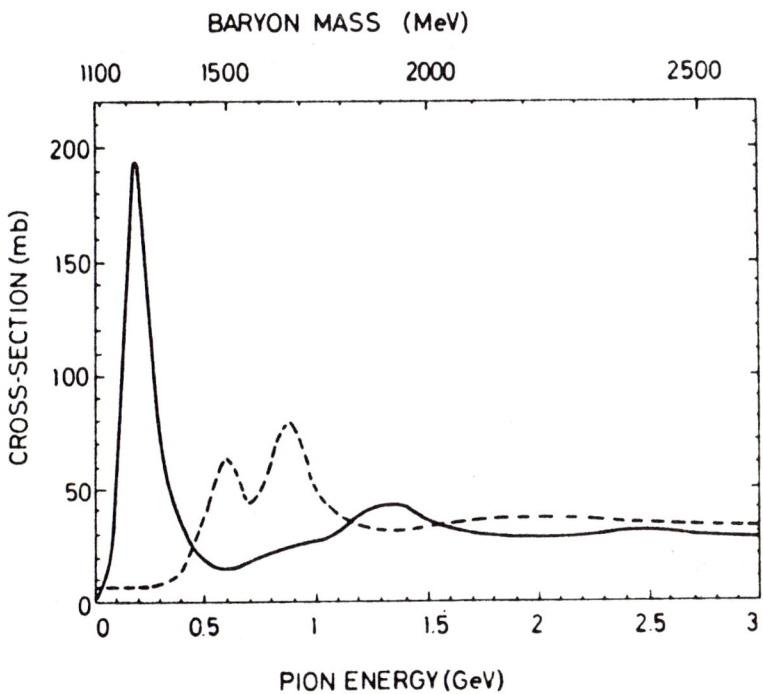

FIGURE 6
Total cross section for the T = 1/2 (dashed curve) and T = 3/2 (solid curve) isospin states of the pion-nucleon interaction.

At energies around the resonance the π^+ interacts essentially with the protons and the π^- with the neutrons in the nucleus. This remarkable isovector sensitiveness has been, indeed, fully exploited in a large number of inelastic scattering experiments, even because near the well known Δ_{33} there are no other resonances, with different properties, to confuse the dynamical picture of the excitations.

Going up with pion energy, say between .5 and 1 GeV, some isovector sensitiveness is still there (Fig.6) but now the increasing inelasticity in the elementary interactions and the large number of overlapping resonances are likely to pose serious problems for the interpretation of the data even if these resonances were all well known.

It is therefore not surprising to hear divergent opinions[8,9] on the usefulness of pion beams for nuclear structure studies in this energy range.

Above 1 GeV the isovector sensitivity is lost in pion-nucleon scattering and so it is very unilikely to have any isovector effect in nuclei.

Another point concerns the relative strength of the central, t_c, and spin-orbit, t_{LS}, parts of the interaction, as only the latter can induce the interesting spin-flip transitions to unnatural parity states. Fig. 7 gives[8] t_c and t_{LS} as a function of momentum transfer for pion energies of 165 and 600 MeV. There could be no doubts that the lower energy is the more interesting one: at $q \simeq 1.5$ fm^{-1}, for instance, $t_{LS}/t_c \simeq 10$ while, for the same momentum transfer but at 600 MeV, $t_{LS}/t_c \simeq 10^{-1}$.

The prospects for proton-nucleus elastic or inelastic scattering studies above 1 GeV are not much more encouraging[10]. In addition to a short mean free path, in fact, Fig. 8 indicates an overall isoscalar behaviour. Moreover, Fig.9 suggests[13] that the relatively uninteresting isoscalar central part of the N-N matrix has the unfortunate tendency to become much larger than the corresponding spin-, isospin-, spin-isospin-flip quantities, at energies above 500 MeV.

These more interesting parts actually have a ratio to the isoscalar part which is more favourable in the 200-500 MeV energy range.

In conclusion, it seems that there is no special need or advantage for conventional IENP to extend at higher energies the elastic or inelastic scattering studies with pion or proton beams.

Nevertheless, there may well be a number of other interesting experiments to do, pertaining to a less conventional nuclear physics (but still quark-free). In the following, some suggestions[8,14] are reported:
- π-N resonances above 300 MeV.

This is a program of primary importance. The existing data are not always of

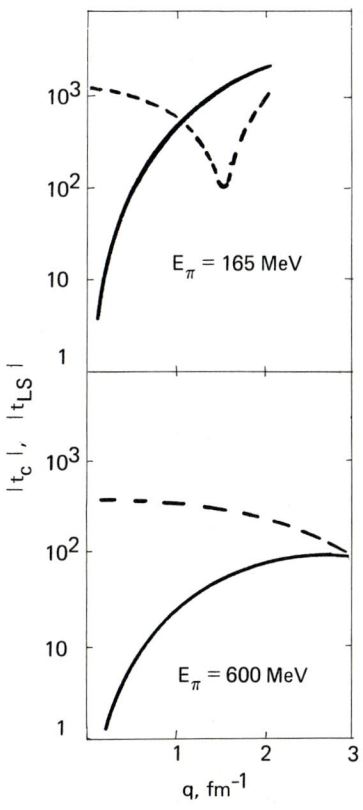

FIGURE 7
Magnitude of t_c (dashed lines) and t_{LS} (solid lines) as a function of momentum transfer q for pion energies of 165 MeV and 600 MeV [8].

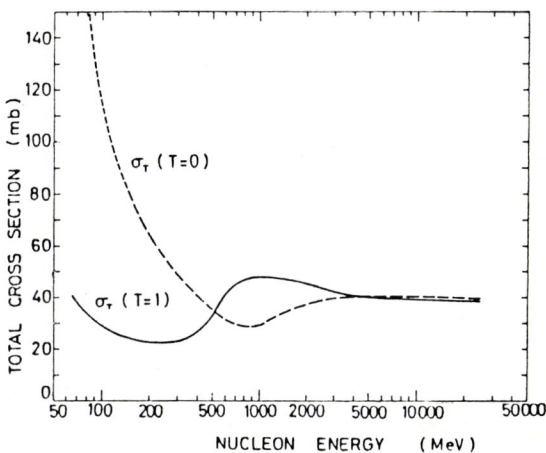

FIGURE 8
N-N total cross section in the isospin channel representation

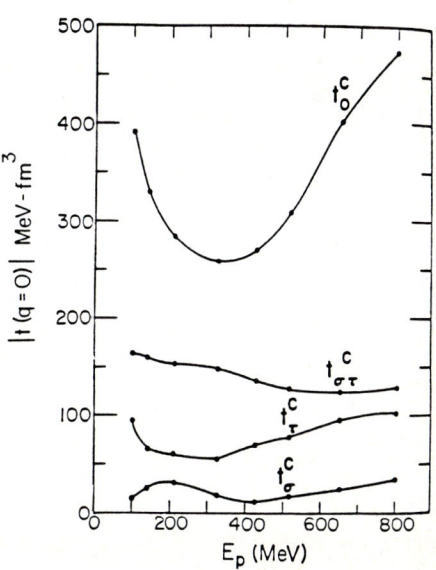

FIGURE 9
Energy dependence of the magnitude of the central parts of the N-N t-matrix [13]

the best quality, and some resonances have still to be confirmed. Of interest would be to see how these survive and propagate in nuclear matter.
- Quasi elastic pion scattering.

The study of deep hole-states through $(\pi,\pi'N)$ could help to solve some still open questions arising from $(e,e'p)$ and $(p,2p)$ reactions.
- Pion-pion interaction.

This interesting subject can be investigated using reactions in which two pions are produced simultaneously.
- Iponuclei, i.e. nuclei with strangeness + 1,

could be formed through N (π^+,Λ) K^+ leaving the K^+ almost at rest (magic momentum) in the nucleus.
- High momentum components in nuclear wave function.

These could be studied by means of high "momentum mismatch" reactions such as (p,π), (π,p)... etc.

Let's now go back to the kaons.

Not being members of the same isodoublet, and therefore not linked by isospin invariance, $K^\pm$ may have different strong interactions, and in fact the $K^\pm$-N total cross sections are markedly different. Not only the K^--N cross section is larger by a factor of about three in the interesting energy range but it shows the existence of several resonances (Y^*) both in the T=0 (Λ) and T=1 (Σ) channels[15]. On the other hand, the K^+-N cross sections are smooth, without evidence of resonances due to the formation of five-quark (Z^*) exotic objects.

Consequently, the K^+-nucleus cross section (which turns out to be proportional to A and not to $A^{2/3}$) is the smallest of any strongly interacting probe, and the amplitudes are comparatively simple, due to a more direct and controllable link between the kaon-nucleon t-matrix and kaon-nucleus optical potential.

Consider the experiment[16] done at BNL with 800 MeV/c kaons and magnetic spectrometer (MOBYDICK), whose results are shown in Fig.10. The fits, obtained with Kisslinger's optical potential, are not terribly bad, but let us just look at the data: they are not very sharply diffractive but nevertheless indicate that the K^- minima i) occur at smaller angles than for K^+ and ii) are deeper than those of K^+, as though K^+ were able to see a nucleus smaller and with a less sharp rim than that seen by K^-. This naive interpretation is in agreement with

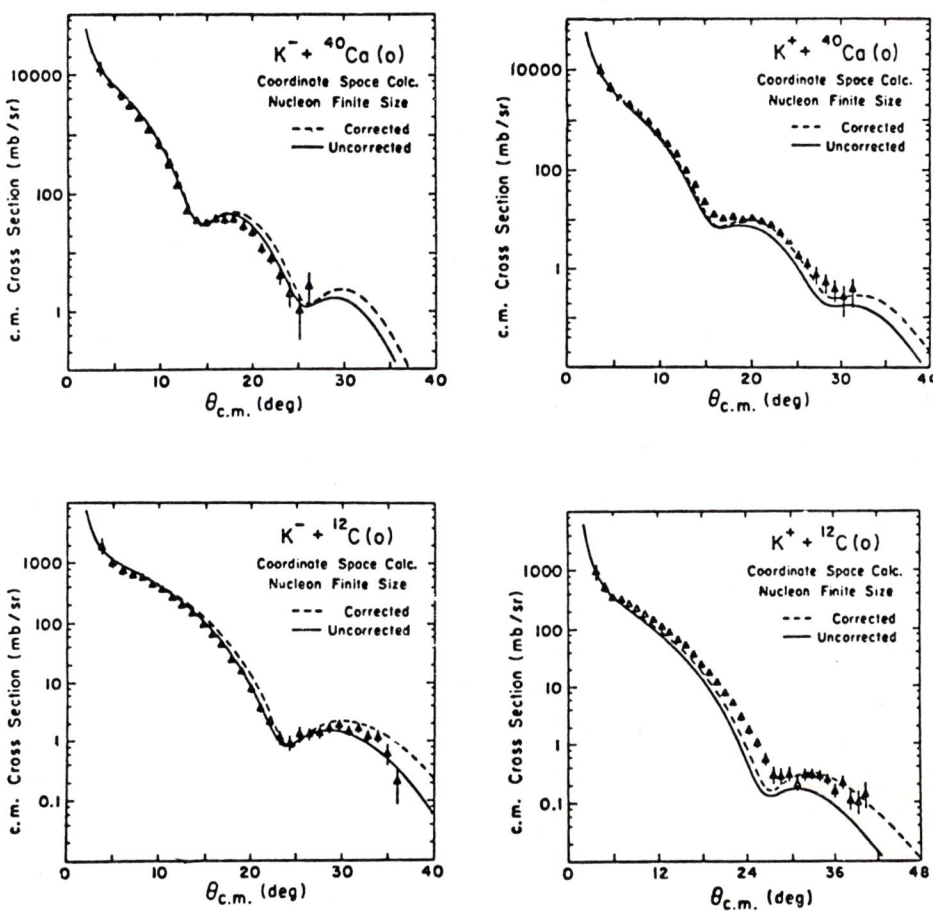

FIGURE 10

Experimental differential cross section for $K^{\pm}$ elastic scattering on ^{12}C and ^{40}Ca. The curves are obtained from the Kisslinger model with e^- scattering nuclear densities corrected (---) or not (——) for finite nucleon size.

the K^- interacting predominantly in the nuclear periphery, and the K^+ accordantly to the "true" density distribution.

L. Ray[18] has done an instructive exercise to evaluate the relative sensitivity of protons and K^+ to the detailed shape of neutron density distribution. He assumes that i) the experimental data can be predicted utilizing first order, local KMT optical potentials with (for protons) or without (for K^+) spin dependence, ii) the neutron density is described by a 2-parameter Fermi distribution

(2pF) plus modulations (j_o) and/or gaussian bumps in the interior and in the surface and tail regions, i.e.

$$\rho_n(r) = \rho_o \left\{ \frac{1}{1+e^{(r-c)/Z}} + A\, e^{-d(r-r_o)^2} j_o(\omega r) + S\, e^{-b(r-r_1)^2} \right\}$$

with various choices of the parameters.

A comparison, Fig.11, of the percent changes in the predicted cross sections

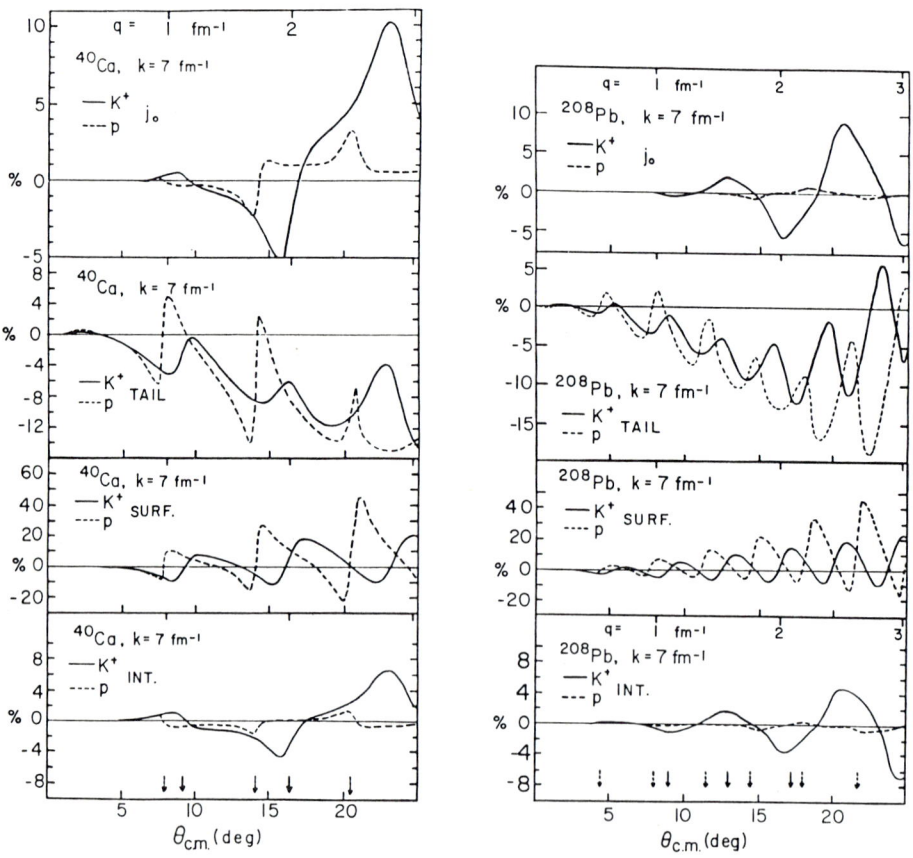

FIGURE 11

Percent change in the K^+ (solid curves) and p (dashed curves) cross sections at c.m. momentum of 7 fm^{-1}, due to the perturbation in the neutron densities of ^{40}Ca and ^{208}Pb (arrows denote minima in the K^+ cross sections).

i.e.

$$\frac{\sigma(\text{perturbed}) - \sigma(\rho = 2pF)}{\sigma(2pF)}$$

between K^+A and pA cases, indicates the relative sensitiveness of these probes to different regions of the neutron density distribution.

As expected, protons can compete with K^+ only in the tail and surface regions. In the nuclear interior the sensitivity of kaons is dramatically better, an effect increasing, obviously, with mass number.

From the above, we see why the K^+ has been claimed to be "the electron of strong interactions", and it may well be so, taking also into account the improvements promised by RIA.

However, the existing data (not so many, indeed) leave some ambiguities on K-N amplitudes. It is therefore imperative to foresee, in primis, a complete and accurate research program on these amplitudes, in order to be able to exploit fully this unique opportunity for nuclear structure studies.

REFERENCES

1) I. Sick et al., Phys. Lett. 88 B (1979) 245.

2) A. K. Kerman, H. McManus and R. M. Thaler, Ann. Phys. (N.Y.) 8 (1959) 551.

3) H. Feshbach, A. Gal and J. Hüfner, Ann. Phys. (N.Y.) 66 (1971) 20.

4) G. W. Hoffmann et al. Phys. Rev. Lett. 47 (1981) 1436.

5) J. A. McNeil et al., Phys. Rev. Lett. 50 (1983) 1439.

6) B. C. Clark et al., Phys. Rev. Lett. 50 (1983) 1644.

7) L. Ray, Phys. Rev. C 19 (1979) 1855.

8) K. Seth, Proc. LAMPF II Workshop, July 19-22 (1982) 367.

9) J. A. Carr, Ibidem pag. 708.

10) J. M. Moss, ibidem, pag. 362.

11) G. E. Walker, ibidem, pag. 202

12) B. Frois et al., Phys. Rev. Lett. 38 (1977) 152

13) W. G. Love and M. A. Franey, Phys. Rev. C 24 (1981) 1073.

14) A. S. Goldhaber, Proc. LAMPF II Workshop, July 19-22 (1982) 171.

15) C. B. Dover and G. E. Walker, Phys. Reports, 89 (1982) 1

16) D. Marlow et al., Phys. Rev. C 25 (1982) 2619.

17) L. S. Kisslinger, Phys. Rev. 98 (1955) 761.

18) L. Ray, Proc. LAMPF II Workshop, July 18-28 (1983) 419.

19) A. Rahbar et al., Phys. Rev. Lett., 47 (1981) 1811.

ANTIPROTON-NUCLEUS INTERACTION: REVIEW OF THE EXPERIMENTAL SITUATION

Guido PIRAGINO

Istituto di Fisica Generale dell'Università di Torino and
I.N.F.N. Sezione di Torino

This paper is a review of the experimental situation of the $\bar{p}$-nucleus interaction studies after the publication of the first results obtained at the LEAR facility of CERN. Before the advent of LEAR in 1983, very little was known about the $\bar{p}$-nucleus interaction at low energy. From the experimental point of view the data were scarce and of rather poor quality, consisting mainly of bubble-chamber data[1], a few reaction cross-sections[2] and level widths and shifts from X-ray studies of antiprotonic atoms[3]. Important progress has been made since intense and pure $\bar{p}$ beams have been available at LEAR. In particular the results obtained in the knowledge of the $\bar{p}$ interaction with the nuclei by the experiment PS184 (devoted to the elastic and inelastic scattering) and PS179 (devoted to the exclusive channels of the non elastic reactions) will be described.

In PS184 (see Fig 1), the incident $\bar{p}$ (with an intensity ranging from $2 \cdot 10^4$ to 10^5 s^{-1}) were monitored by a 0.36 mm thick scintillation counter (S_1) located 25 cm in front of the target. Scattered $\bar{p}$ were momentum-analysed[1] in the Saclay magnetic spectrometer SPESII, which has a momentum resolution of $5 \cdot 10^{-4}$ a solid angle of 30 msr, and momentum acceptance of ±18%. They were detected in three multiwire proportional chambers (MWPCs) and a scintillation hodoscope located near the focal plane. Pions produced by annihilation in the target were discarded by time-of-flight measurement. Information from the MWPCs was used to compute the scattering angle and the excitation energy of the residual nucleus. For elastic and inelastic scattering, the full angular acceptance was divided into 1.67° bins. The energy resolution was about 1 MeV (FWHM), and the overall angular resolution, including multiple scattering in the target, varied from about 2° for the C and Ca targets to about 3° for the Pb target. The uncertainty on the absolute scattering angle was 0.2°. On the absolute normalization, it was 10%. The angular distributions of $\bar{p}$ elastic scattering from ^{12}C,

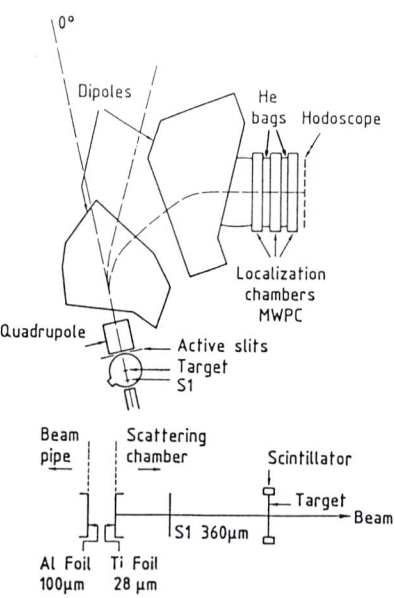

FIGURE 1
The PS184 experimental set-up. SPESII is represented at 0° scattering angle.

^{16,18}O, ^{40}Ca, ^{208}Pb have been measured at about 48 and 179 MeV. The results are described in ref.4,5,6. In Fig.2 the data obtained at KEK[8] are compared with some data of PS184, it is evident the progress due to the high quality beams of the new facility of CERN. These accurate angular distributions have been analysed in the framework of the optical model[5]. Extensive searches on the relative sensitivities to the different potential parameters have been done. Optical potentials with charge-distribution geometries failed in reproducing the experimental data: the finite range of the interaction has to be taken into account. The corresponding optical potentials have shallow real parts 5 MeV $\leq V_o$ $\leq$ 105 MeV, their imaginary depths are at least twice larger than their real depths. The sensitivity of the optical-model analysis of elastic scattering data to the details of the radial potential has been investigated. Target mass and energy evolution of the strong-absorption radius and reaction cross-section confirm that the antiproton penetrates more inside the target nucleus when the incident energy increases, as shown by PS179 experiment (see later). On the other hand in ref.9 has been shown that the elastic and inelastic $\bar{p}$ scattering can

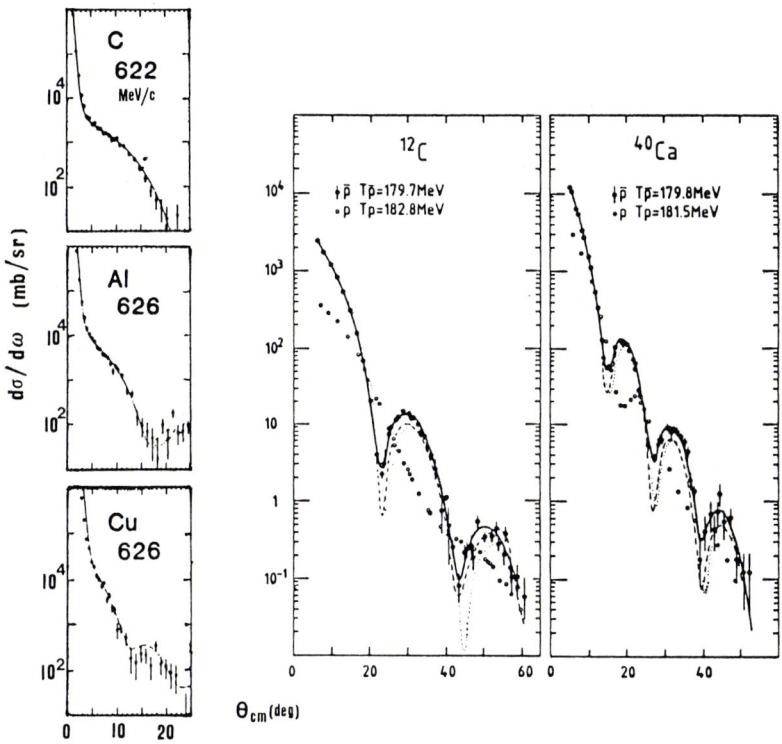

FIGURE 2
Differential cross-section for $\bar{p}$ elastic scattering, at left KEK ref.8, at right PS184. The solid curves result from optical-model fit to the data[6]. Open circles: proton data[7].

be described well in the Glauber approximation[10] also at low energies. This approach has proved succesful in describing corresponding processes in the scattering of protons by nuclei only for energies higher than about 170 MeV. The energies of the $\bar{p}$ the which are discussed here are low (a few tens of MeV), so that it might be expected that the Glauber approximation would not work well in this case. However, the amplitude for the elementary $\bar{p}p$ scattering in this case (in contrast with pp scattering) has a sharply defined forward directionality as shown[11] by the PS173 experiment at LEAR. With decreasing energy the slope of the cone increases (by way of comparison, the slope of the cone in $\bar{p}p$ scattering at 46.8 MeV is 35.6 GeV^{-2}; the cross-section for pp scattering is essentially isotropic at this energy[12], and the slope of the cone for pp scattering at higher energies does not exceed $\lesssim$ 6 GeV^{-2}). Another interesting feature, pointed out in ref.9, is the sensitivity of the depth of the minima of the angular distribution for elastic and inelastic scattering, to the real to imaginary ratio ε of the elementary amplitude. When Coulomb-nuclear interference is taken into account the ε values are in good agreement[4] with direct determinations[11] from $\bar{p}p$ elastic scattering at forward angles. The comparative study[6] of the elastic scattering of $\bar{p}$ by $^{16,18}O$ isotopes shows differences which can be related to neutron excess in ^{18}O.

Inelastic-scattering ($\bar{p},\bar{p}'$) angular distributions measured from ^{12}C at 46,8 MeV and 180 MeV are displayed in Fig.3. For the 4.44 MeV 2^+ state, it is clear that they are typical of a diffractional pattern, the oscillations being out of phase with those of the elastic scattering.

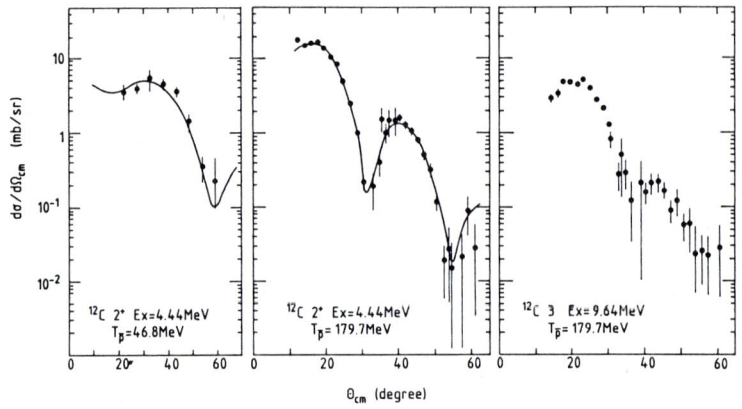

FIGURE 3
Differential cross-section for ($\bar{p},\bar{p}'$) inelastic scattering by ^{12}C. The solid curves result from a coupled channel fit[4].

These angular distributions are also well reproduced by the microscopic calculations of ref.9. A feature that should be pointed out is that the continuous background, corresponding to the carbon break-up, is smaller by a factor of about 10 compared with the case of proton scattering at 46.8 MeV.

Experiment PS184 carried out also a search for narrow $\bar{p}$-nucleus states at about 600 MeV/c using the A($\bar{p}$,p)X reaction at a small angle, detecting protons by reversing the field of the SPESII magnetic spectrometer. There are two main advantages in using the knock-out reaction: a) the outgoing p at Θ=0° carries most of the incoming $\bar{p}$ momentum, leaving the $\bar{p}$ almost "recoilless" in the target, thus favouring the formation of $(\bar{p}-(A-1, Z-1))$ states; b) one can choose the incident $\bar{p}$ momentum close to the maximum of the backward $\bar{p}p$ scattering cross-section, to improve the yield of the $A(\bar{p},p)X$ reaction.

The necessary conditions for $\bar{p}$-nucleus states to exist are that the $\bar{p}$-nucleus interaction is attractive and that the real part $V(r)$ of the $\bar{p}$-nucleus potential must be greater than the imaginary part $W(r)$ near the nuclear surface, i.e. $V(r)>W(r)$. In the pre-LEAR era $\bar{p}$-atom data were consistent with a variety of potentials having, in general a strong absorptive part and a real part ranging from repulsive to strongly attractive with values of several hundreds of MeV. Such a variety of optical potentials could in turn accomodate many resonant or bound $\bar{p}$-nucleus states. Theoretical predictions for the widths of such states range from a few MeV for resonant states to about 100 MeV for bound states. The rather scarce theoretical predictions for the production cross sections of such states are of order of $d^2\sigma/d\Omega dE \sim 0.25$ mb/sr·MeV[13]. The conclusions of the recent LEAR experiments studying the $\bar{p}$-nucleus interaction through elastic and inelastic scattering indicate rather shallow $\bar{p}$-nucleus potentials with central values for the real part $V_o \lesssim 50$ MeV, and that[5] near the surface $W(r) \gtrsim 2V(r)$. The question then arises whether, in such an interaction, $\bar{p}$-nucleus states may be observable or not. In the PS184 targets made of scintillator (CH), ^{12}C, ^{63}Cu and ^{209}Bi were used in a first exploratory experiment, but the most statistically significant results were obtained with ^{6}Li ($2.4 \cdot 10^9$ $\bar{p}$) and with a scintillator target ($2.1 \cdot 10^9$ $\bar{p}$), for outgoing proton energies between 120 and 290 MeV. No evidence for narrow peaks[14] corresponding to bound or resonant $\bar{p}$-nucleus states could be found in the proton spectra. Experimental limits for the production of $(\bar{p}-^{5}He)$ and $(\bar{p}-^{11}B)$ states on ^{6}Li and ^{12}C, respectively, can be deduced for different outgoing proton energies and level widths. Assuming a width of 2 MeV, and considering proton energies close to the incident $\bar{p}$ energy, i.e. states in which the $\bar{p}$ binding energy is equal to the binding energy of the ejected proton, such limits (3σ) are $\sim 12\mu b/sr$ in ^{6}Li and $\sim 40\mu b/sr$ in ^{12}C, about one order of magnitude lower than theoretically predicted[13].

Despite the optical-model ambiguities, the reaction cross-sections ($\sigma_R = \sigma_{tot}-\sigma_{el}$) are well determined[5] in the PS184 (within ±5%). They are displayed in Fig.4 together with those measured by other groups[2,7] and PS179 and compared with those of protons[14]. These reaction cross-sections can be expressed as $\sigma_R = \pi(r_o A^{1/3} + a)^2$; r_o and a taking respectively the values (1.705 ± 0.009) fm and (0.539 ± 0.023) fm at 50 MeV, and (1.49 ± 0.01) fm and (0.65 ± 0.01) fm at 180 MeV. For ^{12}C, the determination of σ_R agrees with that of Nakamura et al.[8], whereas that quoted by Aihara et al[2] is 20% lower. As far as the energy dependence is concerned, like the strong-absorption radius, the reaction cross-sections decrease when the incident energy increases, which is consistent with the energy dependence of the $\bar{p}p$ cross section[11].

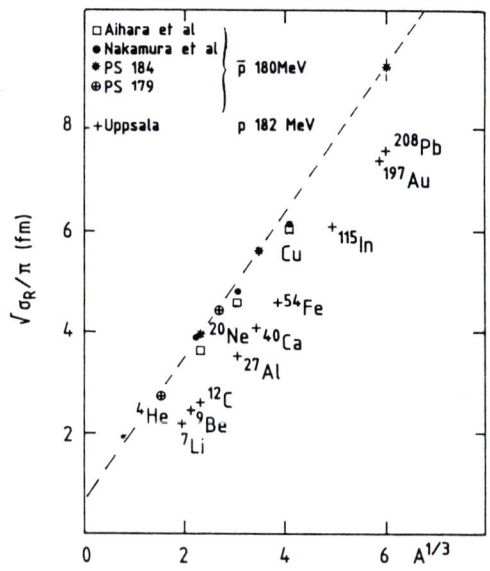

FIGURE 4

The quantity $\sqrt{\sigma_R/\pi}$ is plotted as a function of $A^{1/3}$ for an incident energy of 180 MeV. Data are from ref.2,8 and PS179 for $\bar{p}$, and 7 for protons. The dashed line corresponds to $\sigma_R = \pi(a + r_o A^{1/3})^2$, with $a = 0.65$ fm and $r_o = 1.49$ fm.

The study of the behaviour of the reaction cross-section as a function of the atomic number and of the $\bar{p}$ energy is the common part of the PS184 and PS179 experiments.

In the PS179 (Fig.5) a self-shunted streamer chamber, in a magnetic field, filled at atmospheric pressure with the gas target[15], has been exposed to the low energy $\bar{p}$ beam of LEAR. The volume of the chamber was of 113 litres in the case of ^{3,4}He and Ne targets and of 76 litres for the ^{1,2}H targets. Up today

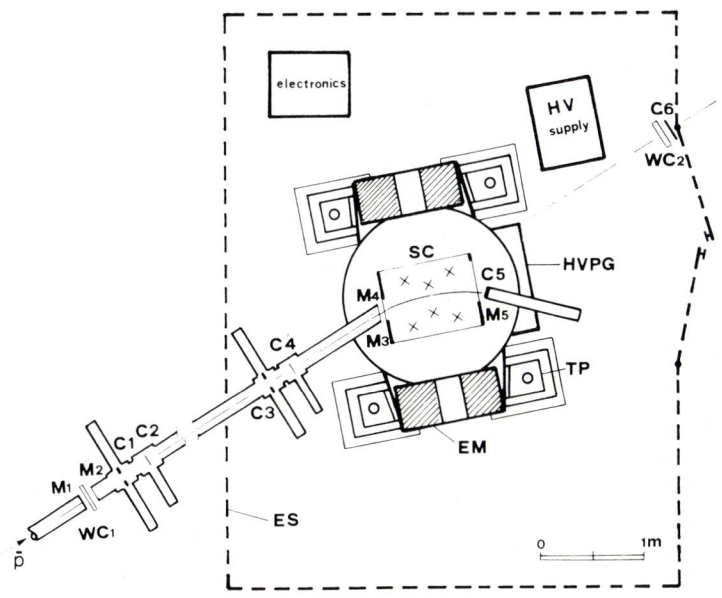

FIGURE 5

The PS179 experimental set-up. EM-electromagnet; SC streamer chamber; HVPG-high voltage pulse generator; TP-travelling platform; ES-electrostatic screening; WC-wire chambers; C-scintillation counters; M-thin walls.

the $\bar{p}$ interaction with ^{4}He and Ne nuclei have been studied at 201, 308 and 609 MeV/c. Exposures with $\bar{p}$ at rest in the center of the chamber have been performed with ^{3,4}He,Ne and the program will be fulfilled with ^{1,2}H. At 201 MeV/c exposures with all these targets are planned before the end of 1986. Also ($\bar{p}$, Ag/Br) annihilation events have been registered in a stack of nuclear emulsions and studied for $\bar{p}$ at rest and in flight at 100,200,300,400 and 500 MeV/c. In Fig.6 the σ_R values, directly measured separating the two prongs events with coplanar tracks (elastic events), are compared with those deduced by PS184 and by other authors[8,11,16]. The σ_R values show an energy behaviour very similar to that of the $\bar{p}p$ interaction, suggesting the hypothesis of $\bar{p}$ annihilations on quasi-free nucleons. Fig.7 shows the σ_R behaviour as a function of $A^{2/3}$, compared with the values obtained in the Glauber approximation[9]. As in the case of the elastic scattering, the agreement with the experimental data is good[4,17].

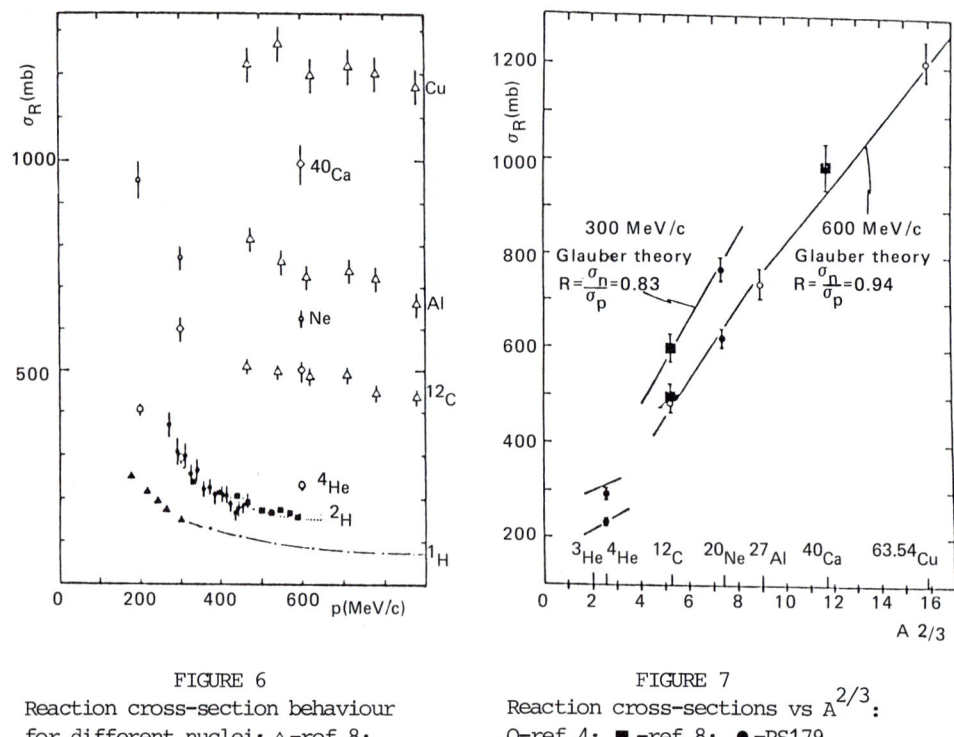

FIGURE 6
Reaction cross-section behaviour
for different nuclei: △-ref.8;
◇-ref.4; -.-.-, ..., ref. 16;
▲-PS173; o-PS179.

FIGURE 7
Reaction cross-sections vs $A^{2/3}$:
O-ref.4; ■-ref.8; ●-PS179

These behaviours of σ_R show that the $\bar{p}$ interaction occurs mainly on the nuclear surface and that the nuclei, in terms of the optical model, are black to $\bar{p}$.

In Fig.8 the charged prong multiplicity distributions obtained[18] for ^{4}He are shown. The even prong events are $\bar{p}$ annihilations on a neutron, with ^{3}He recoil. It must be noted that at 19.6 MeV (where only annihilations are possible) the 2-and 3-prong event percentages are the same as at higher energies, within the experimental errors. This fact shows that the knock-out reaction contributions are small, as in the case of the inelastic scattering[4]. In other words the non-elastic reactions, with $\bar{p}$ present in the final state, occur with less probability than the absorption reactions. The strong suppression of the annihilationless break-up reactions, of $(\bar{p},\bar{p}')$ type, show that their cross-section are by an order of magnitude less than the cross-section of the similar reactions in proton-^{4}He interaction[19]. This effect may be explained by the

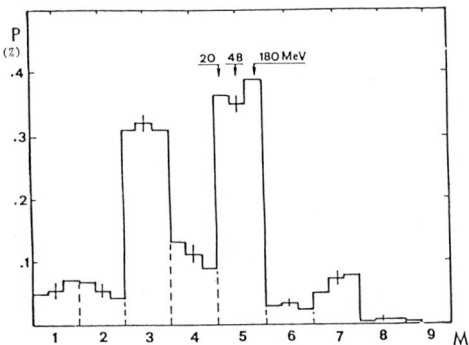

FIGURE 8

Charged prong multiplicity distributions of $\bar{p}$ ^{4}He annihilation events.

specific features of $\bar{p}p$ interaction: the PS173 experiment[11] showed that the $\bar{p}p$ elastic scattering at low energies is characterised by an extremely narrow cone in the differential cross-section for scattering at small angles. Such an angular dependence of the amplitude of the elementary act (scattering on quasi free nucleon) leads to the average momentum transferred to a nucleon in the nucleus, turning out to be too small in $\bar{p}$-nucleus scattering for breaking up the nucleus[19].

Using the ^{3}He production cross-section values and the actual value of the ^{3}He/^{4}He relative abundance in the universe, it is possible[20] to deduce the upper limits on the ratio $n_{\bar{p}}/n_p$ between the amount of antimatter and matter in the early universe. Upper limits of 0.7, 0.9 and $1.1 \cdot 10^{-3}$ at 201, 308 and 609 MeV/c, respectively, have been obtained. These restrictions on the abundance of antimatter in the early universe (after nucleonsynthesis), at time $10^{-3} \leq t \leq 10^{13}$s, exceed by more than 3 order of magnitude the earlier restriction (<1) from thermal spectrum of relic radiation and put new limits on the concentration of gravitinos with mass $\sim 10^2$ GeV, on the temperature of the universe after inflation, in agreement with the limits deduced by Ellis et al[20].

Fig.9 shows the energy behaviour of the ratio $R=\sigma$ ($\bar{p}n$)/σ ($\bar{p}p$) deduced for bound nucleons by the analysis of ($\bar{p}$, ^{4}He) annihilation events. The experimen-

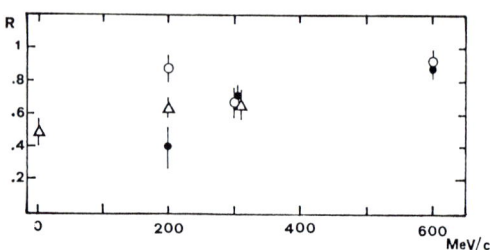

FIGURE 9

$R = \sigma(\bar{p}n)/\sigma(\bar{p}p)$ from $(\bar{p},{}^4\text{He})$ annihilation events: △ ;form Glauber theory; ●-parameters of ref.11 and ○-parameters of ref.21.

tal values are compared with those deduced for free nucleons on the basis of Glauber theory[17] using two different sets of parameters[11,21].

Interesting information has been also deduced by the charged prong multiplicity distributions. Fig.10 shows the behaviour of the cross-section vs the

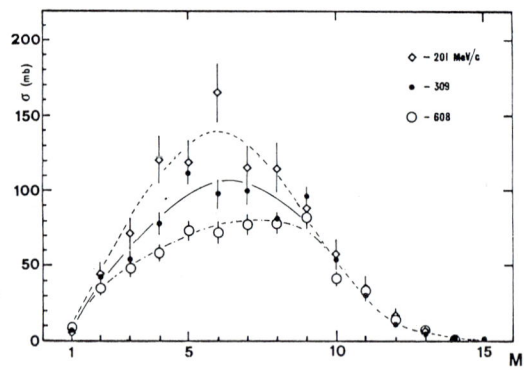

FIGURE 10

Charged prong multiplicity cross-sections for $\bar{p}$Ne non elastic events.

multiplicity M measured for the $(\bar{p},\text{Ne})$ non elastic events at different momenta[17]. One can distinguish two regions, the first where the cross-sections increase with the lowering of the $\bar{p}$ momentum, the second $(M \gtrsim 9)$ where the interaction probability is independent from the $\bar{p}$ kinetic energy. The total percentage of these high multiplicity events increases with the $\bar{p}$ momentum suggesting that these events are result of $\bar{p}$ annihilations inside the nucleus, where the nuclear density is about constant (at these momenta the kinetic energy is negligible in comparison with the annihilation energy). The average multiplicity of these events results nearly constant[17]. Similar behaviour has been found also for the multiplicity distributions[22] of ($\bar{p}$,Ag/Br) interactions. In Fig.11

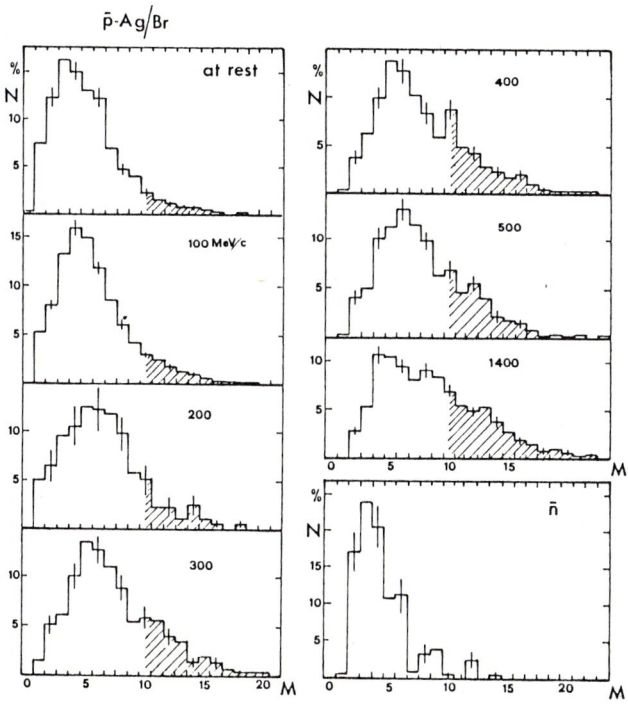

FIGURE 11
Charged prong multiplicity distributions of $\bar{p}$ and $\bar{n}$ annihilation events in nuclear emulsion. The 1400 MeV/c data are from ref.23.

the multiplicity distributions obtained for more than thousand events each are shown. Preliminary result obtained for about 250 annihilations of $\bar{n}$, produced by charge exchange reaction of $\bar{p}$ on nuclear emulsion nuclei, is also reported. As in the case of Ne, the high multiplicity tail (hatched histograms in Fig.11) increases with $\bar{p}$ momenta, suggesting that these events with constant average multiplicity are produced by $\bar{p}$ annihilations deep inside the nuclei[22]. In Fig. 12 the distribution of the nuclear matter density $\rho(r)/\rho(0)$ of Ag and Br nuclei is shown and compared with the $\bar{p}$ absorption probability calculated by Clo-

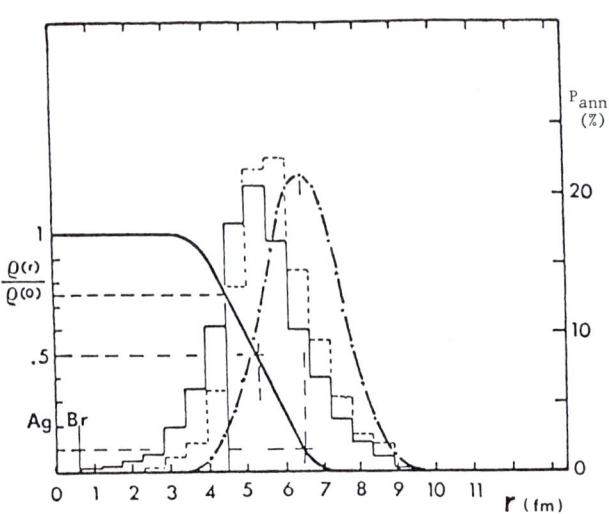

FIGURE 12

Probability of p̄ annihilation in Ag/Br nuclei vs the nuclear radius. The histograms are calculations of Clover et al[24] for 590 MeV/c (the solid curve corresponds to a p̄-nucleus potential V=250 MeV, the dashed curve to V=0). The dot dashed line is a calculation by Iljinov et al[25] for p̄ at rest. The origin of the radius scale corresponding to the center of Ag and to the center of Br nucleus, is indicated[26].

ver et al[23] for 590 MeV/c p̄ (with two values, 0 and 250 MeV, of the real p̄A attractive potential) and by Iljinov et al[24] for p̄ at rest. In the first case, the P_{max} value occurs approximately in the 50% density region and the probability of p̄ annihilation, in the region $\rho(r) > 0.75 \; \rho(0)$, amounts to 14÷27% of the total number of events, in good agreement with the percentage of annihilations in the depth of the nuclei as estimated in Table I at 500 MeV/c. In the case of p̄ at rest, the P_{max} occurs in the $\rho(r) \simeq 0.1 \; \rho(0)$ region and the probability of p̄ annihilation in the region $\rho(r) > 75\%$ is about 5% off all the events. Tentatively a fit has been achieved with the Maxwell distribution.

$$\sigma(M)\,dM \propto \sqrt{M-M_0} \cdot \exp(M-M_0)/T \cdot dM \qquad (1)$$

for $M_0=10$, of the experimental data[22]. The percentages of the high multiplicity events subtended by the (1) are shown in Table I. They are in good agreement with the hypothesis that the high multiplicity events are produced mainly by $\bar{p}$ annihilating in nuclear dense matter ($\rho(r)/\rho(0) \gtrsim 0.75$).

Table I - Percentage of $\bar{p}$ annihilations occurring in the nuclear surface or deep inside the nuclear matter of Ag/Br nuclei. The fit errors with (1) are estimated of about 12%.

$\bar{p}$ momentum (MeV/c)	surface (M<10) (%)	Internal (M>10) (%)
at rest	93	7
100	88	12
200	75	25
300	74	26
400	73	27
500	73	27
1400[23]	63	37

Similar result has been obtained also for $\bar{p}$Ne annihilations[17].

At this point, it seems possible to conclude that two mechanisms can be distinguished in the $\bar{p}$ annihilation in nuclei:

a) $\bar{p}$ annihilation on the nuclear surface, with production of average multiplicity of 5÷6, the relative number of which increases at lower $\bar{p}$ energies. Also the σ_R values increase at lower energies (Fig.6), as in the case of $\bar{p}p$ reaction cross-section, indicating that the main mechanism of $\bar{p}$ absorption is the annihilation on quasi-free nucleon of the nuclear surface (the most important absorption mechanism for $\bar{p}$ at rest).

b) $\bar{p}$ annihilation in the nuclear dense matter, deep inside the nucleus, with production of events with an high average multiplicity of 11÷13, the relative number of which increases with the $\bar{p}$ kinetic energy (penetrability).

The detection of deep annihilations is of interest because theoretical works predict that they might develop through mechanisms different from those included in the "standard physics" of intranuclear cascade models, such as the annihilation on more than one nucleon[27] and the excitation of new degrees of freedom of the nuclear matter[28]. It is a common opinion that typical signatures of these exotic phenomena are an enhancement in hyperon and strange-meson production and the emission of a high number of energetic nuclear fragments. For example, in $\bar{p}p$ annihilation at rest, a $\Lambda\bar{\Lambda}$ pair cannot be created, because of energy conservation (the minimum c.m. energy for the $\Lambda\bar{\Lambda}$ system, is 2231.2 MeV). However, when low energy $\bar{p}$ annihilate in nuclei, $\bar{\Lambda}$-particle production is possible. Mandelkern et al[29] have suggested that Λ production from the reaction $\bar{p}d \to \Lambda + \text{anything}$ seems to be consistent with a double scattering mechanism, in which a K^- is exchanged between two nucleons. This latter conclusion would not appear to be compelling since their model calculation systematically underestimates the experimental cross-section (5 of 6 measurements) for $\bar{p}$ momenta less than 1 GeV/c, as noted in ref.30. On the other hand, studying the same reaction at rest[31], Bizzarri et al[31] have shown that a large fraction of these events cannot be described by K^- rescattering but have to be considered as three-body annihilations.

The Λ emission probability for $\bar{p}$ stopping in ^{2}H is $P_\Lambda \sim 3.6 \cdot 10^{-3}$ of the annihilations[31]. This value decreases, with the increase of $\bar{p}$ energy, down to $P_\Lambda \sim 2.5 \cdot 10^{-3}$, near the threshold of $\Lambda\bar{\Lambda}$ pair production. Afterwards, P_Λ increases with energy. It must be noted that in ^{2}H, the K°_s emission probability is higher ($P_{K^\circ_s} \sim 16 \cdot 10^{-3}$) than that of Λ production[31,32].

In the case of complex nuclei (C,Ti,Ta,Pb), the absorption of low momentum $\bar{p}$ (<300 MeV/c) was found[33] to produce Λ hyperons with a frequency $P_\Lambda = (19\pm 4)10^{-3}$, which is about five times that in ^{2}H. In this low statistic experiment, the K°_s emission has not been measured. The reaction $\bar{p}^{131}\text{Ta} \to V^\circ + \text{anything}$ ($V^\circ = \Lambda, \bar{\Lambda}$ or K°_s) at 4 GeV/c has been studied[34] with good statistics, and a strong enhancement of Λ production was observed. The K°_s production is less frequent ($P_{K^\circ_s} = 50 \cdot 10^{-3}$) than that of Λ's ($P_\Lambda = 118 \cdot 10^{-3}$), the opposite of what occurs in the case of ^{2}H. The Λ and K°_s rapidity distributions[34] show that $(\bar{p}13N)$ and $(\bar{p}3N)$ systems in the c.m.s. are involved respectively in their production. These hot clusters emit the Λ and K°_s particles equally into the forward and backward hemispheres in their $(\bar{p}13N)$ and $(\bar{p}3N)$ c.m. frames, with an evaporation-like mechanism, but at temperatures not high enough to induce a phase transition to quagma[35]. On the other hand, this kind of mechanism is somewhat questionable because the time required for the particle emission is much smaller than that needed for the statistical decay of a thermalized source. In fact the nucleons, in the $\bar{p}$ absorption process, cannot reach statistical equilibrium in a very short time.

An isotropic angular distribution of the reaction products does not necessarily correspond to a slow evaporation-like emission, but the hot spot does not live long, exploding in some 10^{-23} s, with statistical distribution of the number of fragments, as shown by the charged prong multiplicity distributions. Hence the study of Λ and K_s^0 emission and the measurement of the K_s^0/Λ ratio (>1 for ^2H and <1 for heavier nuclei) for $\bar{p}$ annihilation in nuclei is of great interest for understanding the properties of highly excited nuclear matter in order to discover whether quark degrees of freedom of the fireball come into play and to see whether resonant states are formed.

The PS179 studied at 607 MeV/c reaction $\bar{p}Ne \to V^0 + anything$ ($V^0 = \Lambda$ or K_s^0). The total number of $\bar{p}$ considered was $9.4 \cdot 10^6$ and 7622 annihilations in Ne yielded $(147\pm27)\Lambda$ and $(64\pm13)K_s^0$ mesons, which give a final fraction of $(1.9\pm0.4)\%$ Λ and $(0.8\pm0.2)\%$ K_s^0. The neutral strange particles were detected by their charged particle decays: $\Lambda \to p\pi^-$ and $K_s^0 \to \pi^+\pi^-$. The K_s^0/Λ ratio is close to that measured[34] in ^{131}Ta at 4 GeV/c. Also for Ne the rapidity values of Λ and K_s^0, $<y> = (0.08\pm0.03)$ and $<y> = (0.44\pm0.08)$ respectively, show that $(\bar{p}, 4 \div 11N)$ and $(\bar{p}, 1 \div 2N)$ systems in the c.m.s. are involved in their production. Very preliminary measurements of Λ and K_s^0 production in Ne for $\bar{p}$ at rest show that K_s^0/Λ ratio is more than 10 times higher than at 607 MeV/c. It seems possible to conclude that when the $\bar{p}$ annihilate in the nuclear soft surface the K_s^0 production is higher than that of the Λ's. When the $\bar{p}$ annihilation occurs deep inside the nucleus, nucleon-systems are involved and more Λ are produced.

This review of the new results obtained at LEAR shows the great interest of the proposal made at CERN to continue the study of the $\bar{p}$-nucleus interaction, with a large acceptance and high resolution detector as OBELIX[36], in particular in the following directions:

a) wide survey of $\bar{p}$ annihilation channels in nuclear targets as $^2H, ^{3,4}He, N, Ne, Ar, Kr, Xe$ up to the highest $\bar{p}$ momenta ($1 \div 2$ GeV/c);

b) study of quark substructures and manifestations of quark-gluon plasma in nuclei[37]. Such "exotic" effects could be detected studying:

1) pionless annihilations $\bar{p}A \to p(A-2)N$ as, for example, $\bar{p}\,^3He \to n\,p$;

2) single pion annihilation $\bar{p}A \to \pi(A-1)N$ as, for example, $\bar{p}\,^2H \to \pi^-p$, or quasi deuteron $\bar{p}$ absorption;

in these reactions the annihilation occurs on very short ($0.2 \div 0.3$ fm) correlated nucleons.

3) neutral strange partiche production $\bar{p}A \to V° +$ anything ($V° = \Lambda, \bar{\Lambda}, K_s$) and $\bar{p}A \to \Lambda\Lambda + KK +$ anything (A>3) in the search for bound (H dibaryon) or unbound $\Lambda\Lambda$ states[30].

REFERENCES

1) L. E. Agnew et al. Phys. Rev. 108 (1957) 1545.

2) H. Aihara et al. Nucl. Phys. A306 (1981) 291, and refs. quoted therein.

3) C. J. Batty, Nucl. Phys. A372 (1981) 433, and refs. quoted therein.

4) D. Garreta, Antinucleon-and Nucleon-Nucleus Interaction, eds. G. E. Walker, C. D. Goodman and C. Olmer (Plenum, 1985) p.49, and refs. quoted therein.

5) S. Janouin et al. Nucl. Phys. 451 (1986) 541.

6) C. Bruge et al. Phys. Lett. B169 (1986) 14.

7) A. Johansson et al. Ark. Fys. 19 (1961) 541; J. H. Menet et al. Phys. Rev. C4 (1971) 1114; R. F. Carlson et al. Phys. Rev. C12 (1975) 1167.

8) K. Nakamura et al. Phys. Rev. Lett. 52 (1984) 731.

9) O. D. Dalkarov and V. A. Karmanov, Nucl. Phys. A445 (1985) 578 and Lebedev Phys. Inst. preprint 87 (1986).

10) R. Glauber and G. Matthiae, Nucl. Phys. B21 (1970) 135.

11) W. Brukner et al. Phys. Lett. B158 (1985) 180 and B166 (1986) 113.

12) A. K. Kerman et al. Ann. Phys. 8 (1959) 551.

13) H. Heiselberg et al. Phys. Lett. B132 (1983) 279.

14) D. Garreta et al. Phys. Lett. B150 (1985) 95.

15) F. Balestra et al. Nucl. Instr. and Meth. A234 (1985) 30.

16) T. E. Kalogeropoulos et al. Phys. Rev. D22 (1980) 2585; R. Bizzarri et al. Nuovo Cim. A22 (1974) 225; R. P. Hamilton et al. Phys. Rev. Lett. 44 (1980) 1182; V. Flaminio et al. CERN-HERA 84-01 (1984).

17) F. Balestra et al. Nucl. Phys. A452 (1986) 573.

18) F. Balestra et al. Phys. Lett. B165 (1985) 265.

19) Yu. A. Batusov et al. JINR Rap. Comm. 12 (1985) 6.

20) I. V. Falomkin et al. Nuovo Cim. A79 (1984); Yu. A. Batusov et al. Nuovo Cim. Lett. 41 (1984) 223 and JINR Rap. Comm. 6 (1985) 11; M. Yu. Khlopov et al. Phys. Lett. B138 (1984); J. Ellis et al. Phys. Lett. B145 (1984) 181 and Nucl. Phys. B259 (1985) 175.

21) H. Iwasaki et al. Nucl. Phys. A433 (1985) 580.

22) F. Balestra et al. CERN-EP/85-122 and Europhys. Lett. (in press.).

23) F. O. Breivik et al. Phys. Scripta 28 (1983) 362.

24) M. R. Clover et al. Phys. Rev. C36 (1982) 2138.

25) A. S. Iljinov et al. Nucl. Phys. A382 (1982) 378.

26) L. R. B. Elton, Nuclear Sizes (Oxford University Press, 1961).

27) J. Cugnon et al. Phys. Lett. B146 (1984) 16; C. Derreth et al. Phys. Rev. C31 (1985) 1360.

28) J. Rafelski, Phys. Lett. B91 (1980) 281; J. Rafelski et al. Phys. Rev. Lett. 48 (1982) 1066; S. C. Phatak et al. (in press).

29) M. A. Mandelkern et al. Phys. Rev. D27 (1983) 19.

30) G. T. Condo et al. Phys. Lett. B144 (1984) 27; M. Locher et al. Adv. Nucl. Phys. (in press).

31) R. Bizzarri et al. Nuovo Cim. Lett. 9 (1969) 431.

32) B. Y. Oh et al. Nucl. Phys. B51 (1973) 57.

33) G. T. Condo et al. Phys. Rev. C59 (1984) 1531.

34) K. Miyano et al. Phys. Rev. Lett. 53 (1984) 1725.

35) E. V. Shurvak, Phys. Rep. C61 (1980) 71.

36) R. Armenteros et al. CERN/PSCC/86-4 (1986).

37) D. S. Koltun et al. Phys. Lett. B172 (1986) 267.

EXPLORING QUARK-GLUON DEGREES OF FREEDOM IN NUCLEI

Emilio CHIAVASSA

Istituto di Fisica Superiore dell'Università di Torino
Istituto Nazionale di Fisica Nucleare - Sezione di Torino.

Nuclear Drell-Yan process and heavy mesons production in nuclei are briefly discussed as possible experiments for exploring quark-gluon degrees of freedom at a hadron facility.

1. INTRODUCTION

In the last years an increasing amount of work has been dedicated to look for experiments that can evidentiate quark and gluon degrees of freedom in nuclei. Both theoretical and experimental re sults seem to indicate the usefulness of describing nuclei in term of quarks and gluons.

A good discussion of the problem may be found in reference 1). From the point of view of nuclear physics this will be one of the principal motivations for building a hadron facility producing high quality beams of protons and mesons at energies greater than 30 GeV. Many suggestions concerning nuclear physic experiments can be found in the literature; see for example the ref.1) and the proposal for LAMPF II[2].

In this school many arguments as the EMC effect, the hypernucle ar physics, the spin effects will be treated elsewhere so that my task will be to present other experiments which I judge important as well to explore quark and gluon degrees of freedom in nuclei.

The hadron facility, when done, will be a "meeting point" among intermediate energy nuclear physicists and particle physicists;for this reason I choose to present here two experiments that can be considered an "extrapolation" of typical experiments in these two different research fields; I mean the nuclear Drell-Yan process and the heavy meson production in nuclei.

2. THE NUCLEAR DRELL-YAN PROCESS.

The possibility of using the Drell-Yan process to study the quark and antiquark distributions in nuclei has been suggested by R.F.Bickerstaff et al.[3] and has been considered in the LAMPF II proposal[2]. The authors suggested that the Drell-Yan process would give information about the role of quarks in nuclei that may complement what we could learn by means of the EMC effect. As it is well known the Drell-Yan process consists in the production of a lepton pair of opposite charge from a hadron-hadron collision:

$$A + B \rightarrow \ell^+ + \ell^- + \text{anything}$$

A complete description of this process has been given by I.R.Kenyon[4]. In general, the detection of dimuons has been preferred to that of e^-e^+ pairs. In fig.1 (taken from ref. 4) is reported the

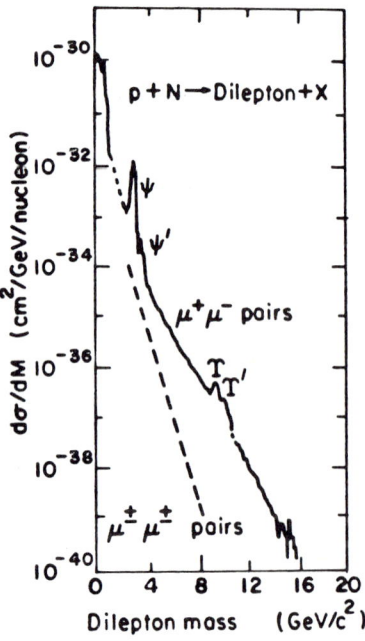

FIGURE 1
Dimuon spectra for 400 GeV/c protons incident on a Pt target.

dimuon spectra for 400 GeV/c proton on a Pt target.

The continuum of $\mu^+\mu^-$ has been explained by Drell and Yan[5] by means of the process pictured in fig.2.

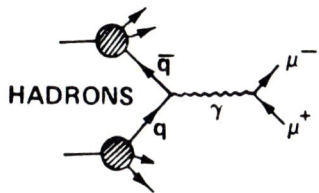

FIGURE 2
Feynmann diagram for the Drell-Yan process.

In this process a quark and an antiquark annihilate to give a virtual photon that subsequently converts to a pair of opposite charge leptons of mass M. The process is calculable and a factor M^{-4} appears in the cross section that explains the rapid fall with the dilepton mass (see fig.1).

In order to calculate the differential cross section the longitudinal momentum distribution of quark and antiquark in the hadrons must be known; for this reason the Drell-Yan process has so far been utilized for measuring the structure functions of nucleons and mesons. Miller et al.[3] proposed to use this process to measure the nuclear quark and antiquark distribution in nuclei: they found for the double differential cross section of a Drell-Yan process induced by proton in nuclei the following expression:

$$\frac{d^2\sigma}{d^2MdX_f} = K \frac{4\pi\alpha^2}{9M^2} \left(\frac{X_1 X_2}{X_1+X_2}\right) \{\Sigma_f Q_f^2 v_f^P(X_1)\bar{q}_f^T(X_2) + \frac{\bar{q}^P(X_1)}{X_2} \frac{F^T(X_2)}{2}\}$$

where:

K   is a constant approximately equal to 2

α is the fine structure constant

X_1, X_2 are the fractional momenta of the quark and antiquark in the projectile and target.

Q_f is the charge of quark with flavor f.

$v_f^P(X_1)$ is the proton valence quark distribution

$\bar{q}_f^T(X_2)$ is the antiquark distribution in the target

$F_2^T(X_2)$ is the target structure function defined by

$$F_2(X_2) = X_2 \Sigma_f Q_f^2 \{q_f^T(X_2) + \bar{q}_f^T(X_2)\}$$

$\bar{q}^P(X_2)$ is the antiquark proton distribution taken independent of flavor.

From the known value of v_f^P the first term in the bracket is dominant for $X_1 > 0.3$ while the second term acquires relevance at small

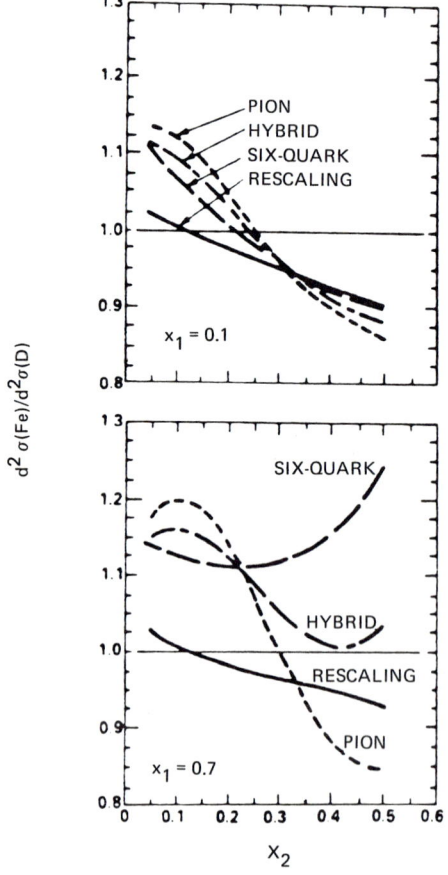

FIGURE 3
Ratio of double differential cross sections for Drell-Yan process on Fe and D.

value of X_1. This means that at low X_1 values the Drell-Yan process gives information similar to deep inelastic scattering, while at large X_1 it rappresents a unique possibility to study the antiquark distribution of the target nucleus. In fig.3, taken from ref.3), the ratio of the double differential Drell-Yan cross sections on Fe and D is presented for different X_1 values and models.

The figure demonstrates the high sensitivity of the Drell-Yan process at high X_1 to discriminate between the proposed nuclear models. The question now arises if this appealing process could be measured at a hadron facility.

A typical apparatus for studying the Drell-Yan process is sketched in fig.4 taken from ref.4.

The figure 4 rapresent the CERN-OMEGA spectrometer that has been used to study the Drell-Yan process for an unseparated beam of 40 GeV/c.

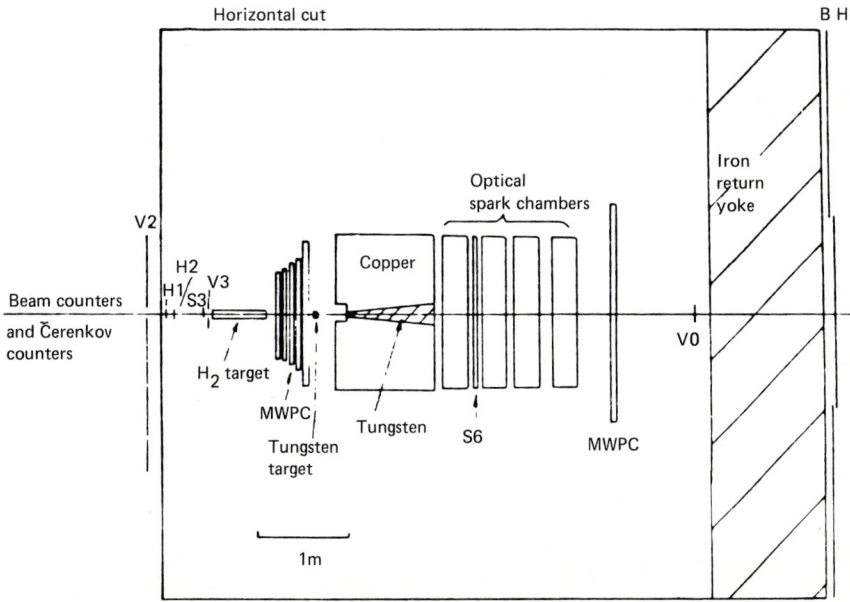

FIGURE 4

The CERN-OMEGA spectrometer.

The beam particles enter in the target that is followed by a thick absorber. The emerging charged particles are analysed in a magnetic spectrometer of 1.8T field. The magnet yoke provides a further iron absorber for the residual hadrons that penetrated the first one. The mass resolution of the spectrometer is 6% and the accepted values for $\sqrt{X_1 X_2}$ range from 0.22 to 0.55, with $X_F = X_1 - X_2$ greater than -0.5. The requirements for a useful measurement of a nuclear Drell-Yan process are[3]:

a) mass of the dilepton greater than 2 GeV/c^2;
b) measurement of the double differential cross section $\frac{d^2\sigma}{d^2 M dX_f}$ in different nuclei in the region of $\sqrt{X_1 X_2} \sim 0.4-0.5$ with rapidity lower than zero.
c) measurement of dimuon pairs in a region free of resonances;that is, looking at fig.1, around 2 GeV/c^2 or, much better, between 4 and 8 GeV/c^2.

All these conditions would be satisfied by a hadron beam of microamperes intensity and energy of 30 or better 40 GeV. In ref.2)it is reported an evaluation of a counting rate of 10 dilepton pairs per second when a microampere proton beam impinges on a target of one radiation length.

Finally I want to recall that, in the same report, very interesting tests on nuclear structure functions using Drell-Yan processes induced by $K^{\pm}$ are proposed.

The $q\bar{q}$ annihilation with K^+ valence quarks will be produced by target sea quarks, since the K^+ contains $\bar{s}$ u quarks, while the $\bar{u}$ of the K^- will annihilate with a quark of the target.

It is superfluous to recall that producing very good K beams is one of the main goals of a hadron facility.

3. HEAVY MESON PRODUCTION.

Studying short living heavy meson production in nuclei could be one of the most interesting subjects to be pursued in a hadron facility. Production of $\eta, \eta', \rho, \omega$ and φ mesons will provide extreme conditions where nuclear properties must be described using the constituents of the nucleons. It must be recalled that heavy mesons are responsible for nuclear forces at short distances just where QCD considers explicitly quark and gluon contributions.

Excellent reviews of "hadronic meson production" and "η meson production" are given respectively by R.J.Comfort[6] and J.C.Peng[7].

In this talk I will refer about the studies concerning η and η' mesons. Experiments of η production by means of π interactions are in progress at LAMPF[7] and in a near future the (p,η) reaction will be studied at SATURNE[8]. In both cases the η detection will be done using a two arms, high resolution spectrometer in which are measured energy and position of the γ rays produced in the 39% of η decays.

Such a spectrometer has been firstly built in Los Alamos[9] to be used for π^0 detection. A similar apparatus is now under construction at Turin[10] and in fig.5 is presented the exploded view of one arm of this spectrometer.

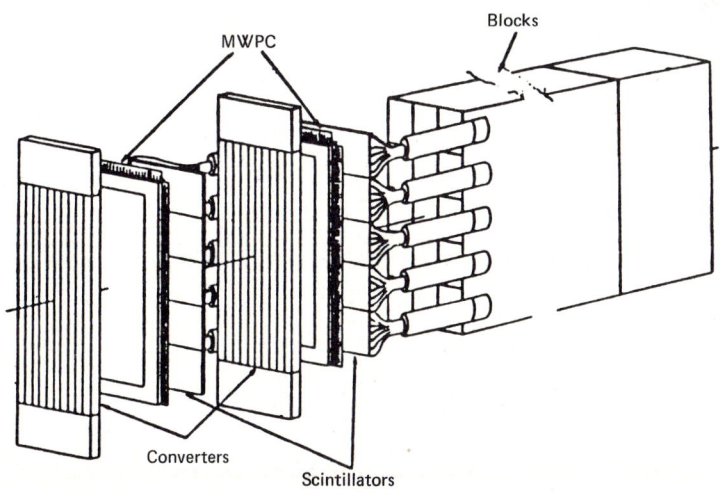

FIGURE 5
Exploded view of one arm of the Turin π^0,η spectrometer.

It consists of two scintillator glass converters, in which γ rays originate a shower, followed by MWPC'S, where the charged particles emerging from the converters are detected, and finally by blocks of scintillator glass where the shower stops. The main features of the spectrometer are:

π^0 energy range	0-500 MeV
η energy range	0-100 MeV
Angular resolution	< 3°
π^0 detection efficiency	50%
Effective solide angle	10^{-3} sr
Maximum rate per arm	10^6 pps

With this apparatus, and the existing beams, it will be certainly improved our knowledge of the η-nucleus interaction, but more complete studies could be done only with more energetic and intense beams of pions and protons. With good pion beams of more than 1 GeV we could measure η' production by means of (π⁻p) reactions that present integrated cross section of the order of μb[11].

It will also be possible to produce heavy mesons almost at rest, as suggested for example by K. Kilian[12], profiting of the reaction kinematics. We want here just to recall that when a neutron is emitted at zero degree we will have an η at rest in the following cases:

$\pi + p \rightarrow n + \eta$ $p_\eta \rightarrow 0$ when $p_{lab}(\pi) \rightarrow 900$ MeV/c

$p + n \rightarrow d + \eta$ $p_\eta \rightarrow 0$ when $p_{lab}(p) \rightarrow 3040$ MeV/c

Detecting heavy mesons produced in nuclei will pose experimental problems and spectrometers with greater acceptance should be constructed, especially for what concerns η' that decays in two γ's only in 2% of the cases; for revealing this meson one must probabily profite of other decays, for example the ργ that has a 30% branching ratio. Before concluding I want to recall an argument given by F.Lenz[12] and F.Cannata[13] to illustrate the relevance of studying heavy mesons for understanding quark-gluon degrees of freedom in nuclei.

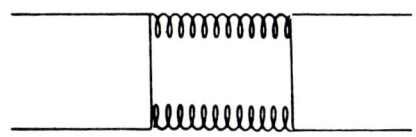

FIGURE 6

The annihilation diagram which is responsibile for the larger mass of the η'.

It is generally accepted that the η has a quark wave function that could be described as:

$$\eta \to \frac{u\bar{u}+d\bar{d}-2s\bar{s}}{\sqrt{6}}$$

while the η' will contain an appreciable two gluons pseudoscaler configuration due to an annihilation diagram as in fig. 6.

This means that studying η in nuclei will be an useful tool for understanding the behaviour of the strange quarks in nuclei while η' production could give information concerning gluons in nuclei. In fact the gluon of η' may couple with the gluons of the nucleus affecting considerably the production of this meson.

REFERENCES

1) Quark/gluon phenomena in Nuclear Physics. "Hadronic Probes and Nuclear Interactions". Eds.
J.R. Comfort, W. R. Gibbs, B.G. Ritchie
A.I.P. (1985) pag.110.

2) A proposal to extend the intensity frontier of nuclear and particle physics to 45 GeV. LA-UR-84-3982

3) R. P. Bickerstaff et al.,"Hadronic Probes and Nuclear Interactions",A.I.P. (1985) pag.144.

4) I. R. Kenyon, Rep.Prog. Phys. 45 (1982) 1261.

5) S. D. Drell and T. M. Yan, Ann. Phys. 66 (1971) 578.

6) J. R. Comfort, Hadronic Probes and Nuclear Interactions, A.I.P. N.Y. 1985 pag.242.

7) J. C. Peng, Hadronic Probes and Nuclear Interactions, A.I.P., N. Y. 1985, pag. 255.

8) G. C. Bonazzola et al., Proposition d'experience, LNS 125/1985.

9) H. W. Baer et al. Nucl. Inst. and Meth. 180 (1981) 445.

10) E. Chiavassa et al., Nuclear and Particle Physics at Intermediate Energies with Hadrons, Eds T. Bressani and G. Pauli, Conf. Proc. vol.3 S.I.F. 1986 pag. 347.

11) V. Flaminio et al. CERN-HERA 79-01 (1979).

12) K. Kilian, CERN EP/85-17.

13) F. Lenz, The future of medium and high energy physics in Switzerland, Les Rasses, May 1985. Proc. SIN.

14) F. Cannata, Nuclear and Particles Physics at Intermediate Energies with Hadrons. Eds. T. Bressani and G. Pauli, Conf. Proc. Vol.3 S.I.F. 1986 pag. 127.

15) F. Cannata, Color degrees of Freedom in Nuclear Physics. This school.

16) J. F. Donoghue et al., Phys. Rev. D28 (1983) 2800.

17) V. P. Efrosinin et al., Z. Phys. C 28 (1985) 211.

HIGH-ENERGY NUCLEUS-NUCLEUS COLLISIONS AND NUCLEAR MATTER

Renato A. RICCI

Department of Physics, University of Padova and Laboratori Nazionali di Legnaro - INFN, Padova

1. INTRODUCTION

Figure 1 shows a representation of the possible common inte-

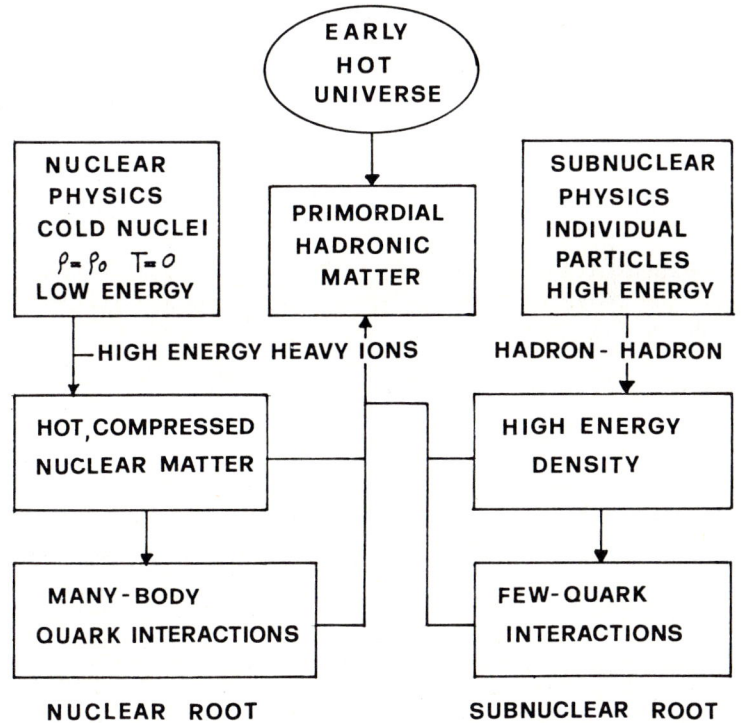

Fig. 1. Fundamental problems and corresponding rootes to approach them for nuclear and subnuclear physics, respectively. ρ_B, T and ε are the baryon density, temperature and energy-density, respectively.

rests and fundamental problems of *nuclear* and *subnuclear* physics as far as the intimate structure of the matter of the universe is concerned.

The known "subnuclear root" to approach the primordial matter is related to the addition of ever more energy to regions of ever decreasing sizes; this is done via interaction between individual hadrons, with a *restricted number of quarks*. The major problem is the "confinement" of the elementary constituents.

On the other hand, the "nuclear root" to the primordial hadronic matter, which can be pursued by nucleus-nucleus collisions, is related to the possible observation of *many-body quark interactions*. Ever higher energy is added to volumes containing many nucleons and, bringing together a large number of quarks and gluons, at ever higher density (compression), they can be released not as a collection of individual particles but as a new form of matter, the "quark-gluon plasma". This is the form of matter corresponding to the earliest universe, just after the big-bang, from which elementary particles and subsequently atomic nuclei have been generated.

Moreover the common *interface* between nuclear and elementary-particle physics as arising from such a scheme, is, on one hand, that the nucleus itself can be used as a laboratory to evidentiate some sub-nuclear peculiarities and, on the other, that one can identify sub-nuclear effects in the nuclear behaviour due to the more elementary degrees of freedom (mesons, quark, gluons)

2. THE "HOT" AND "DENSE" NUCLEAR MATTER

We have already mentioned that, starting at certain energies ($\sim$100 MeV/A) one can observe, in nucleus-nucleus collisions, some *phase-transition* (liquid to vapor). This indicates that, increasing the energy deposition into the nuclear matter and compressing it to some extent, the cohesive nuclear forces are overcome by nucleons and a new nuclear stage is reached which develops in scattering fragments.

Such events clearly demonstrate that nuclear matter is dramatically sensitive to changes in temperature (*heating*), pressure and density (*compression*) and would justify the search of an "equation of state" for atomic nuclei, so as for ordinary substances.

A further demonstration is given by other experimental observations. For instance in the high-energy (1.8 GeV/A) collision between ^{40}Ar (projectile) and ^{208}Pb (target), as reported in a streamer-chamber experiment at the University of California[1], more than 130 charged-debris emerging from the site of the collision were found, more than the 100 proton charges belonging to the original nuclei. This means that 30 *new particles* (mainly pions) have been created in such a violent process.

Here we are already in the relativistic domain and the creation of a *new hadronic matter* is achieved by the *transformation of nucleons into more elementary particles* or into *higher-mass hadrons* (mesons and barions). We shall briefly summarize, in this lecture, the main phenomenological aspects of the possible "phase evolution" of nuclear matter and the corresponding expected behaviour and significance.

First of all we should recall that our present knowledge of nuclear matter is limited to nuclear excitations close to the ground-state (*cold nuclei*). This "normal nuclear matter" can be compared to a liquid drop where nucleons (protons and neutrons) move freely at zero temperature and pressure (p=0 since the nucleus is *unconfined* otherwise is would expand, for p>0, or it would collapse for

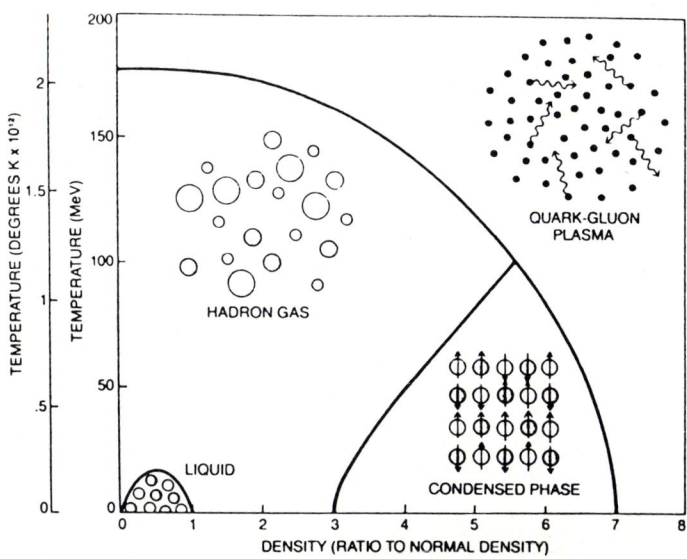

Fig. 2. Phase diagram (T,ρ) of nuclear matter with the different speculative fundamental states (from ref. 9).

p<0). The corresponding normal density is $\rho_0 = 2.8 \cdot 10^{14}$ g/cm² (0.16 fm⁻³) while the energy-density is $\varepsilon_0 = 0.13$ GeV·fm⁻³.

To reach higher temperatures and densities one needs to heat and compress nuclei even to *extreme conditions* (T∿100 MeV, ρ∿5÷10ρ_0) which are only possible in astrophysical bodies or in ultrarelativistic heavy-ion reactions.

Fig. 2 shows a (speculative) "phase diagram" for nuclear matter which is commonly used in the actual theoretical and phenomenological investigations. Until now our experimental knowledge is limited to the region around ($\rho/\rho_0 \cong 1$, T≅0). However some results, as the ones already mentioned, at CM energies between 0.1 and 2 GeV per nucleon would already give useful indications on the trend of nucleus-nucleus collision as a function of the excitation energy and baryon density

The main conjectures one can make in following the different

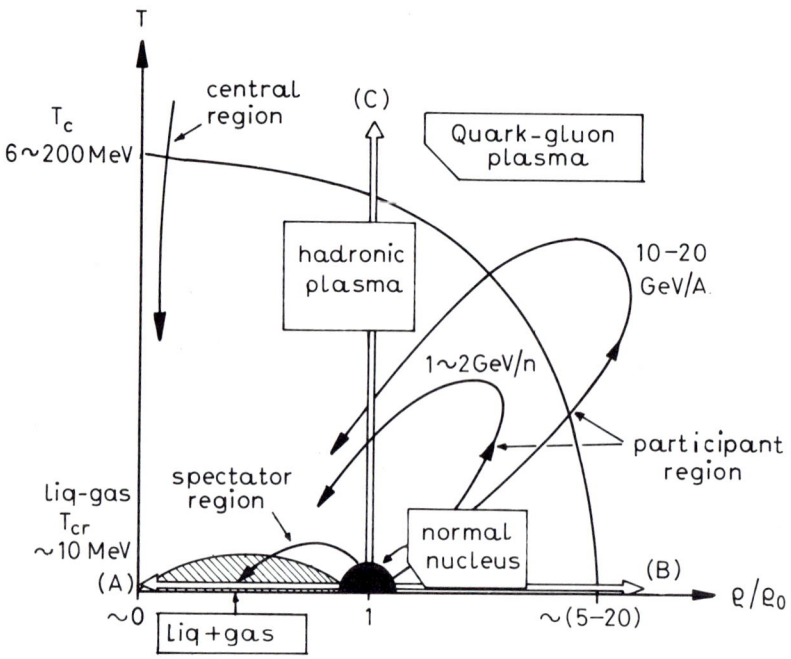

Fig. 3. The different roots to new phases of nuclear matter (from ref. 3).

roots as presented by the phase diagram are (see Fig. 3):
a) decreasing density at low temperatures(T<20 MeV)(*cold expansion*) for $\rho<\rho_0$ one has the "liquid-vapor phase transition", (*boiling of nuclear matter*), as already mentioned.

In such a case the internucleon distance increases until it becomes so large, for a certain limit ($\rho<<\rho_0$), that a nucleonic gas will be formed. The intermediate region would correspond to a *liquid-vapor mixed stage*. As for the ordinary matter, this occurs below a certain "critical temperature" $T_{cr} \cong 10$ MeV (see Fig. 2 root A).

b) increasing density ($\rho \cong 3 \div 5\rho_0$) at still low temperature (*cold compression*, root B) one could enter into a special phase of "abnormal nuclear matter" (*density isomers* and *pion condensate*). The "condensate phase" corresponds to "freezing" the nuclear matter into a *lattice structure* similar to a crystalline solid, where protons and neutrons have alternate opposite spin orientations. This expectations is based on a special binding arising from the interaction between pions and nucleons at such densities[4].

On the other hand, the "density isomers" as discussed by different authors, among whom T.D.Lee and G.Wick[5] and A.Migdal[4], correspond to the fact that in compressing nucleons closer together, the curve described by the "equation of state" would not necessarily continue to rise indefinitely. Secondary minima could appear corresponding to *stable high-density configurations*. Such abnormal states can also be introduced in the relativistic mean-field theory of normal nuclear matter taking into account Δ-resonances[6].

c) At higher temperatures (T>50 MeV) the way to "pionization" is open (root C) provided the impact energy is sufficient to give rise to internal transformation of nucleons. In that case a real "chemical" change will occurr: nucleons "create" hadrons, among which pions (i.e. the lowest-mass mesons) are the first to appear. Pion emission has been observed even at low-energy (25 MeV/A) in nucleus-nucleus collisions, indicating the presence of large cooperative effects of many-nucleons. The creation of the "hadronic gas-phase" is not only limited to pions but will contain other excited baryons i.e. hadrons heavier than pions will be created at energies above 1 GeV/A, in which the constituent quarks have taken on additional energy and angular momentum. Increasing energy would contribute in producing heavier and heavier hadrons until a *highest mass hadron* is produced. However, since this might not exist, there should be an absolute limiting temperature (the Hagedorn limit T_c)[7].

This means that, at a certain point, the nucleon character is destroyed giving rise to an assembly of quarks and gluons, since their mutual distance becomes smaller than the nucleon and pion size (∼0.8 fm) ("quark-gluon plasma)[8].

d) The Hagedorn limit on the temperature side marks a new phase-transition from hadronic to quark-gluon matter. This transition could also be reached following the root B, exploiting more and more the cold compression (limiting density at $\rho/\rho_0 \cong 5 \div 10$). In that case nucleons start to overlap and are expected to melt into quarks and gluons[8], which, therefore, will be *deconfined*.

e) Of course this "deconfinement" phase transition can occur following intermediate roots between the limits of very high temperatures (T=150-250 MeV) and baryon densities ($\rho \cong 5 \div 10 \rho_0$).

This is what happens (or could happen) in the high-energy heavy-ion reactions as shown in Fig. 3 by the curved roots.

3. PHENOMENOLOGICAL MODELS AND NUCLEAR EQUATION OF STATE

A phenomenological model which takes into account the various features leading to phase transitions inside and outside the deconfinement limit (see Fig. 3) is, for instance, the "participant-spectator" representation as shown in Fig. 4, taken from Nagamiya[3].

Here the overlapping fraction of the two colliding nuclei *(participant zone)* is separated from the two *spectator partners* and gives off different species of hadronic matter (nucleons, pions, massive hadrons ...) immediately after a short period during which temperature and density are both largely increased due to the violent nucleon-nucleon collisions. That situation corresponds to the system starting to expand and to be cooled down (see the lower right part of Fig. 4).

If the energy deposition is sufficiently high, not only an hadronic plasma, but also a quark-gluon plasma (deconfined) can be created.

The spectator region, on the contrary, corresponds to the nucleons passing through each other *(transparency)* since no very hard collisions occur. Density will decrease without increasing

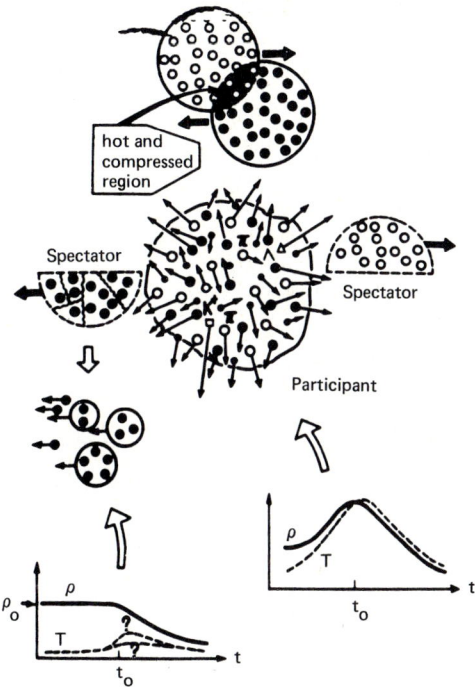

Fig. 4. The participant-spectator model of high-energy nuclear collision as a function of time (from ref. 3).

significantly temperature (see the lower left part of Fig. 4) and, therefore, together with projectile and target-like fragment emission, one could expect a transition to a liquid-gas mixed phase (see again Fig. 3). One should keep in mind that the chief difference regarding the nucleus-nucleus collisions at high energies as compared with low-energy experiments is that, in the last case ($\sim$10 MeV per nucleon), the basic costituent is the *nucleus* as a whole, since the De Broglie wavelength λ is of the order of the nuclear radius R($\sim$3 fm) and the interaction takes place with the *nuclear mean field* (mean free path Λ>R); in the former case ($\geq$1 GeV per nucleon) since $\lambda\ll d$, where d is the internucleon distance the *nucleon* is the basic constituent and the *nucleon-nucleon interactions* dominate (mean free path $\Lambda\leq$R).

Such aspects are taken into account by two other models, but in different ways: the "hidrodinamical model" introduced by the Frankfurt school[9] and the "intra-cascade model" developed at the Weizman Institute[10], at Dubna[11] and at the University of Liège[12].

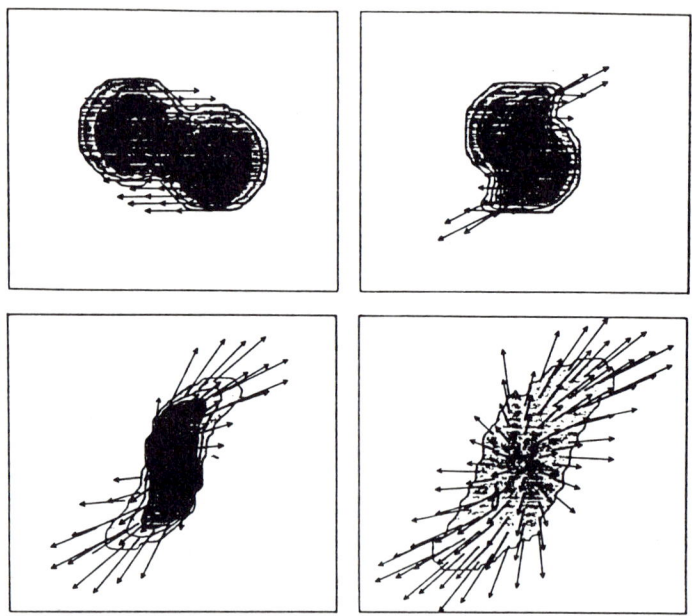

Fig. 5. Colliding nuclei (center of mass system) in the hydrodinamical representation stopping each other and giving rise to an important "side-flow" of the emitted fragments. (from ref. 9).

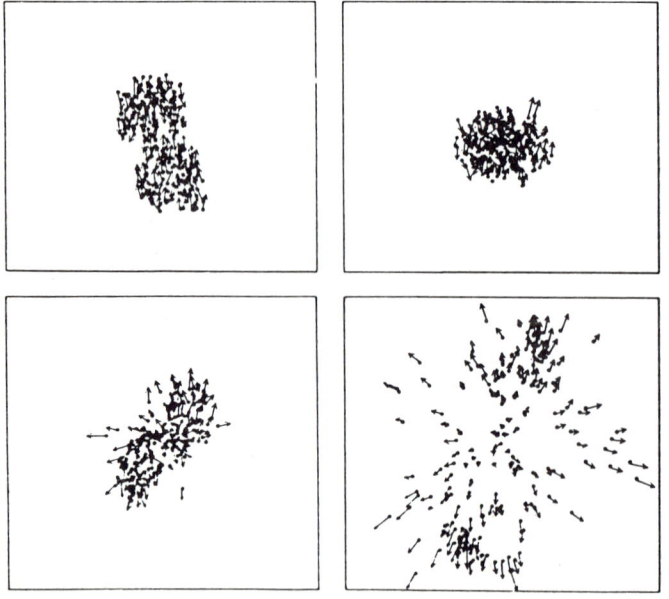

Fig. 6. Colliding nuclei and fragment distribution as given by the cascade model (from ref. 9).

The first takes into account the short mean free path without any other individual particle consideration and since the two colliding nuclei are treated as liquid drops *stopping each other*, local equilibrium is needed and the thermodinamical concepts have to be used. The second deals explicitely with the short De Broglie wave-length and with the nucleon-nucleon collisions; therefore the two colliding nculei behave as if they would be *transparent*.

The two extreme representation are illustrated in Fig. 5 and 6, respectively, taken from W.Greiner and H.Stöcher[3]. One of the major difference is the presence of an important component of the matter-flow perpendicular to the beam axis in the hydrodinamical picture, whereas the cascade model gives rise to an almost uniform distribution.

In both models, so as in the participant-spectator representation,there is an initial stage of *compression*, followed by an *expansion* and a *disintegration*.

The *compression stage* corresponds to the *interpretation* of the two colliding nuclei, mainly in the participant zone. This stage, in the hydrodinamical picture, would be related to the appearance and development of "shock-compression waves" at a density $\rho \cong 3$ to 5 ρ_0. The subsequent *high-energy-density stage* corresponding to highly compressed nuclear matter means, in the language of the participant representation, the creation of the central "fire-ball" (hot nuclear matter).

Expansion should be followed by *cooling down* or *freezing out* of the system, which terminates the whole cycle of nuclear collisions.

The bombarding-energy interval corresponding to such different stages can vary from 50 MeV/A (sound velocity in the nuclear matter) which is the lower limit for the occurrence of shock-waves in central collisions to hundreeds of GeV/A (225 GeV/A is the actual top energy at the CERN-SPS) and could reach the TeV/A domain for cosmic-ray events.

Of course, there are some major problems to be faced. First, the use of thermodinamical concepts for phase-transition models, is not very clear when the expansion time is of the order of the time between nucleon-nucleon collisions and the meaning of thermal equilibrium remains an open question.

Since, for a nuclear system, the transition occurs at a speed which is only a factor of few unities smaller than the average nucleon speed, the thermodinamic quantitites cannot be well defined as for infinitely slow transformations.

Second, the only messengers carrying out the available information on the initial dense and/or hot nuclear system, are the nuclear fragments which, before detection, have passed through the expansion and freezing-out stages. Expansion, which is due to the high internal pressure, leads to so low density ($\sim 0.5\rho_0$) that the particles in the nuclear fluid do no interact anymore and a transition to a free gas of nuclear fragments travelling towards detectors, occurs. To calculate their composition one can use quantum-statistical procedures which need limiting restrictions (for instance, in the hydrodynamical model, the conservation of the baryon number and of the energy for interacting particle).

Third, in the extreme picture where only the nucleon degrees of freedom are considered and the fire-ball is produced in a "cold process" where the projectile drills a hole through the target, one expects that the emitted-nucleon multiplicity would be given by the participant nucleons. This is in contradiction with experimental data, which show that the multiplicity is increasing with the energy per nucleon[13] (see Fig. 7).

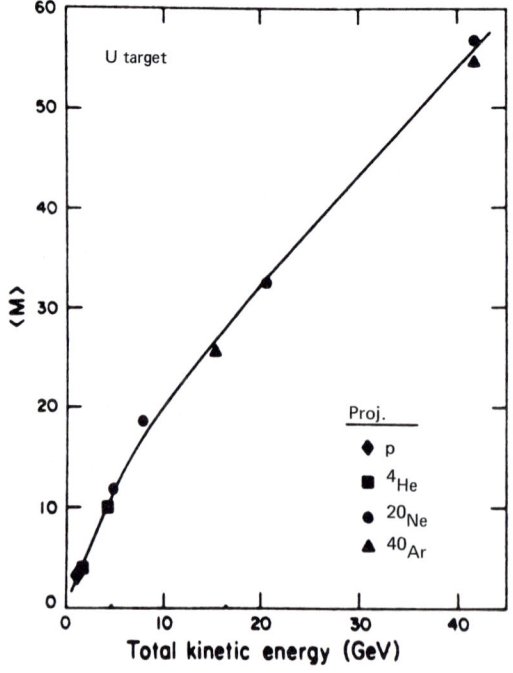

Fig. 7. Particle multiplicity versus the bombarding energy for reactions induced by different projectiles on Uranium target (ref. 13).

However, in the last case, the explanation seems to be found in a fire-ball expanding into the spectator region. A striking feature of the emitted particles is the different slope of the experimental spectra of K^+, protons and pions as obtained from the interaction area with 2.1 GeV/A Ne beams on NaF and Pb targets (see ref. 3). Such a slope which can be related to the "temperature" of the particles indicated that pions have the lowest, protons the intermediate and K^+-mesons the highest temperature as shown in Fig. 8, in the representation used by Faessler[13].

As pointed out by Nagamiya[3], this can be simply related to the mean free path of the different particles. After the initial compression stage when the system starts to expand and to cool down, high-energy protons and secondary particles, π and K^+, would be created copiously. Among them, those with larger $\Lambda(K^+)$ would escape more easily and carry the highest kinetic energy (temperature): protons with intermediate values of Λ and pions with the lowest would experience more interactions (rescattering) and escape subsequently, reflecting later cold stages of the collision.

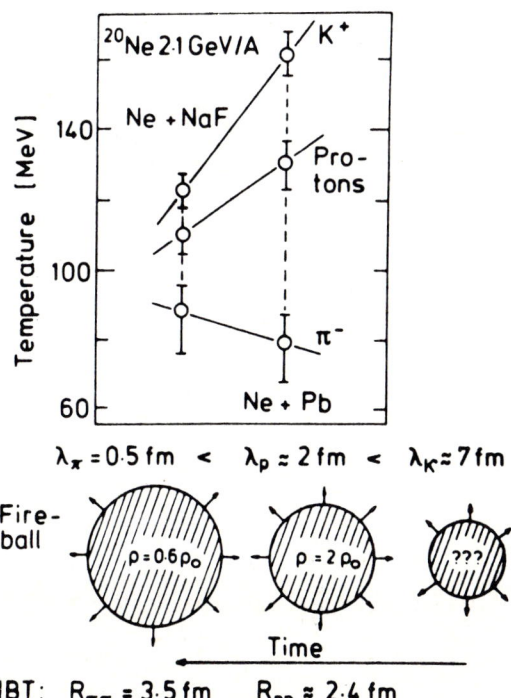

Fig. 8. Time evolution of the "temperature" of different particles as observed in violent nucleus-nucleus collisions (see ref. 3 and 13).

In spite of the different and incomplete validities of the various models, one tries to build a *nuclear equation of state* from semiempirical arguments. For instance, increasing the baryon density, one is able to find a T=0 equation of state (cold nuclear matter)[14] i.e. the relationship between the energy per nucleon W and the density ρ as illustrated in Fig. 9, where $W(\rho,T=0)$ is plotted as a function of the density ratio ρ/ρ_0' for two conventional theories: the mean-field approach and the variational method. The empirical hatched region is taken from pion-yield data. It should be remembered that pion emission has been found down to 25 MeV/A bombarding energies.

The main remarks one can make are the following:
- a certain *minimum energy* is required around ρ_0 in order to get compression; here the parabolic behaviour dominates
- the transient high-density stage ($\rho>3\rho_0$) is easier to reach, with its *stationary density* ($d\rho/dt\cong 0$) than the high $d\rho/dt$ expansion (*stiff equation of state*)

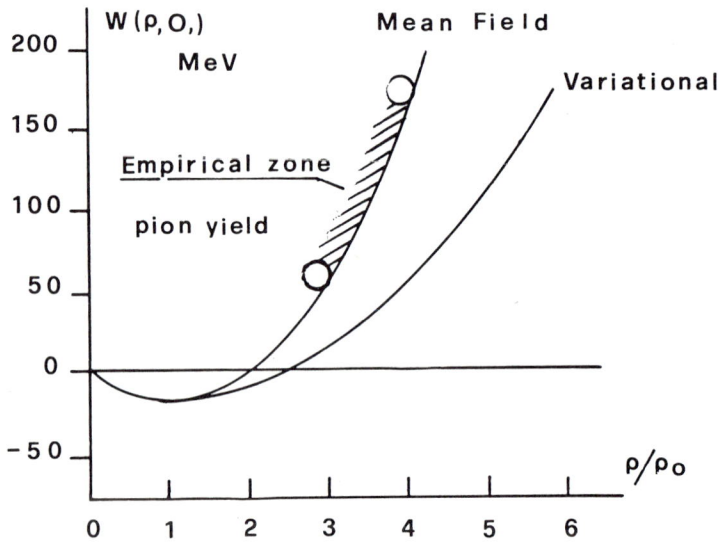

Fig. 9. Ground-state $W(\rho,T=0)$ of nuclear matter (from ref. 14).

- the interval from ρ_0 to $2\rho_0$ is not easy to be studied and no simple functional connection can be obtained between the ρ_0 domain and the high ρ region.
- to relate $W(\rho,T)$ with $W(\rho,T=0)$ is a complicated theoretical problem; on the other hand, shock compressions need high entropy and temperature, which could rise steeply with ρ in the interval $2.5\rho_0$-$4\rho_0$.[14]

4. EXPERIMENTAL OBSERVATIONS AND PERSPECTIVES

The experimental techniques now commonly designed for very-high energy heavy ion experiments are dealing with the measurement of *high-multiplicity final states* i.e. particle-yield and spectra as well as energy and flow distributions. Such data could be related to some "primordial observables", as pointed out by Stock[14], which have survived the various stages of the "compression-expansion-freezing out" sequence of the violent nucleus-nucleus collisions. The usual experimental devices are 4π-detectors, streamer chambers and plastic balls, while the available accelerator facilities are the Bevalac-LBL which can accelerate all ions up to Uranium from 100 MeV/A to 2 GeV/A, the Sincrophasotron of Dubna (Neon beam up to 4 GeV/A) and quite soon the SPS at CERN (225 GeV/A of ^{16}O and 60 GeV/A of ^{32}S).

The first experimental indication of the possibility of heating and compressing the nuclear matter in violent central collisions with $A \geq 100$ systems, was the discovery of *shock waves* by means of a momentum-flow analysis of emerging particles in the Plastc Ball Collaboration (GSI-LBL-Heidelberg)[2]. This is shown in Fig. 10, where the distribution of flow angles $dN/cos\theta$ for the reaction ^{93}Nb+^{93}Nb at 400 MeV/A is compared with the theoretical expectations of the hydrodinamic and cascade modes. The *sideward peaking* of the emitted fragments is in agreement, with the hydrodynamical predictions indicating the appearance of shock-waves, as already mentioned (see Fig. 5).

There are various experimental data on proton and pion-yields and spectra; others are coming from Kaons, which could give information on the time evolution of nuclear collisions producing hadronic plasma[3].

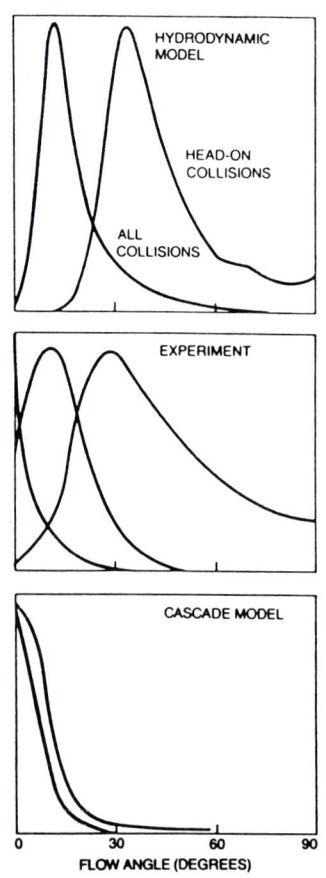

Fig. 10. Flow distributions of emitted fragment for the Nb+Nb reaction at 400 MeV/A as compared with the predictions of the hydrodynamical and cascade models (ref. 9).

This primordial observables can be related to model quantities in some phenomenological way. For instance the *temperature* T, is derived from the *slope parameter* E_0 of the nearly exponential shape of the proton and pion spectra observed experimentally, which are of the form[3]:

$$E(d^3\sigma/dp^3) \alpha \exp(-E_{CM}/E_0)$$

In spite of the nuclear validity of thermodynamic concepts such parameter E_0 is connected to the kinetic energy of the emerging particles which depends on the heating of the nuclear system emitting them.

If one plots $T=E_0$ as a function of the bombarding energies of various reactions experimentally investigated, one finds the behaviour illustrated in Fig. 11 (refs.9 and 13). Here the dashed line FG indicates the temperature dependence of a free nucleon gas,

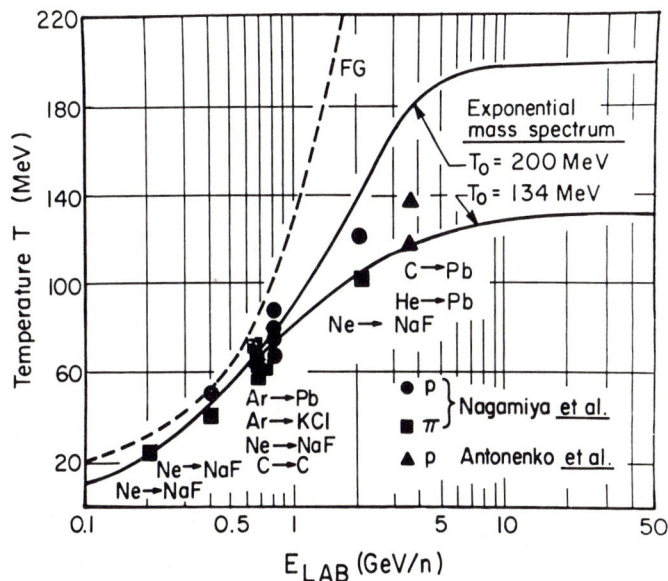

Fig. 11. Experimental "temperatures" (slope-factors E_0) attained at various bombarding energies for different heavy-ion reactions, as compared with the energy dependence of theoretical ones, as calculated in different models (see text).

which would give rise to a continuous increasing. On the contrary the data for energies above 0.8 GeV/A show a different behaviour, which could be approached by a model taking into account a mixture of non interacting hadron systems with an exponentially growing spectrum leading to a Hagedorn *limiting temperature*[9,13].

This limiting temperature, which could be of the order of 134 or 200 MeV, is not yet indicated by the available experimental data.

Other "measurable" quantities are the "thermal energy" which an be related to the observed ratio, of pion to baryon yields and the "fire-ball entropy" derived from nucleon to cluster yield ratio[14]. Another important information is the asymmetry in the pressure distribution that is related to flow-direction i.e. to the ratio between parallel and transverse particle momenta, as we have mentioned for the observation of shock-waves.

When more and more energy could be brought into the interaction region the "deconfinement" of the quark-gluon plasma is the major goal of the next planned experiments (CERN-SPS [15], Brookhaven-AGS[3]).

There are theoretical speculations[9,16] on the possibility of creating such quark-gluon plasma at two energy regions. The first is at 10-20 GeV per nucleon, where the two colliding nuclei stop each other increasing both temperature and density until nucleons are melt into quarks and gluons. The second is of the order of a TeV per nucleon or 10 GeV/A in a colliding mode; at this energy nucleons pass through each other but a hot baryon-free zone can be formed with a so high density that a quark-gluon plasma would be generated.

An illustration of the energy-density which could be attained as a function of the bombarding energy (GeV per nucleon in the Laboratory system) is shown in Fig. 12 (taken from Stöcker and Greiner, ref. 9), as obtained from simplified hydrodinamical calculations. Such energy-densities depend of the equation of state used as indicated on the figure. The major point to be emphasized is that, for heavy-nuclei collisions (Pb+Pb, for instance) it seems that not very high energy is required ($E_{Lab} \cong 5$ GeV/A) to reach the baryon density needed for the deconfinement-limit to be overcome.

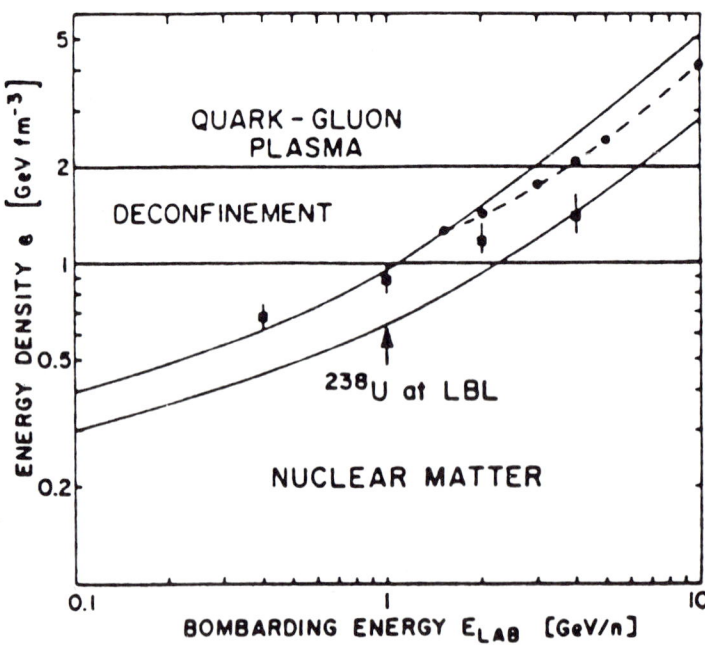

Fig. 12. Energy density as a function of the bombarding energy, calculated in the hydrodinamical model. The upper curve corresponds to a Hagedorn gas, the lower to a free-nucleon gas, the dashed-dotted curve to deconfinement (from ref. 9).

What is still uncertain is the clear experimental fingerprint of the quark-gluon plasma and the corresponding detection.

Current arguments are dealing with the possible decay of the "quagma" which would correspond to regrouping quarks and antiquarks into hadrons (strange mesons) or clusters which should be detected and recognized[3,15].

REFERENCES

1) R.E.Renfordt et al., Phys. Rev. Lett. 53 (1984) 763.

2) H.A.Gustafsson et al., Phys. Rev. Lett. 52 (1984)1590; and Phys. Lett. 142B (1984) 141.

3) S.Nagamiya, From reaction mechanism to new forms of matter, in: Proc. in Particle and Nucleon Physics, Vol. 15, ed. A.Faessler (Pergamon Press, Oxford 1985) 363-402.

4) A.B.Migdal, JETP 34 (1972) 1184; W.Weise and G.E.Brown, Phys. Reports 2TC (1976) 1.

5) A.R.Bodmer, Phys. Rev. D4 (1971) 1601; T.D.Lee and G.Wick, Phys. Rev. D9 (1974) 2201.

6) J.Boguta, Phys. Lett. 109B (1982) 251.

7) R.Hagedorn,Proceedings of Quark matter, Vol. 221, ed. K.Kajantie (Springer-Verlag, Berlin 1985) 53.

8) M.Jacob and H.Satz (ed.) Quark matter formation and Heavy Ion Collisions (World Scient. Pub. Singapore 1982); T.W.Ludlam and H.E.Weneger, Proc. of Quark matter, Nucl. Phys. A418 (1984); K.Kajantie, Proc. of Quark matter, Lecture Notes in Physics, Vol. 221 (Springer-Verlag 1981).

9) see for instance: W.Scheid, R.Ligensa and W.Greiner, Phys. Rev. Lett. 21(1968)1479; H.Stöcker and W.Greiner, Scientific American, January 1985; see also: G.F.Chapline,M.H.Johnson, E.Teller and M.S.Weiss, Phys. Rev. D8 (1973) 4302.

10) Y.Yariv and Z.Fraenkel, Phys. Rev. C20 (1979) 2277 and C24 (1981) 488; Z.Fraenkel, Nucl. Phys. A428 (1984) 373c.

11) K.K.Gudima and V.D.Toneev, Yad. Fiz. 27 (1978) 658; K.K.Gudima, H.Iwe and V.D.Toneev, J.Phys. 65 (1979) 229.

12) J.Cugnon, Phys. Rev. C22 (1980) 1885; J.Cugnon, D.Kinet and J. Van der Meulen, Nucl. Phys. A379 (1982) 553.

13) A.Faessler: Proc. Int. Conf. on Nucleus-Nucleus Collisions, eds.: G.F.Bertsch, C.K.Gelbke and D.K.Scott (North Holland, 1983) 565c.

14) See R.Stock, Progress in Particle and Nucl. Phys. Vol. 15, ed. A.Faessler (Pergamon Press, 1985), 455-478 and references quoted herein.

15) H.J.Specht, Progress in Particle and Nucl. Phys. Vol. 15, ed. A.Faessler (Pergamon Press, 1985), 479-481 and references quoted herein.

16) M.Gyulassy, Progress in Particle and Nucl. Phys. Vol. 15, ed. A.Faessler (Pergamon Press, 1985), 403-442 and references quoted herein.

IV
INSTRUMENTAL ASPECTS

HIGH INTENSITY PROTON SYNCHROTRONS AND THE EHF PROJECT

Franco BRADAMANTE
Dipartimento di Fisica, Università di Trieste, Italy and
Istituto Nazionale di Fisica Nucleare, Sezione di Trieste, Italy

1. HISTORICAL PERSPECTIVE

The origins of the Proton synchrotrons can be traced back to 1930, when E.O. Lawrence proposed the first circular accelerator, its most celebrated cyclotron. A resonance condition between the revolution frequency of the charged particles, moving in a uniform magnetic field, and the frequency of the RF voltage in an accelerating gap, would allow an energy gain much higher than that corresponding to the highest voltage in the gap.

It is well known that this resonance condition occurs at the cyclotron frequency ω_c given by

$$\omega_c = \frac{qB}{m}$$

where q is the charge of the particle, m its relativistic mass and B the (uniform) magnetic field. And it is also very well known that, for a fixed RF frequency, this condition is lost when the particle becomes relativistic. The way out was to modulate the RF to keep it in step with the cyclotron frequency, and Mac Millan[1] and Veksler[2] demonstrated indipendently in 1945 that stability in longitudinal phase space was guaranteed (principle of phase stability). Synchro-cyclotrons accelerating proton beams well into the relativistic domain could thus be built, the most powerful ones in operation being the 600 MeV SC at CERN and the 680 MeV machine in Dubna (USSR). I must stress, however, that the gain in energy was obtained at the expense of the beam intensity, since the synchronous accelerator has to be pulsed, while the cyclotron is a continuous beam accelerator.

The basic ideas for the synchrotron were proposed by Mac Millan and, independently, by Veksler in the papers I already quoted in which they demonstrated phase stability. In a synchrotron the beam circles on an orbit with fixed radius, and acceleration is obtained by an RF voltage which usually has a frequency which is an integer multiple (harmonic) of the revolution frequency. As the particle energy increases, the field is also increased, in order to keep the particle orbit the same at all energies, and the RF frequency is changed in synchronism with the revolution frequency.

The first proton synchrotron to enter into operation was the 3 GeV Cosmotron at Brookhaven, which was completed in 1952. As everybody knows a decisive push to the understanding of elementary particle physics was given by the experimental work carried on at the 28 GeV PS at CERN and the 32 GeV AGS (Alternating Gradient Synchrotron) at Brookhaven, which entered into operation in the early 60's. The largest proton synchrotrons existing today are the 500 GeV machine at the Fermi National Laboratory (Batavia, Ill., USA) and the 400 GeV SPS at CERN, but plans are already put forward for much more powerful machines, capable of accelerating a proton beam into the 20-50 TeV range, and requiring circumferences of several hundreds of kilometers.

To complete this historical introduction, I'd like to mention also that the last fifteen years had witnessed the development of proton-proton and proton-antiproton colliders, i.e. 2 x 31 GeV CERN ISR (Intersecting Storage Rings) in 1971 and the 2 x 270 GeV CERN Sp$\bar{\text{p}}$S in 1983. At the end of 1985 also Fermilab proton-antiproton colliders had the first test runs at 2 x 800 GeV. Colliders make the "best" use of the energy of the beams, which is all available for the interaction at variance with "fixed target" accelerators, where the centre of mass energy is only a fraction of the beam energy. The price one has to pay in a colliding beam machine as compared to a "fixed-target" machine is a reduction in the interaction rate, but today development in accelerators have allowed such increase in the intensity and quality of the primary beams to give quite acceptable values for the luminosity.

2. THE INTENSITY FRONTIER

In the previous paragraph I stressed, in the development of accelerators over the last 50 years, the growth in the maximum energy, which has increased roughly by one and a half orders of magnitude in energy per decade. Correspondingly, the higher and higher energies available for the collisions have allowed fundamental advances in the field of elementary particles, as best witnessed by the most recent discovery of the weak bosons at the CERN Sp$\bar{\text{p}}$S collider.

One has to remember, however, that our progress in the physics of elementary particles has been marked also by discoveries which have been made thank to the use of very high intensity beams and of sophisticated experimental techniques, and as examples I'd like to remind the discovery of the weak neutral current at the CERN PS and the finding of the J particle at the AGS. These discoveries were not done using the highest energy accelerators, and still are considered as the most spectacular advances in this field in the last twenty years.

Generally speaking, the increase in energy has been accompained by an equivalent increase in beam intensity and quality, which has allowed to push further and further the frontier of precision measurements, namely to look into subtler and subtler experimental effects.

The history of particle accelerators is marked by a continuous progress in beam handling and in understanding beam dynamics, and by a few break-through ideas, like strong focusing and beam cooling. Excellent treatments of all these topics exist in the literature, and I really would feel it inappropriate to repeat these subjects in these written notes. In particular I would like to remind the interested reader that most recently both the U.S. comunity and the European comunity have set up many Accelerator Schools and written Proceedings exist with detailed treatments of both elementary and advanced topics.[3]

In the following pragraph I would just like to give the essence of strong focusing because it represented a fundamental step in the construction of higher intensity and higher energy machines.

3. ALTERNATING GRADIENT SYNCHROTRONS

In all circular machines the beam undergoes oscillations around the equilibrium orbit, defined as the periodic solution of the equation of motion, and which usually lies in the horizontal plane. These oscillation take place both in the radial and in the vertical directions, and are usually referred to as betatron oscillations. Oscillations around the equilibrium values occur also in energy and phase, but their periods are much longer than the periods of the betatron oscillations, so generally the particle motion can be studied independently in transverse and longitudinal phase space.

In a machine like the cyclotron, transverse stability is guaranteed by the so called "weak focusing" property of the magnet. The guide field is not constant, but decreases slightly with increasing radius, and its gradient is constant along the circumference of the magnet. Very tight limits are necessary for the field index, $n = -\frac{r}{B}\frac{dB}{dr}$, namely

$$0 < n < 1$$

in order to have simultaneosly restoring forces in both radial and axial direction.

The gradient n must be positive to assure vertical stability, while it should be smaller than 1 for stability in the radial direction. When both conditions are satisfied, transverse stability is guaranteed, and the particles undergo slow oscillations around the equilibrium orbit. The ratio of the frequency of the betatron oscillation and of the revolution frequency is known as the betatrone tune, and is equal to $\sqrt{n}$ for the vertical oscillations and to $\sqrt{1-n}$ for the radial oscillations. Since the intensity of the accelerated beam depends crucially on the angular divergence of the beam accepted by the magnets system, a weak focusing synchrotron required larger and larger magnet apertures to go to larger and larger energies, so that these accelerators soon became prohibitively expensive, and the practical limit was given by the 12 GeV Argonne ZGS (Zero-Gradient Synchrotron).

The only way to keep the magnet apertures to reasonable values was to increase the restoring forces, thus increasing the betatron tune and correspondingly decreasing the wave length and tha maximum swing of the betatron oscillations. This could not be obtained simultaneously in the axial and radial directions, because in one case one needed a large and positive value of n, in the other a large and negative value. The breakthrough solution[4,5] was to satisfy these conflicting requirements <u>alternatively</u> along the circumference of a synchrotron, namely to provide a succession of high gradients, ($|n| \sim 300$) alternatively positive and negative. The net effect is focusing in both horizontal and vertical plane, with betatron tunes roughly equal to $\sqrt{\frac{|n|}{2}}$. This could be achieved either by using bending magnets in which the field was not uniform, but had alternatively a large positive and a large negative field gradient, or by inserting in between the (uniform) bending magnets special elements with strong magnetic field gradients, alternatively positive or negative, the so-called quadrupoles. In the first case one talks of combined functions synchrotrons (like the AGS at Brookhaven), in the second case of separated functions machines (like the CERN PS).

Strong focussing provided the key to the development of the present very high energy synchrotrons, which all can handle the same number of particles, about 2×10^{13} ppp (proton per pulse), roughly independent of energy. This number cannot be easily increased, because the accelerators are normally operating near to the limits given by the space charge forces. On the other hand, precision experiments by now demand to push further away the intensity frontier and a substantial factor can be gained only by going to fast-cycling synchrotrons, i.e. to accelerators which are pulsed many times per second. The rest of my talk will be devoted to a project which foresees the use of several fast-cycling synchrotrons, the European Hadron Facility.

4. THE EHF PROJECT

The European Hadron Facility (EHF) is a high current proton synchrotron in the 30-40 GeV energy range that a large community of nuclear physicists and elementary particle physicists is proposing to construct in Western Europe. The starting action of this community has been the "Workshop on the Future of Intermediate Energy Physics in Europe[6], which was held in Freiburg im Breisgau fro 10 to 13 April 1984, and the next one will be the "International Conference on a European Hadron Facility", which will take place in Mainz, from 10 to 14 March 1986. In between, this community has taken many actions to promote this project, and a proposal is being prepared to be submitted to the relevant authorities possibly within 1986 or early 1987.

We are very well aware of the existence of similar projects[7,8] somewhere else in the world, but we believe that given the size and competence of the community, and the long and sucessful tradition of particle and nuclear physics at intermediate energy in Europe, the construction of such Facility in Europe is fully iustified[9]. The physics case is very well known and very well agreed upon, and in a situation in which particle physics, concentrating on high energy colliders, and nuclear physics, investing primarily in heavy ion colliders, drift irreversibly apart, EHF offers the unique possibility of bridging over these two fields and investigating with a common language the central issue of hadronic interactions, namely confinement.

A preliminary conceptual design for the European Hadron Facility has been worked out and in this report I will give a short description of the various accelerators involved and of the criteria adopted. The work has been carried through by an international team made up of the accelerator physicists from Europe and from oversea listed in Table 1. The activities of the group have been sponsored in so far by Germany, Italy and Switzerland, and have had as milestones two dedicated workshops, in Trieste (Oct. 9 - Oct. 16, 1985), at SIN (Dec. 9 - Dec. 11, 1985), and a third one is about to take place in Karlsruhe (March 3 - March 5, 1986).

The terms of reference for the EHF asked for an accelerator complex to accelerate a 100 µAmp proton beam ($.6 \times 10^{15}$ protons/sec) to an energy of 30 Gev. Further constraints regarded
i) the capability to produce polarized proton beams;
ii) the presence of both fast and slow extraction systems (duty factors 10^{-4} and 1 respectively) to produce neutrino beams and a spectrum of intense, high quality μ, π, k and $\bar{p}$ beams;
iii) easy upgrade of the designed maximum energy to~40 GeV.

No existing injector was suggested, since no existing site or laboratory was to be assumed for the moment, and in this sense the project has been usually referred to as the "siteless EHF"[6]. Being a European project clearly a Swiss option using the 590 MeV SIN isochronous cyclotron as injector has been kept in mind from the very beginning (option A in ref. 6) and is still being investigated by the SIN staff[10], but has not been studied by the group in Table 1 and I will not report about it. The possibility of building the EHF in the framework of CERN suggested a further constraint, i.e.

iv) 960 m for the length of the 30 GeV proton-synchrotron so that eventually it could fit into the ISR tunnel.

Actually, given the energy of the machine, this last requirement turned out to be no constraint at all.

TABLE 1: EHF Design Group.

R. BAARTMAN	TRIUMF	G. MACKENZIE	TRIUMF
P. BLÜM	Karlsruhe	A. MASSAROTTI	Trieste
D. BÖHNE	GSI Darmstadt	D. MÖHL	CERN
K. BONGARDT	KFA Jülich	F. PILAT	Trieste
A. CITRON	Karlsruhe	M. PUSTERLA	Padua
E. COLTON	Los Alamos	G. REES	Rutheford
M. CONTE	Genoa	A. RUGGIERO	Argonne
M. CORNACCHIA	Berkeley	G. SCHAFFER	Karlsruhe
E. COURANT	BNL	H. SCHÖNAUER	CERN
M. CRADDOCK	TRIUMF	A. THIESSEN	Los Alamos
J. CRAWFORD	SIN Villingen	C. TSCHALAR	SIN Villingen
J. GRIFFIN	FNLA	V. VACCARO	Naples
W. JOHO	SIN Villingen	M. WEISS	CERN
H. KLEIN	Frankfurt	C. WIEDNER	MPI Heidelberg
P. LAPOSTOLLE	France		

5. ACCELERATORS PARAMETERS

The proposed EHF is the complex of accelerators schematically illustrated in Fig. 1, whose main components are a high-energy LINAC, accelerating a H$^-$ beam to 1.2 GeV, and two synchrotrons, a 9 GeV Booster Ring and a 30 GeV Main Ring, with radii and repetition rates of ratios 1:2 and 2:1 respectively. In order to cope with the required beam intensity the synchrotrons have to be fast cycling. The repetition rates of the LINAC and of the Booster are the same, 25 Hz. The H$^-$ beam pulse coming from the LINAC is stripped into a proton beam by passing through a thin foil and injected directly into the Booster over 200 turns.

Two more rings complement the system, a 9 GeV Accumulator, with the same

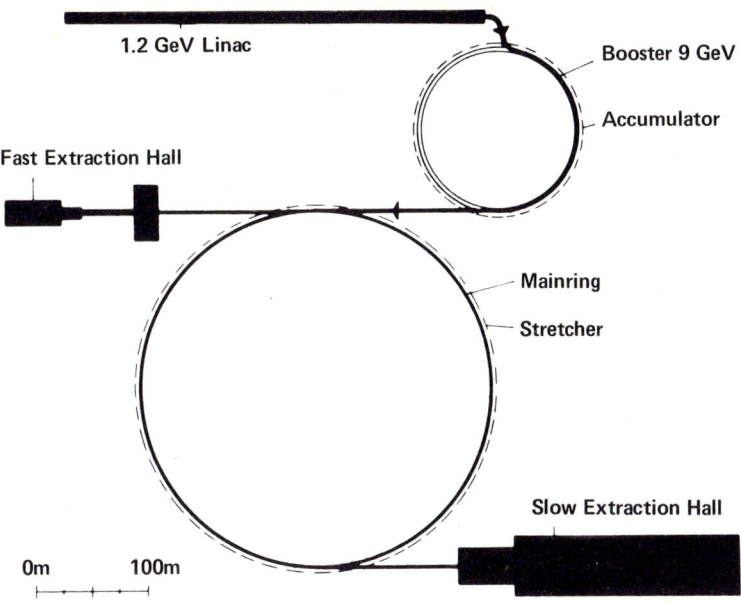

Fig. 1 - Schematic layout of the European Hadron Facility.

radius as the Booster, and where the Booster pulses are stored before being transferred to the Main Ring, and a 30 GeV Stretcher Ring, having the same radius as the Main Ring synchrotron, where the fast extracted 30 GeV beam from the Main Ring is stored and then slowly extracted to produce 100% duty factor secondary beams. The 1:2 ratio between the repetition rates of the Main Ring and the Booster allows us to have an Accumulator of the same size as the Booster, rather than of the same size as the Main Ring as customary. Only one Booster pulse is thus stored in the Accumulator, the second one passing through the just emptied Accumulator and going directly to the Main Ring. The use of these two relatively low-cost storage rings allows us to continuously run acceleration cycles in the Booster and Main Ring without the need to "flat-top" or "flat-bottom" the magnet cycles. The net advantages are less strain on the rf system and a 100% duty factor for the slowly extracted proton beam.

The operation of the complex can be understood by looking at the time diagram in Fig. 2. Acceleration cycles of the Booster and of the Main Ring are shown, as well as the beam transfers between the various rings. Bucket-to-bucket transfer of the beam from one machine to the next minimizes beam losses, as will be explained in Section 8.

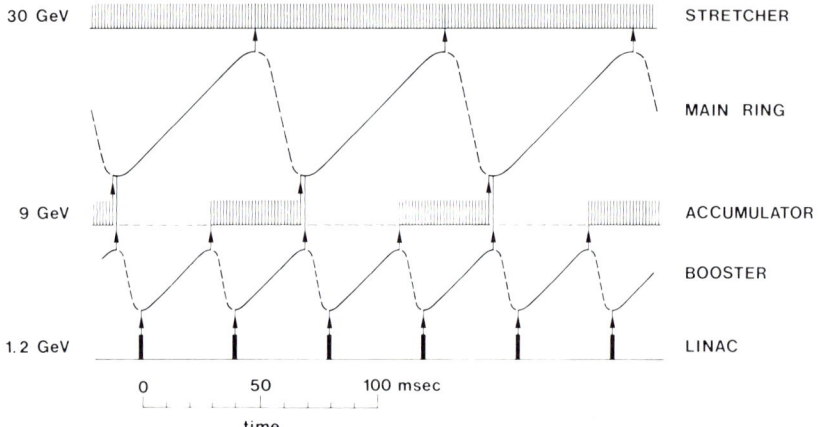

Fig. 2 - Time diagram for the EHF. The acceleration cycles of the Booster and of the Main Ring, and are shown, as well as the beam transfers in the various stages.

6. BEAM LOSSES

A crucial issue in the design of every single part of the EHF has been the minimization of beam losses.

A 1% beam loss at EHF would correspond to the loss of the full beam at the CERN PS or at the Brookhaven AGS, a perspective clearly unacceptable and which would put excessive requirements on shielding and remote control of the machine components and seriously question the reliability of the entire complex. It turns out, however, that by careful design and parameter optimization beam handling can be effectively lossless, so that the general radioactivity level will be equivalent to that of an existing accelerator of the same energy. Of course, the machine must be adequately shielded (underground tunnels, ~20 mt deep, are needed to house the accelerators) but mainly as a precaution against accidents, and no particular radiation damage of the various components is foreseen.

Minimization of beam losses has been achieved by (i) properly choosing the main accelerator parameters which are of relevance for the beam instabilities, (ii) phase locking all the machines, so that the beam is always transferred bunch-to-bucket (iii) proposing a new beam injection technique for entering into the Booster, as described in sect. 8, and (iv) placing very safe margins on the magnets apertures.

As regard point (i), it is important to stress that crossing of transition energy is avoided both in the Booster and in the Main Ring. At transition longitudinal stability is lost (the stable phase goes through $90°$) and, although the crossing of transition energy is considered as a standard procedure in existing synchrotrons, we believe that in presence of space charge limitations transition crossing could be a source of considerable beam losses.

As regard point (iv), we have taken an acceptance which is four times the beam transverse emittance at injection, and includes allowances for closed orbit distorsions and dispersion.

Although a precise evaluation of the beam losses has not yet been done and will require the running of long simulation programs, we are confident that they can be kept within the 10^{-4} level. Losses will be significant, at the 10^{-2} level, only at two points, namely at injection into the Booster, where the H^- beam passes through the stripping foil, and in the septum magnet for the slow extraction from the Stretcher, where again beam losses are unavoidable. In both cases, however, beam losses are localized, and will be dealt with by using collimators and beam dumps, and, where necessary, particular pieces of equipment will be handled by remote control.

7. DESIGN CRITERIA

Some basic choices in the design are straightforward, and in particular this is the case for the building blocks in Fig. 1. Also the repetition rate of the Main Ring, 12.5 Hz, follows directly from the EHF current requirements (100 µAmp) and from the limit to the maximum number of protons per pulse given by current experience with existing similar machines ($N_p \simeq 5 \times 10^{13}$). On the other hand, the definition of the parameters of the various machines is the result of a long and careful optimization procedure aimed at maximizing the reliability and minimizing beam losses, within the boundary conditions set by the EHF Study Group and within a reasonable economic frame. A characteristic feature of the complex we propose is the injection energy of the Main Ring, 9 GeV, which is rather high as compared with that of similar projects (3 GeV for TRIUMF[7], 6 GeV for LAMPF II[8]), and which in turn demands a rather high energy LINAC.

Such a choice for the injection energy into the Main Ring presents several advantages with respect to somewhat cheaper options with a 4-30 GeV Main Ring, namely

i) the possibility of using Siberian Snakes[11] in the Main Ring so as to preserve the beam polarization. Since the magnets in the snakes work at constant field and some 15 T.m are needed, at least 9 GeV are required to keep the magnet apertures within reasonable limits.

ii) More flexibility with lattice design. It is easy to avoid transition crossing by placing the transition energy well outside the energy range of interest. Also, since polarization is guaranteed by the Siberian Snakes, low super-periodicity lattices may be used for the Main Ring, thus allowing long dispersion-free straight sections for RF cavities, injection and extraction.

iii) Owing to the rather high injection energy into the Booster one can increase the number of particles per pulse in the Booster and consequently decrease the repetition rates of the LINAC and Booster to 25 Hz, namely twice that of the Main Ring. As a nice consequence of this ratio for the repetition rates, as already seen in Section 5, all four rings are evently distributed into the two tunnels, Main Ring and Stretcher in the big tunnel and Booster and Accumulator in the small tunnel.

iv) The RF-requirements in the Booster are released, since the RF swing is rather small (11%) due to the high injection energy (1.2 GeV).

v) An upgrade of the energy of the Main Ring to 40 GeV is quite easy, due to the low packing factor.

vi) There is a possibility for a staged construction where the LINAC, Booster and Accumulator are built first and the Main Ring and Stretcher added later. It is indeed possible to carry out a useful physics program at 9 GeV, especially with low energy neutrinos, and for that reason the Fast Extraction Hall in Fig. 1 can be fed directly from the Booster. Eventually even polarized proton physics or kaon physics can be carried on by operating the Accumulator as a Stretcher for the Booster pulses.

The energy of injection into the Booster ring has been fixed essentially on the basis of space-charge considerations, and it turns out to be 1.2 GeV. The complex therefore requires a rather long and expensive LINAC, which however is well within present technology, as will be shown in Section 9.

Other Scenarios have been considered by the Design Group, but the one I have outlined here has been found to be, for the moment, the most effective, although somewhat more expensive. Another Scenario which has been examined centred around 4 GeV Booster with a circumference four times smaller than the Main Ring circumference. In this Scenario the Booster has a repetition rate of 50 Hz, and four pulses are transferred into the Main Ring at a time, via an Accumulator which has to be located in the Main Ring tunnel. The injection energy into the Booster had to be 0.6 GeV. Though this Scenario is ceaper than the 9 GeV Scenario, it has been nevertheless disfavoured because of several other technical drawbacks.

8. INJECTION SCHEME FROM LINAC INTO BOOSTER

LINAC's and synchrotrons have conflicting requirements as regards the RF systems, since LINAC's want high frequencies, even a few GHz, to reduce beam losses and the length of the structure, whilst synchrotron cavities prefer a "low" frequency because of the frequency swing (11% in our Booster). The final choices have been made to minimize costs, 50 MHz in the Booster and 400 MHz in the LINAC. The beam pulse coming from the LINAC is thus a train of bunches at a frequency which is 8 times the RF frequency at injection of the Booster.

To cope with the required intensity a beam pulse equivalent to the length of 200 turns has to be injected into the Booster. The only way to do this is by H^- stripping, i.e. inject a beam of H^- which will eventually lose their orbiting electrons by hitting a foil during the first time they go around into the Booster. This method is not phase space area preserving and therefore allows reasonably small beam emittances. In each RF buckets the bunches coming from the LINAC will add up every turn to the ones previously injected, by using a special "painting" technique in both longitudinal and transverse phase space, which has been first suggested to us by the Los Alamos and TRIUMF groups. Such a technique will allow to fill the bucket area uniformily, thus reducing the the space-charge forces.

Detailed calculations are still needed to evaluate precisely the losses of the proposed scheme, but preliminary estimates already make us confident that the system has a very high efficiency and well justify our claim that beam handling will not be a serious problem for EHF.

9. THE 1.2 GEV LINAC

The linear accelerator is made up of a source of negative hydrogen ions, followed by a combination of a chopper and two RFQ's (Radio Frequency Quadrupoles), operating at 50 MHz and 400 MHz respectively, one drift tube LINAC (DTL), operating at 400 MHz and finally a side coupled LINAC (SCL) running at 1200 MHz. The final energies of the various stages are 0.2, 2, 150 and 1200 MeV respectively.

The beam pulse will be 360 μsec long. The desired time structure for optimal injection into the Booster is generated in the two RFQ's, as illustrated in Fig. 3. The first RFQ captures and bunches the dc beam from the ion source to a certain extent. The compressed phase structure of that beam is then taken over by the second RFQ, which will further accelerate and bunch the beam to match it to the 400 MHz time structure of the Alvarez section. In between the

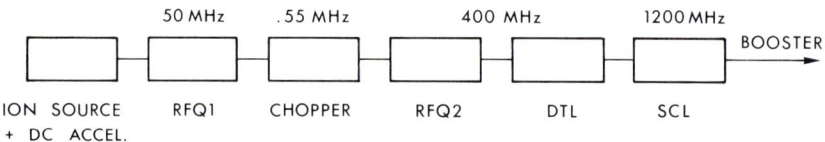

Fig. 3 - Schematic diagram of the Linear Accelerators.

two RFQ's, a 0.55 MHz chopper will create a 100 nsec hole (5 empty buckets out of a total of 90 buckets) in the Booster to allow the operation of the kicker and a lossless fast extraction of the beam from the Booster into the Accumulator.

After coming out from RFQ2, the beam is injected into the 400 MHz Alvarez Structure. No problems are foreseen in handling the effective beam current of 50 mA. From the DTL the beam is injected into the SCL, where there will be a frequency jump of a factor of three to increase the LINAC efficiency. Space charge problems are even less important than in the DTL due to the increased particle energy.

The design of both the DTL and of the SCL are on the safe side, and beam dynamic calculations showed no problems. In particular, the DTL is very similar to the CERN 50 MeV DTL, being somewhat more efficient because of the higher frequency (400 MHz rather than 200 MHz), which in turn is possible since the peak current is here 3 times less. The SCL is more demanding, and it will be the highest energy proton LINAC in the world. Its structure is similar to that of the Los Alamos 800 MeV LINAC, but again it will be more efficient because of the higher frequency and of the higher gradients, which are possible thanks to the present existence of more powerful klystrons. The DTL will accelerate with an average gradient of 2 MV/m and the SCL with an average gradient of 3.2 MV/m, resulting in a total length of about 400m for the whole LINAC, but it is possible that these gradients be increased in the final project to reduce somewhat the investment cost.

10. BEAM POLARIZATION IN THE SYNCHROTRONS

A very clear message from the EHF study group demanded for this machine capability of accelerating polarized proton beams of the same intensity as the unpolarized beams. Even if today polarized H^- sources cannot deliver such intensities, there is confidence that technological improvements will fill in this gap by the time EHF is in operation. Whilst the maintenance of polarization is guaranteed in the LINAC, the horizontal components of the magnetic fields in the four Rings, required for the strong focusing, induce rotations of the proton spins which tend to destroy the vertical polarization. This situation is particularly dangerous when depolarizing resonances occur[11], namely when the precession frequency of the spin $f = G\gamma$ (G is the usual proton gyromagnetic ratio $G = (g-2)/2 = 1.79$) coincides with the frequency of a disturbing horizontal magnetic field as seen by the circulating beam (either due to an imperfection in the lattice, or to the intrinsic periodicity of the strong focusing forces). Such resonance condition is met when $G \cdot \gamma = \ell$ (imperfection resonances) or when $G \cdot \gamma = kP \pm Q_y$ (intrinsic resonances), where ℓ and k are any integers and P and Q_y are the number of superperiods and the vertical betatron tune of the lattice respectively. One superperiod is that part of the lattice which repeats itself identically along the circumference of the synchrotron. When these resonance conditions are met, depolarization of the beam takes place at a rate which depends on δ, the intrinsic strength of the resonance and on α, the speed with which the resonance is crossed during the acceleration cycle, according to the Froissart - Stora formula $P_f/P_i = 2 \cdot \exp[-\pi|\delta|^2/2\alpha] - 1$. These factors, as well as the number of depolarizing resonances themselves, depend crucially on the choice of the lattice. For this reason great care and effort have been devoted to choose lattices for the booster and the Main Ring which would minimize the depolarization of the beam.

There exist several ways to preserve the polarization,

i) choose a high periodicity lattice to avoid intrinsic resonances,
ii) cross resonances with a fast Q-jump or an adiabatical spin flip. These are methods already used in the AGS and in Saturne,
iii) introduce spin trasparency into the lattice, which makes the strength of the resonance small when the resonance is crossed,
iv) use a pair of Siberian Snakes diametrically located in the lattice in order to make the spin tune energy independent.

Solution (i) was eventually ruled out because it demanded for unpractically large values for P or Q, incompatible with other requirements, such as long straigth sections for injection and extraction.

Solution (iii) has been kept in mind, but the proposed lattices could not reduce the strength of all the resonances down to the desired level, so that it could not solve the problem by itself alone. At the end we decided to propose method (ii) for the Booster and method (iv) for the Main Ring.

11. THE BOOSTER SYNCHROTRON

The Booster Ring is the most challenging part of the project, therefore particular care has been devoted to the choice of the lattice.

Since it was impossible to avoid the 0^+ resonance ($G \cdot \gamma = Q_y$), some depolarizing resonances had to be accepted in the lattice, but they have been chosen to be not too strong and will be crossed by fast Q-jump. A separated function lattice has been adopted to have more flexibility to adjust the tune. As the beam at injection is close to the space-charge limit, a doublet has been taken as focusing structure. Transition energy has been pushed well above the maximum energy of the Booster. There are six superperiods, providing six long (2x6.5 m) straight dispersion-free sections for installation of the RF cavities and for injection and extraction. Table 2 summarizes the Booster parameters and Fig. 5 shows a possible schematic layout.

TABLE 2: Booster parameters.

energy range	1.2 - 9.0 GeV
circumference	480 m
repetition rate	25 Hz
particles/pulse	2.5 10^{13} ppp
number of superperiodes	6
number of cells	54
cell structure	separated function magnets, DFO
phase advance/cell	$\sim \pi/2$
tune: horizontal	13.4
vertical	10.2
γ_{TR}	12.7
β_{max}: horizontal	12.8 m
vertical	14.3 m
D_{max} : horizontal	1.9 m
depolarizing resonances (kP)$^\pm$	18^- $\gamma = 4.4$ $\delta = 0.0025$
	0^+ $\gamma = 5.7$ $\delta = 0.0181$
	24^- $\gamma = 7.7$ $\delta = 0.0076$
	6^+ $\gamma = 9.0$ $\delta = 0.0136$

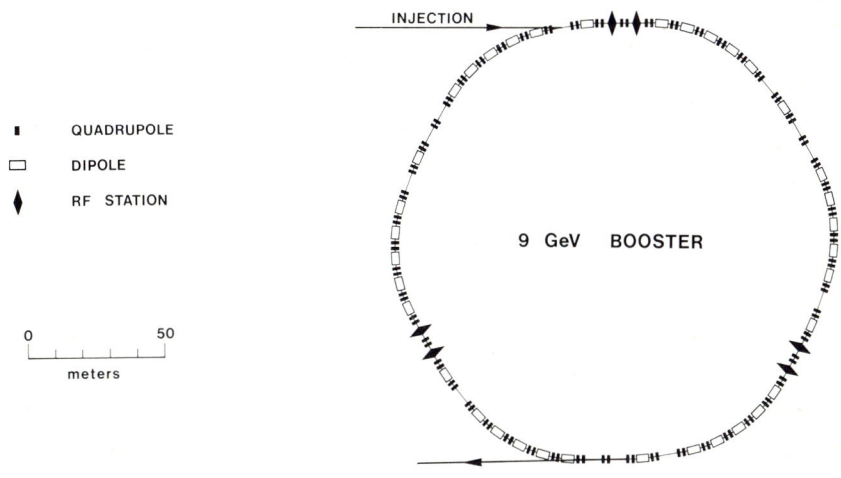

Fig. 4 - Schematic layout of the Booster Ring, showing a possible arrangement of magnets and RF cavities.

The beam power is 1.6 M Watt, and so as not to have beam loading problems we decided for an rf system delivering 3.2 M Watt total power. To provide the required 11% frequency swing, ferrite-loaded cavities will be used, similar to the ones developed at Fermilab. Twenty cavities will be needed, with a peak voltage of 50kV per cavity.

12. THE MAIN RING SYNCHROTRON

The Main Ring lattice has been chosen essentially to provide long (3x7.8 m) dispersion-free straight sections for insertion of the Siberian snakes (20 m) magnets and for extraction. Slow Extraction in this energy range requires straight sections about 20 m long. Although the slow extraction system will be installed in the Stretcher, it still represents a constraint for the Main Ring, since it will be convenient to install the two machines one on top of the other, so clearly they must have a similar lattice.

Like for the Booster, we have selected for the Main Ring a separated function lattice, with a regular FODO cell structure and a superperiodicity of 8. All the relevant parameters of the lattice are summarized in Table 3.

As the frequency swing is only 0.4%, a higher beam-loading ratio of 2:1 has been adopted. The total power needed is then 6.6 MWatt for a beam power of 4.4 MWatt. Twenty-four cavities, again of the type developed at FNAL, will be used,

TABLE 3: Main Ring parameters.

energy range	9.0 - 30.0 GeV
circumference	960 m
repetition rate	12.5 Hz
particles/pulse	5×10^{13} ppp
number of superperiodes	8
number of cells	56
cell structure	separated function, FODO
phase advance/cell	$\pi/3$
tune: horizontal	9.2
vertical	9.2
γ_{TR}	8.2
β_{max}: horizontal	30.2 m
vertical	29.5 m
D_{max}: horizontal	6.9
Siberian Snakes	yes

at a peak voltage of 85 kV/cavity, and will be arranged in four straight sections. Two more sections will contain the Siberian Snakes, one more section will house the injection and extraction systems and one section is left free for future options. A possible arrangement of magnets and RF cavities is shown in Fig. 5.

No detailed designs have been worked out at this stage of the project for the Accumulator and the Stretcher ring, but they are simple DC rings and do not pose any particular problem.

13. EXPERIMENTAL AREAS AND BEAM INTENSITY

As shown in Fig. 1, both slow extracted and fast extracted beam will be used for experiments.

The intensity gain of two orders of magnitude in the primary proton beam of the EHF as compared with the beam of the CERN PS or of the AGS will guarantee the same gain in the intensities of the secondary beams, the quality factor of capital importance to the physicists who will come to carry out experiments. Clearly, this gain in intensity can in many cases be usefully converted into a gain in beam quality, that is in beam purity, momentum resolution, and phase space. It turns out, however, that by improved target systems and secondary beam design, larger gains can actually be expected. This is the case, for instance, when adopting the MAXIM system[12] for the slowly extracted primary proton beam,

Fig. 5 - Arrangement of magnets and RF cavities for the Main Ring.

a system for Multiple Achromatic Extraction of Independent Momentum beams. Such a system allows extraction of several secondary high-energy charged-particle beams from a production target at zero angle to the primary beam. The beam intensity and quality is thus improved because generally the production cross sections are peaked in the forward direction and because long targets can be used (tipically one interaction length), without increasing the transverse dimension of the beam source.

The system consists of a pair of sectors of concentric circular bending magnets of opposite polarities, centred on the production target. As the magnetic field is rotationally symmetric with respect to a vertical axis through the target, any charged particle emitted from the target centre will emerge from the system travelling on a radial plane through the axis. Although all momenta are focused into any radial direction, there is a one-to-one correspondence between any given direction and the sign and momentum of the forward going-secondaries.

A possible beam layout for the Slow Extraction Area using two MAXIM schemes and three production targets to eat up essentially all the primary beam is shown in Fig.6. The first target is half an interaction length, while the other two are one interaction length. The layout is based on previous work done at SIN by the secondary beam group for HIPS (High Intensity Proton Synchrotron). The parameters of the beam lines are given in Table 4. The angular acceptance for the unseparated beam lines are given in parenthesis. The π, k and $\bar{p}$ intensities which can be obtained in the various lines are given in Fig. 7 and are really impressive.

The Fast Extraction Area in Fig. 1 is situated so that it can be fed either from the Main Ring or from the Booster. This possibility is quite attractive in a staged construction of the EHF because the 9 GeV Booster can be an excellent source of low-energy neutrino beams and fast-pulsed muon beams. The facility will consist of the usual production target, a magnetic horn as focusing device, a long decay section, and a huge shielding block in front of the fast extraction hall.

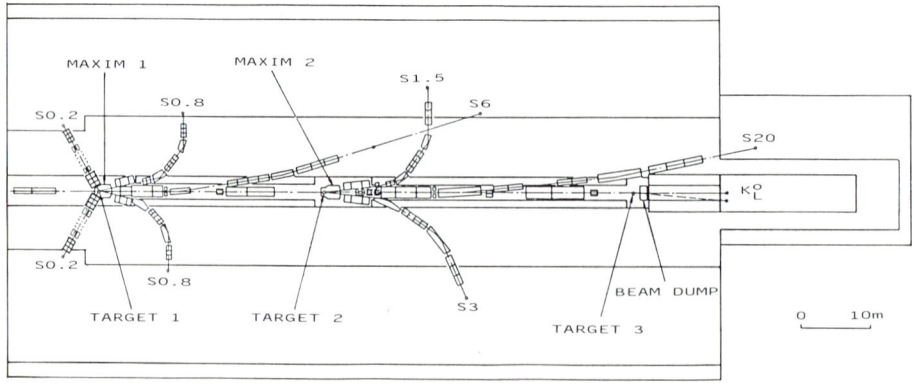

Fig. 6 - Possible layout for the Slow Extraction experimental area.

TABLE 4: Secondary beam characteristics for external W-targets (slow extraction).

BEAM	LENGTH m	l_{tg} mm	INTER. PER PROTON	ABSORPTION FACTOR	α_s mr	β_s mr	a_s mm	b_s mm	Ω msr	$\Delta p/p$	EXTRACT. ANGLE	PROD. ANGLE
S 0.2	20	50	.4	.9(π, μ)	80	200	1.1	2.1	50	.15	120°	120°
S 0.8	18	50	.4	.76(K,$\bar{p}$)	22	71	0.9	1.8	5(8)	.05	11.5°	0 - 5°
S 1.5	25	100	.4	.64(K,$\bar{p}$)	11	56	.74	2.8	2(5)	.03	5.7°	0 - 4°
S 3	35	100	.4	.64(K,$\bar{p}$)	5.5	56	.57	2.8	1(5)	.05	5.7°	0 - 3°
S 6	75	50	.4	.75(K,$\bar{p}$)	0.6	36	.50	1.0	0.07(3)	.05	0°	0 - 3°
S 20	130	100	.4	.55(K,$\bar{p}$)	0.6	25	.50	1.35	0.05(.5)	.05	0°	0°
K_L^0	15	100	.14	.7 (K^0)	10	2.5	.70	.52	0.1 wide band		0°	0°

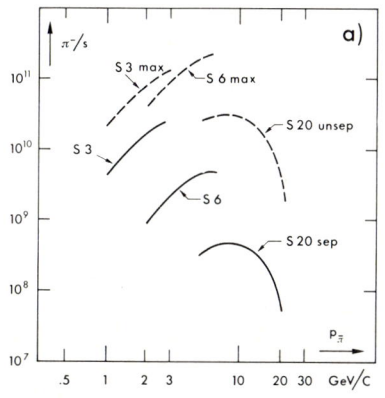

Fig. 7 -

Particle Intensities vs momentum in the beam lines shown in Fig. 6, for pions (a), for kaons (b) and antiprotons (c). Dashed curves of full curves refer to unseparated or separated beam lines respectively.

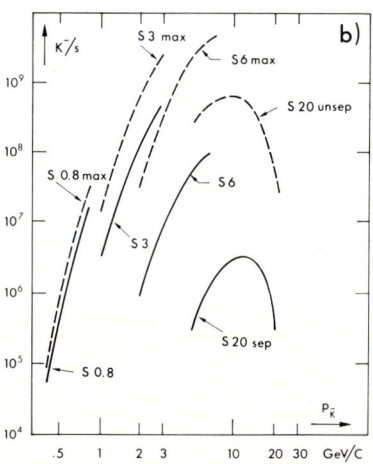

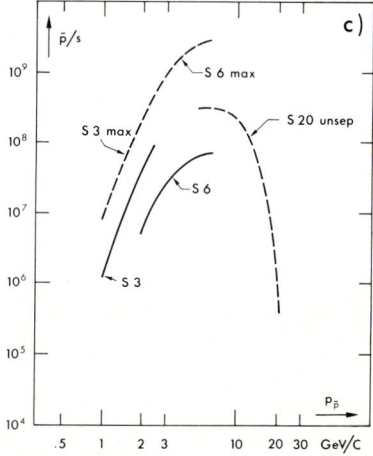

14. COST

A cost estimate for a "green pasture" laboratory including machines, tunnels, buildings, basic infrastructures, and two experimental areas, namely the Fast Extraction and the Slow Extraction ones, is given in Table 5. The estimates were mainly based on the comparable facilities proposed at TRIUMF and Los Alamos, but were adjusted to European prices by using the experience of CERN, DESY, GSI and SIN. The table does not include minor items such as the control system, probably 5% of the cost of the accelerators, and it has no allowance for inflation or contingency. Also, the running cost of the Facility has not yet been estimated.

As mentioned in Section 7, some cost reduction can be obtained by lowering the energy of the Booster to ~ 4 GeV and correspondingly adopting a shorter LINAC, accelerating the beam to about 600 MeV. Eventually this scenario was disregarded. Reliability of the design and excellence of performances, essential prerequisites for an ambitious project like EHF, were considered more important than just straightforward cost considerations.

TABLE 5: (in Million of Deutsch Marks).

1.2 GeV LINAC	149
1.2-9 GeV BOOSTER	104
9 GeV ACCUMULATOR	42
9-30 GeV MAIN RING	130
30 GeV STRETCHER	55
BUILDINGS	71
CENTRAL FUNCTION	10
COMMON SERVICES AND UTILITIES	60
FAST EXTRACTION HALL	46
SLOW EXTRACTION HALL	190
ARCHIT./ENG.	10
TOTAL	867

15. CONCLUSIONS

I have tried to summarize in this report the results of a Feasibility Study for a European Hadron Facility[13]. Clearly, a lot of work has still to be done to complete the design: the DC rings have not yet been studied, the SCL structure has to be calculated and tested, the Booster and Main Ring lattices have to be checked and eventually improved, items such as painting, or precise evaluation of beam losses and beam depolarization demand for computer tracking with multiparticle simulation codes. Further studies may also lead to improvements in the design, possibly resulting in better performance and lower cost of the Facility. Still, we believe the project we have worked out to be rather satisfactory and the price estimate reasonable and fairly accurate. The next step therefore is to the comunity of the physicists who want the Facility, and as a member of this comunity I would like to conclude by saying that the physics case for EHF is perfectly well demonstrated and has been agreed upon over the last two or three years, so EHF could really be the first move towards a new International Laboratory for Research in Nuclear Physics and low energy Elementary Particle Physics in Europe.

16. ACKNOWLEDGEMENT

The description I've given of the EHF is based on a Feasibility Study which is about to be concluded, and I wouild like to thank the Design Group listed in Table 1 for their enthusiastic contributions and for permission to quote the present material prior to the pubblication of the Feasibility Study Report.

REFERENCES

1. E.M. McMillan, Phys. Rev. **68**, 143 (1945).
2. V. Veksler, J. of Phys. USSR, **9**, 153 (1945).
3.
4. N.C. Christofilos, unpublished (1950).
5. E.D. Courant, M.S. Livingston and H.S. Snyder, Phys. Rev. **88**, 1190 (1952).
6. Proceedings of the "Workshop on the Future of Intermediate Energy Physics in Europe", Freiburg im Breisgau, 10 - 13 April 1984, Edited by S. Galster, KfK Karlsruhe.
7. TRIUMF "KAON Factory Proposal", September 1985.
8. "A proposal to Extend the Intensity Frontier of Nuclear and Particle Physics to 45 GeV (LAMPF II)", December 1984, LA-UR-84-3982.
9. See, for instance, F. Scheck's Report to Restricted ECFA meeting of October 23, 1985 at Rutherford, UK.
10. W. Joho and J. Crawford, "Proposal for a High Intensity Proton Synchrotron at SIN", Proceedings of the Workshop on Nuclear and Particle Physics at Intermediate Energies with Hadrons, Miramare - Trieste, Italy, 1 - 3 April 1985, Ed. by T. Bressani and G. Pauli, p. 197.
11. See, for instance, R. D. Ruth, "A Review of polarized proton Beam Techniques", Proceedings of the 6th International Symposium on High Energy Spin Physics, Sept. 12 - 19, 1984, Marseille (France), ed. by Journal de Physique, Colloque C2, supplement au n° 2, Février 1985, p. 611.
12. C. Tschalar, "Multiple Achromatic Extraction System", 1986, to be published in Nucl. Instr. and Methods.
13. "Feasibility Study for a European Hadron Facility" to appear in June 1986.

USE OF SEMICONDUCTORS IN EXPERIMENTAL PHYSICS

F. FORTI, M.A. GIORGI and G. TRIGGIANI
INFN sez. Pisa, Scuola Normale Superiore e Universita' di Pisa, Italy.

The working principle of semiconductor detectors is described. The detectors currently used in physics experiments are revised, as live targets for energy sampling, strip detectors, semiconductor drift chambers and CCD's for precise position measurements. Radiation damage effects and applications in high intensity environments are shortly discussed.

1. INTRODUCTION

Before 1980 almost no semiconductor devices were used in experimental physics, apart for nuclear spectroscopy. Afterwards, high energy physicists have used semiconductors (essentially Silicon and Germanium) to build dE/dx and tracking devices and by now projects are in progress for silicon readout of calorimetry. The high spatial resolution obtainable with strip tracking devices and active targets (3÷100 μm), makes these detector suitable for precise measurement of lifetime of shortlived particles, such as charmed and bottomed particles.

With the possibility of measuring decay lengths from about 10 μm to a few mm, corresponding to lifetimes in the range 10^{-14} to 10^{-12} sec, the semiconductor detectors offer the chance of making heavy flavour tagging through the reconstruction of the decay vertex, besides the traditional high transverse momentum lepton identification. Moreover, a detailed study of the event decay chain is possible, allowing, for instance in events containing bottom, the determination of the $b\bar{b}$ mixing.

Examples of physics events containing secondary vertices can be seen in Fig.1 for fixed target and collider experiments.

Semiconductor devices show many other advantages over the more conventional detectors. First of all they are compact, solid and operate properly inside a magnetic field or in vacuum, allowing their use in the core of big collider experiments, even inside the accelerator pipes. In general their signal to noise ratio is rather high (about 10) also with very cheap and fast electronics. High rates are affordable (1÷10 MHz), thanks to the fast readout. Silicon detectors show also a good radiation damage hardness, certainly much better than that of solid state electronic and other conventional detectors, and this feature makes it possible to use these detectors in future high rate,

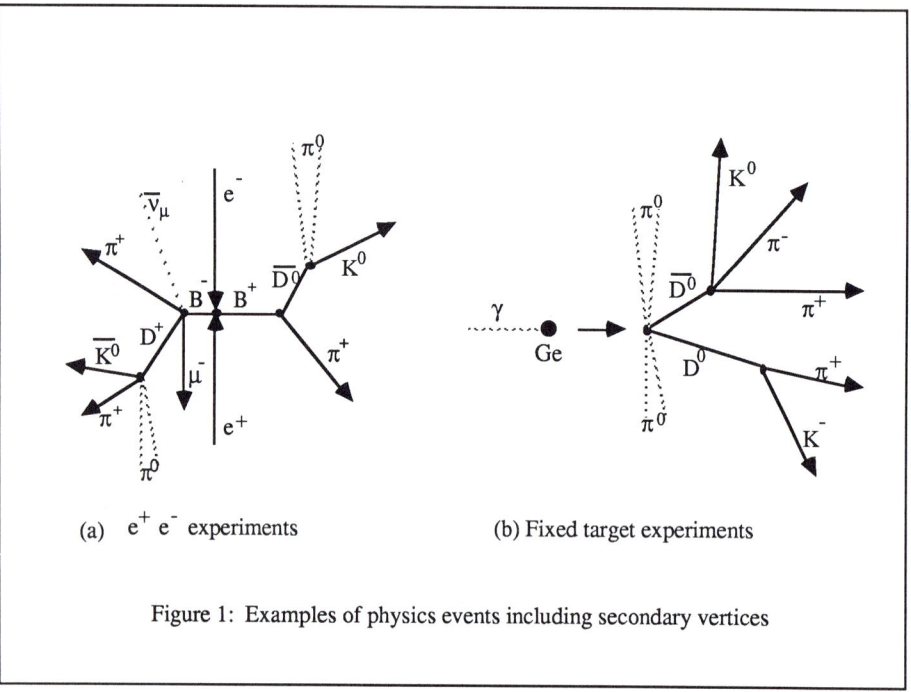

Figure 1: Examples of physics events including secondary vertices

high luminosity machines. Finally, no complicated cooling system is required, except for Charged Coupled Devices (CCD) that need low temperature operation.

In what follows, application of semiconductors to experimental physics will be discussed in many respects, starting with the working principle of detectors, going through the construction methods, and ending with a short discussion of position sensitive detectors. Both active targets and tracking devices will be examined, including telescope and monolithic active targets for lifetime measurement, microstrip detectors with or without charge division read out, silicon drift chambers and pixel detectors like CCD's.

2. WORKING MECHANISM OF SOLID STATE DETECTORS

2.1 Solid state ionization chambers.

The ionization mechanism in solid crystals has been deeply studied and extensively reported in the literature[1)-4)]. It can be summarized as follows.

In a crystal, the atomic energy levels are mixed in a rather complicated way, forming two continuos energy bands called valence and conduction bands. The energy difference between the bottom of the conduction band and the top of the valence band is called band gap, and its temperature dependent magnitude is of the order of 1 eV for semiconductor materials like Silicon and Germanium and a few eV for diamond.

Electrical and optical properties of the crystal depend strongly on band structure and band filling.

An important parameter to determine crystal characteristics is the Fermi level ε_F, defined at 0°K as the energy level at which all higher levels are free and all lower levels are filled. If the Fermi level lies in the conduction band the material is an electrical conductor, if ε_F is in the valence band it is an insulator, if ε_F lies in the band gap it is a semiconductor.

In these latter materials, at very low temperature (at least in the limit T=0°K), all electrons are confined in the valence band and the electric conduction is forbidden. If the crystal sandwiched

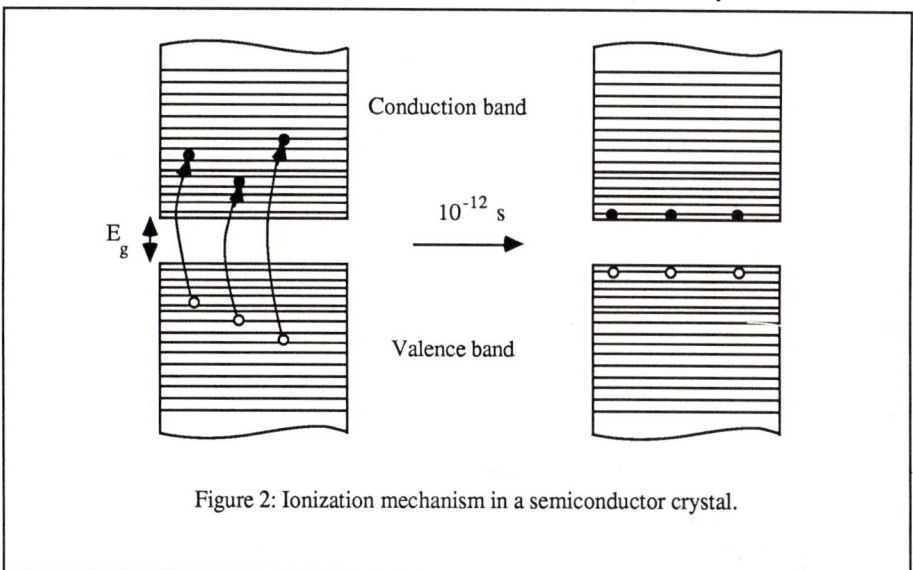

Figure 2: Ionization mechanism in a semiconductor crystal.

between two electrodes is electrically polarized, it behaves as a plane capacitor with no leakage current. A schematic behaviour of an ideal crystal detector is therefore very similar to the gas filled ionization chamber, the main difference being in the ionization energy E_I lower than 30 eV, the typical ionization energy of the gaseous mixtures. Once a charged particle passes through such a device, the energy released by the particle (E_{loss}) contributes to the creation of electron-hole pairs. The electrons jump in the conduction band, exchanging soon afterwards their excess energy with the lattice (electron-phonon scattering) and after about 10^{-12} s fall down to the lowest available level while the holes reach the highest available level in the valence band (see Fig. 2). This mechanism accounts for the difference between ionization energy and band gap observed in semiconductor crystals (see Table 1).

Once a crystal is polarized between two conductive surfaces as sketched in Fig. 3, the electrons in the conduction band are collected by the positively biased electrode, while the electron motion in the valence band results in a net transport of holes to the cathode.

The collection time τ_c, which is in fact the observed rise time of the detected current signal, is determined by the average drift velocity of the free electrons in the conduction band along the electric field E:

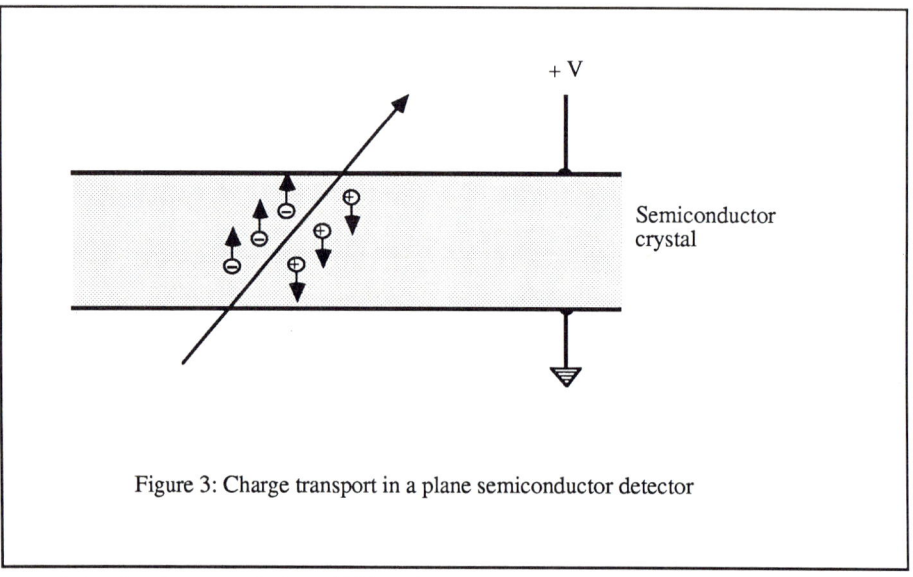

Figure 3: Charge transport in a plane semiconductor detector

$\mu_{e,h}$ being the electron or hole mobilities, that are usually not equal (see Table 1) becuase of the different transport mechanisms of electrons and holes. At finite temperature the conduction band is no longer completely empty because of thermal excitation, the electron density being given by:

$$n_e \propto T^{3/2} e^{-E_g/2kT}$$

where E_g is the energy gap, k is the Boltzman constant and T is the absolute temperature.

Both electrons and holes contribute to electrical conduction, and if n is the density of free carriers inside a crystal, the equivalent resistivity of the material is given by:

$$\rho = \frac{1}{en(\mu_e + \mu_h)}$$

In our example at 0°K, in the absence of the ionization mechanism, n≈0 and ρ→∞. The correct description of the ionization mechanism in real crystals is more complicated and several important parameters have to be taken into account in designing a solid state detector.

1. The ionization energy E_I needed to produce a hole-electron pair has to be as low as possible in order to enhance the signal. E_I is correlated to the band gap, although the vibration modes in the lattice play an important role in the energy losses.
2. The density n_f of free carriers inside the device in the experimental operating conditions has to be as low as possible to reduce the leakage current.
3. The density n_t of trapping centers has to be low as well. Traps and metastable levels that can be present inside the gap between valence and conduction bands produce many undesirable effects:
    a. Reduction of signal produced in the ionization due to deep pit trapping and recombination mechanisms.
    b. Increase of the collection time τ_c due to the electron-hole recombination and

successive reemission mechanism. This is determined mainly by intermediate energy levels between conduction and valence band.

c. Increase of the leakage current determined by the enhancement of the thermal excitation process. In the presence of intermediate energy levels, thermal excitation can proceed via a two or multi-step mechanism where an electron, previously thermally excited from the valence band into an intermediate level, is then re-excited into the conduction band even if the thermal energy kT is much less than the energy gap E_g (see Fig. 4).

4. Finally, as in every other detector, density ρ, atomic number Z and atomic weight A of the crystal, as long as radiation and collision lengths must be evaluated in choosing the material for the detectors.

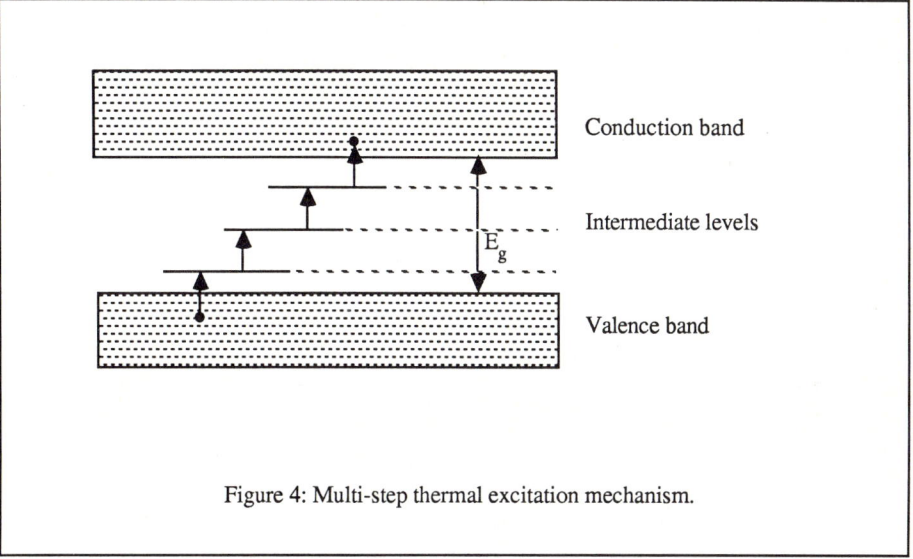

Figure 4: Multi-step thermal excitation mechanism.

A compromise solution must be chosen between points 1. and 2., since a very low E_I is associated with a low E_g and therefore a high density of free carriers due to thermal excitation. Moreover, very pure crystals must be used to reduce trapping and recombination phenomena. It is clear that very few natural crystals satisfy the above requirements of low E_g, <10 eV, low E_I, <30 eV, low density of free carriers at room temperature, high carrier mobility and high carrier lifetime, >100 μsec (i.e. low trapping and recombination effects). The best crystal available is natural, highly pure diamond[5] with E_g = 6 eV, an empty conduction band at room temperature and very few traps and metastable levels thanks to the high purity of the material. Unfortunately, due to the high cost of large pure diamond crystals, they haven't found wide applications in experimental physics.

Semiconductor crystals, such as Silicon and Germanium, have structures rather similar to diamond and show similar characteristics. Each atom has valence 4 and is bonded in tetrahedral lattice with simple covalent bonds. As can be seen in Table 1, where the main physical parameters

TABLE 1
Intrinsic Silicon and Germanium physical parameters

	Silicon	Germanium
Atomic number Z	14	32
Atomic weight A	28.06	72.60
Density (300°K)	2.33 g/cm^3	5.33 g/cm^3
Dielectric constant ε_r	12	16
Energy gap (300°K)	1.106 eV	0.67 eV
Energy gap (77°K)	1.183 eV	0.755 eV
Ionization energy E_I	3.66 eV	2.96 eV
Electron mobility μ_e (300°K)	1350 cm^2 V s	3900 cm^2 V s
Hole mobility μ_h (300°K)	480 cm^2 V s	1900 cm^2 V s
μ_e (77°K)	4×10^4 cm^2 V s	3.6×10^4 cm^2 V s
μ_h (77°K)	1.0×10^4 cm^2 V s	4.2×10^4 cm^2 V s
Radiation length	9.8 cm	2.2 cm
High energy π interaction length	24 cm	10.5 cm
Most probable energy loss for M.I.P.	285 keV/mm	584 keV/mm

of intrinsic Silicon and Germanium are reported, both materials are suitable for making particle detectors, the main problem being the relatively low resistivity shown by these materials at room temperature due to their small band gap. In fact, even in intrinsic materials, that is those materials where conduction is dominated by thermal excitation and not by impurities, at 300°K the Si and Ge resistivities are ρ_{Si} = 230 kΩ·cm and ρ_{Ge} = 45 Ω·cm. These values may be easily calculated from the known density of free carriers as a function of temperature (expressed in °K) :

- for Silicon $n_f = 3.9 \times 10^{16} \, T^{3/2} \cdot e^{-7014/T}$ cm^{-3}
- for Germanium $n_f = 9.2 \times 10^{15} \, T^{3/2} \cdot e^{-4350/T}$ cm^{-3}

Therefore Silicon at 300°K may be defined intrinsic if the density of impurities is lower than 1.5×10^{10} cm^{-3}, which correspond to the above resistivity. Analogously, intrinsic Germanium must have less than 2.4×10^{13} impurities/cm^3. Such levels of purity are reachable for Germanium, which is the purest material presently obtainable, whereas for Silicon it is hard to reach purities better than 3÷4×10^{10} cm^{-3} corresponding to ρ>100 kΩ·cm and normally the purest material commercialy available has $\rho \leq$10 kΩ·cm.

With these low resistivity, a semiconductor crystal simply sandwiched between two polarizing electrodes will never work as a detector, because a huge leakage current i_L would flow, and the random noise, proportional to $\sqrt{i_L}$, would completely mask the particle signals.

2.2 Doped semiconductors

In order to reduce the leakage current, the impurity density and type must be controlled. A doping procedure can be used to compensate not hyperpure crystal and reach high resistivity. There are two kinds of active impurities usually employed in semiconductor technology:

1. *Donors* are atoms with valence 5, such as P, As, Sb, that can substitute a Si or Ge atom. Four electrons establish a covalent bond with the Si or Ge atoms in the lattice, while the fifth electron is only loosely bound by residual Coulomb forces, resulting in a quasi-free electron with ionization energies of about 5×10^{-2} eV in Si and 10^{-2} eV in Ge. At room temperature these electrons are normally free, contributing to the overall conduction in the crystal and leaving positive charges well fixed in lattice sites. Also Li inpurities, with valence 1, show a similar behaviour, even though the charge fixing in the lattice is due to interstitial phenomena and not to covalent bonding.

2. *Acceptors* are atoms with valence 3, such as B, Al, Ga, In. When an atom of this kind substitutes a Si or Ge atom, one of the covalent bonds of the four closest atoms of Si or Ge, is left uncoupled, producing a quasi-free hole level and fixed negative charges.

From the band point of view, donors impurities, or n-type impurities, correspond to filled levels close to the conduction band, while acceptor impurities, or p-type impurities, correspond to empty levels close to the valence band. At thermodynamical equilibrium, the product $n_e \cdot n_h$ may be assumed constant and equal to its value for an intrinsic crystal, i.e. n_f^2.

So in principle it is possible to compensate a low purity crystal doping it with impurities of the right sign obtaining an average condition $n_e \equiv n_h$ equivalent to an intrinsic material.

Impurities of different signs combine via Coulomb attractions and form electric dipoles that do not contribute to the overall current flow. However, this compensating procedure introduces a great deal of attraction centers inside the crystal reducing the free carrier mobility. In any case, in an intrinsic or compensated crystal, the leakage current is too high to detect small ionization signals.

2.3 Junction operations.

To circumvent all these difficulties, diode structures are usually made diffusing, for instance, a p-type dopant in a thin layer of an n-doped crystal. In a n-doped region electrical conduction is due to electrons, and they are therefore called the majority carriers, while holes are called minority carriers. The reverse nomenclature applies obviously to p-doped regions. Near the junction, n and p impurities are in contact and can recombine, leaving a depleted region where no majority carriers are available and where a space charge is present. These fixed space charge produces a voltage drop a cross the junction (contact potential) as can be seen qualitatively in Fig. 5.

When an external reverse bias is applied, i.e. the n side is brought to a higher potential than the p side, the majority carriers are swept away from the crystal and the depletion region extends until,

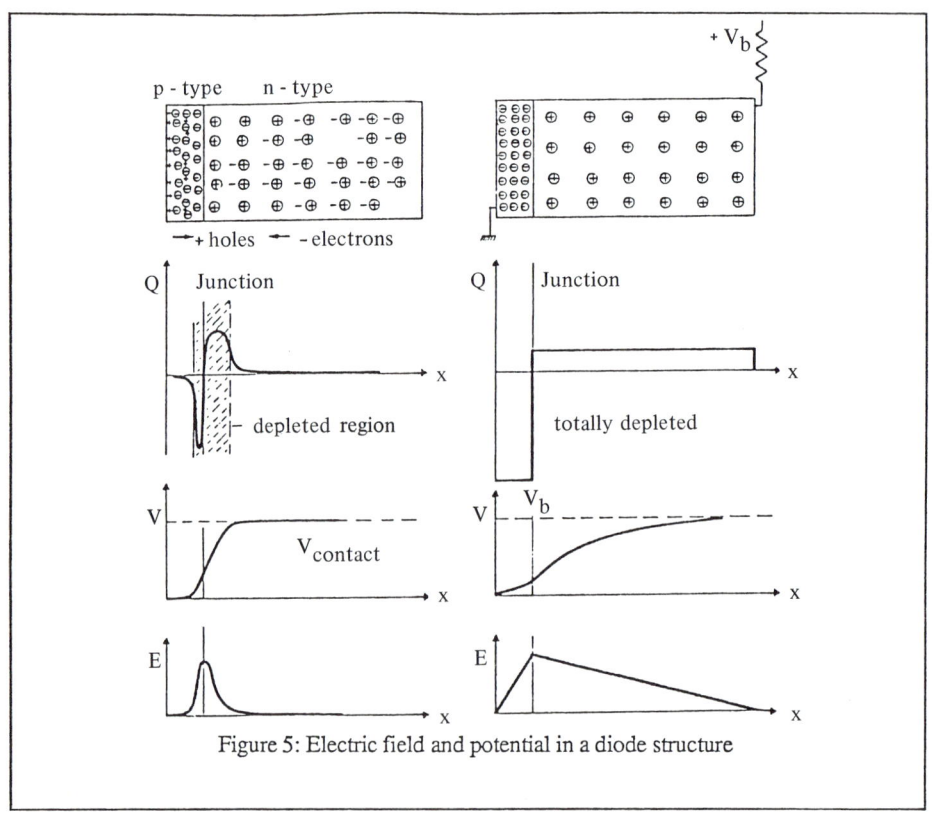

Figure 5: Electric field and potential in a diode structure

at a certain voltage V_b, it covers the full thickness of the device (total depletion). In this condition, electrical conduction must rely only upon minority carriers and the effective resistivity of the crystal is increased for Si by 3-5 orders of magnitude and for cooled Ge (77^0K) even by 9 orders of magnitude. Detailed discussion of diode operation can be found in many textbooks of electronics[6].

For a given applied voltage V, depletion thickness turns out to be $d=(2\varepsilon\mu\rho V)^{1/2}$, where μ is the mobility of the majority carriers. It should be noted, however, that the electric field is linear in the x-coordinate rather than constant and therefore one expects to find differences respect to the conventional ionization chambers. First of all the electrical field is zero outside the depleted region and ionization charge produced in this zone cannot be collected. Then, also in a totally depleted crystal, the electric field tends to zero on at least one electrode and, strictly speaking, the charge collection time is ∞. In practice, because of charge motion, an induced signal appears on the electrodes:

$$q(t) = Q(1 - x_0/d)(1 - e^{-t/\varepsilon\rho})$$

where d is the detector thickness and x_0 is the distance of the produced charge Q from the junction. About 86 % of the charge is collected in time $\tau=2\rho\varepsilon$, which is typically 40 nsec for a 20kΩ·cm silicon detector, as used in the NA1 experiment [7]. Holes collection time is connected to

the electrons collection time through the ratio of mobilities:
$$\tau_h = \tau_e (\mu_e/\mu_h)$$
For room temperature Si, the ratio $\mu_e/\mu_h \approx 3$ and collection is dominated by electrons. In Ge cooled at 77^0K, electrons and holes have the same mobility and contribute almost equally to collection time. Collection time is also affect ed by carriers trapping and reemission, which can considerably lengthen the pulse, as in heavily doped lithium compensated crystals.

In an overdepleted crystal, collection time is determined mainly by drift velocity $v = \mu E_0$ where E_0 is the electrical field at the electrode (that is non zero if the detector is overpolarized). Collection time must be kept low also to limit lateral diffusion of the ionization cloud that can be embrassing for fine-grained detectors with the aim of high space resolution. The width of the cloud after a transit time t is given by $\sigma=(2Dt)^{1/2}$ where the diffusion coefficient D can be expressed as $D=(kT/e)\times \mu$.

Even though this description does not take into account more complex phenomena such as recombination and scattering, the predicted value is in good agreement with measurements giving at 300^0K, $D=35$ cm^2/s for Si. For instance, in a fully depleted 20 kΩ·cm Silicon detector $\sigma = 17$ μm. In a 5 mm thick Germanium detector with an impurity density less than 3×10^{10} atoms/cm^3 overpolarized at 2000 V and operated at 77^0K, $\sigma = 13$ μm.

The energy resolution of a semiconductor detector is affected by different sources of error, which can be summarized in the following way:
1. Intrinsic noise, due essentialy to leakage and stray currents:
    a. Bulk leakage current. This is due free carriers produced by thermal excitation specially in not perfectly pure materials.
    b. Surface leakage current. It is correlated to the formation of surface conduction channels due to a bad passivation process or to a lack of cleaness in the manipulation.
    c. Series resistance produced by a bad ele ctrical contact on the electrodes. This resistance is particulary danger ous because it reduces signal and increases noise (Johnson noise)
    d. Electromagnetic pick-ups and experimental hall noise may be also an important source of noise for high energy detectors usually operated in noisy environments.
2. Noise of preamplifier. It is dominated by the first stage of the head amplifier [8]. This noise grows linearly with the detector capacitance and is inversely proportional to the square root of the pulse duration. This kind of noise is particulary important in designing low signal charge division position sensitive detectors.
3. Fluctuations in the ionization charge. This is more relevant for Minimum Ionizing Particles and in thin detectors (Landau fluctuations).

The pulse height distribution for different numbers of crossing particles are shown in Fig. 6 for a 200 μm thick detector, 350 mm^2 in area [9].

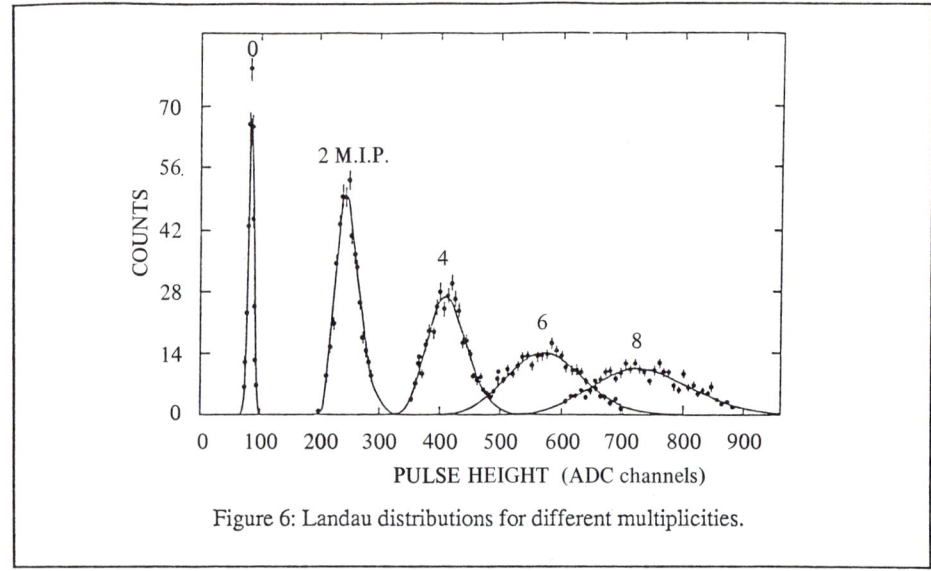

Figure 6: Landau distributions for different multiplicities.

2.4 Radiation damage

Future high luminosity high rate machines will need detectors capable of working also in extreemely intense fluxes of particles. For instance at SSC, assuming a luminosity $L = 10^{33}$ cm^{-2} s^{-1}, a center of mass energy of 40 TeV, a cross section of about 150 mb and a 33 ns bunch distance, at 10 cm from the interaction point detectors have to work in radiation fluxes of about 10^6 rad/year at 90°, 10^7 rad/year at 20°, 10^8 rad/year at 10° (1 rad = 10^2 erg/g = 6.24×10^7 MeV/g of deposited energy). Typical radiation effects in semiconductor detectors are crystal micro-damaging and atomic displacements and a consequent increase of both free carriers and trapping centers. In addition a formation of a heavy space charge at the interface Si-SiO$_2$ is observed. These two combined effects procuce a general deterioration of detector operation, increasing leakage current and reducing carriers lifetime and collected charge [10].

These effects are usually reversible and can be partially eliminated with a medium temperature (400°K) annealing. In any case the disassembly of a working detector for annealing can be unpleasant. Radiation damage tests show that:

- Charge Coupled Devices stand in about 10^5 rads.
- Strip detectors can absorb 10^5-10^6 rads of M.I.P. with at most a 10% reduction of collected charge.
- JFET electronics stands up to 10^8 rads.
- N-MOS electronics, like the CERN-HAWAII-SANTA CRUZ-STANFORD 128-channels μ-plex [11] cannot absorb more than 10^5 rads.

Present detectors and electronics do not seem to comp letely satisfy the needs of experiments at future machines as SSC or LHC, at least as far as radiation damage is concerned. A possible way out for electronics is to go to solutions with thinner oxide layer (see chapter 3).

3. CONSTRUCTION METHODS OF SEMICONDUCTOR DETECTORS.

3.1 Junction detectors.

Junctions are semiconductor crystals in which at least two zones with different doping impurities sign can be distinguished. As discussed in the previous paragraph, this kind of structure is suitable for use as particle detectors and allow the employment of crystals without extreme purity (1 kΩ·cm<ρ<10 kΩ·cm). Construction techniques are those of the electronics industry and are therefore rather defined and secure. By the way these techniques require high temperature processes and a great care in crystal handling obtainable with expensive equipements very difficult to be found in University Departments and Research Institutes.

Moreover, a great experience of laboratory operators is required for the high temperature processes that can easily produce micro-melting or thermal shocks in the crystal, thus altering the lattice regularity. These standard processes needed in detector making can be summarized as follows [12]:

1. Crystal preparation: lapping and etching.
2. Passivation: a 500÷3000 Å layer of Silicon dioxide (SiO_2) or Silicon nitride (Si_3N_4) is growth at high temperature on the crystal surface. Basically the oxide or nitride molecules fill the surface free bonds of the monocrystal that otherwise would contribute to surface leakage currents.
3. Photolithography: the geometrical structure of the detector is determined by windows opened in the passivation layer with a photochemical process. The part of the passivation layer that is not removed by the selective etching acts as a mask for the following processes.
4. Doping: impurities of the correct sign are introduced in the selected zone of the crystal to obtain the junction. Doping is usually made in one of the following two ways:
    a. Ion implantation: the crystal is used as a target for a ion beam, such as P or B, that stops inside the crystal at calculable depth. After implantation, a thermal annealing is required to allow doping atoms to settle in the lattice sites (activation).
    b. Thermal diffusion: the crystal is brought in contact at high temperature (900÷1200°C) with a source of dopant that diffuse into the crystal. The doping source can be either gaseous or solid. Liquid sources are also employed (spin-on dopants), but must be glassified at low temperature before the high temperature process.
5. Metallization: a metal layer, usually Al or Ag, is evaporated or sputtered on the crystal surface to make electric contacts and to protect the doped regions.
6. Photolithography: a final step is required to get the definitive electrodes geometry, by removing the eccess metal from the surface.

An electric contact must be also made on the back side of the wafer, either by direct metal deposition (see next section about metal- semiconductor contacts), or by doping the crystal, for

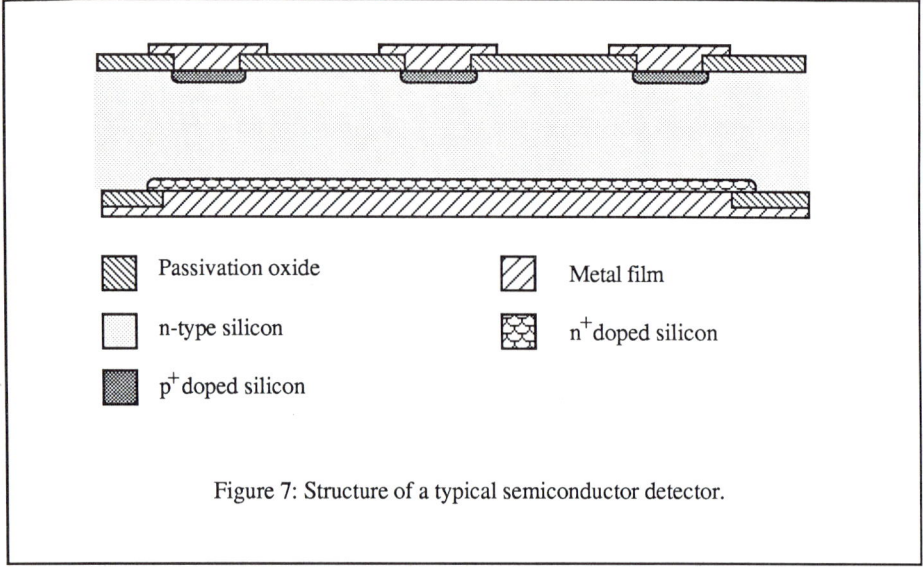

Figure 7: Structure of a typical semiconductor detector.

instance an n type Si crystal, with impurities of the same sign, i.e. n type. The final structure can be seen in Fig.7 for a strip detector.

3.2 Surface barrier detectors.

Diode structures can also be obtained through a direct metal- semiconductor contact. The metal-semiconductor contact acts as a junction or as an ohmic contact, or as none of them. The behaviour depends both on metal nature and on semiconductor type and purity. The phenomenon is not completely understood yet, but it is clear that it originates from differences in the extraction potentials of metal and semiconductor, or better by the related so called work functions. At the contact a potential barrier appears due to the difference between the work functions, given by $\phi_{MS} = \phi_M - \phi_{Si}$. ϕ_{MS} determines whether the contact is rectifying or ohmic.

As can be seen in Table 2, for instance Al can be used as a (not so good) ohmic contact onto an n type Silicon, but it is rectifying on p Silicon. In the same way, gold is rectifying on n Si, and is in fact often used, but it is almost ohmic if used on p Si.

TABLE 2

Potential barrier at metal-semiconductor interface.

	$q\phi_{MS}(n)$	$q\phi_{MS}(p)$
Al	-0.5 eV	0.58 eV
Au	-0.81 eV	0.35 eV
Cu	-0.69 eV	0.46 eV
Ni	-0.67 eV	0.46 eV

The goodness of the contact depends also on the surface characteristics and surface electronic states due to dirt or impurities can often prevent the device from properly working. For this reason the construction processes of surface barrier detectors always include a heavy chemical etching step to remove surface layers possibly damaged and access the even inner lattice structure. The next processing steps are similar to those used in the construction of junction detectors, namely metalization and final electrodes photolithography. Surface barrier detectors don't require high temperature processes and the implied technology is rather simple and accessible by all laboratories. On the other hand the use of higher purity ($\rho > 5000$ $\Omega \cdot$cm) and thus more expensive materials is needed to guarantee stability and reproducibility of the devices.

Moreover surface leakage currents are usually higher than in doped detectors because no surface passivation step is usually performed. In conclusion, surface barrier detectors are relatively easy and fast to produce and can be successfully employed for prototypes and tests. Implanted and diffused detectors offer instead more reliable and possibly precise devices, to be used in large and complex experiment. They can now be produced by the industry at relatively low price.

4. DETECTORS IN ACCELERATOR EXPERIMENTS.

Detectors with whole electrodes, like those described in section 2, are used for precise measurement of the energy loss of ionizing particles. In the past ten years a great boost has also been given to the development of solid state detectors for high precision position measurements. Integrated circuit technology has opened the way to very fine-grained electrode geometries thus allowing microstrip detectors with resolutions down to 5 µm.

A part from the use in test beams by several groups, microstrip silicon detectors have been employed as tracking devices for vertex reconstruction in high multiplicity events (NA32 exp. at CERN [13]). At the moment most of collider experiments plan to have or have already a silicon vertex detector installed inside the apparatus around the pipe in the collision region.

4.1 Live targets.

Basically two kinds of semiconductor targets have been used in high energy physics experiments so far: Silicon telescopes and monolithic Germanium targets. The aim of this kind of live targets, which sample the ionization signal proportional to the number of particles travelling along the beam direction, is to measure the decay path of long-lived heavy particles through the charged multiplicity variation to obtain their meanlife [14].

In this case Landau fluctuations of ionization energy loss in thin wafers play a crucial role, and can completely mask the signal. For this reason charged multiplicity must be kept small (<8÷10) where the differences between Landau curves referring to different multiplicities are more evident. Moreover, noise can be a limiting factor, increased as it is by:
- the high capacitance of the relatively large detectors required;
- the finer the sampling and the higher the resolution, the smaller the signal.

Telescope detectors, made of thin semiconductor wafers (about 200÷300 µm) assembled along the beam axis with small gaps between each other (about 200 µm), can be used only for a relatively coarse sampling, with resolutions not better than 200 µm. The limiting factors can be summarized as follows:

1. Very thin wafers (less than 300 µm) are very difficult to handle and assemble. Gaps smaller than 100 µm are not conceivable for the present technology.
2. The signal is proportional to the detector thickness d while the noise is roughly proportional to the capacitance, i.e. 1/d. S/N $\propto d^2$ and too thin detectors show a poor S/N ratio. This problem can be partially overcome using cooled Ge instead of Si. In fact, due to the different atomic properties, the energy loss in Ge is twice as large as in Si, while the ionization energy of Ge is 1.3 times smaller than in Si, giving altogether a favourable factor 2.6 in signal. Cooling is needed for Ge operation.

One way out of these problems is represented by monolithic Ge targets, such as the one used in the NA1 experiment [15)-16)](see Fig.8). In such a target the active zone is sampled with the aid of readout strips present on the crystal top and bottom which divide the crystal in cells along the beam axis. Granularity can be pushed down to less than 50 µm and is limited only by the S/N ratio, but not by mechanical problems. The advantage of this configuration is that each cell offers a small capacitance to the front-end amplifier and the noise is proportionally small.

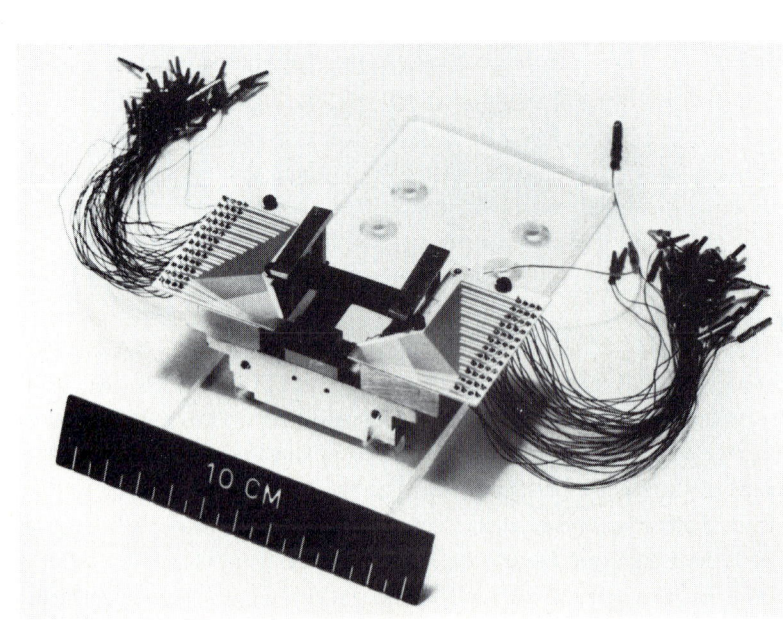

Figure 8: Ge monolithic target for the NA1 experiment.

4.2 Microstrip detectors.

Microstrip detectors are made out of thin semiconductor wafers on which strip-shaped junction zones are built. The crystal is subdivided in cells that refer to different strips, to locally measure the ionization signal. When a charged particle crosses the layer, the ionization charge is collected only on the strips involved by the particle crossing [17].

However, different kinds of wafer polarization and readout configuration are used, leading to different types of detectors. If the crystal is overdepleted, the inter-strip resistance is very high (>100 MΩ) thanks to the lack of free carriers. A charge produced in a single point inside the crystal is collected on a single strip which gives the point coordinate. In this configuration no precise pulse height measurement is required to know the particle impinging point, a part for better cleaning of signal, as the information is essentially digital. Charge released in all points of a cell gives the signal on the same strip and if s is the strip pitch, the corresponding resolution is $\sigma = s/\sqrt{12}$ [18].

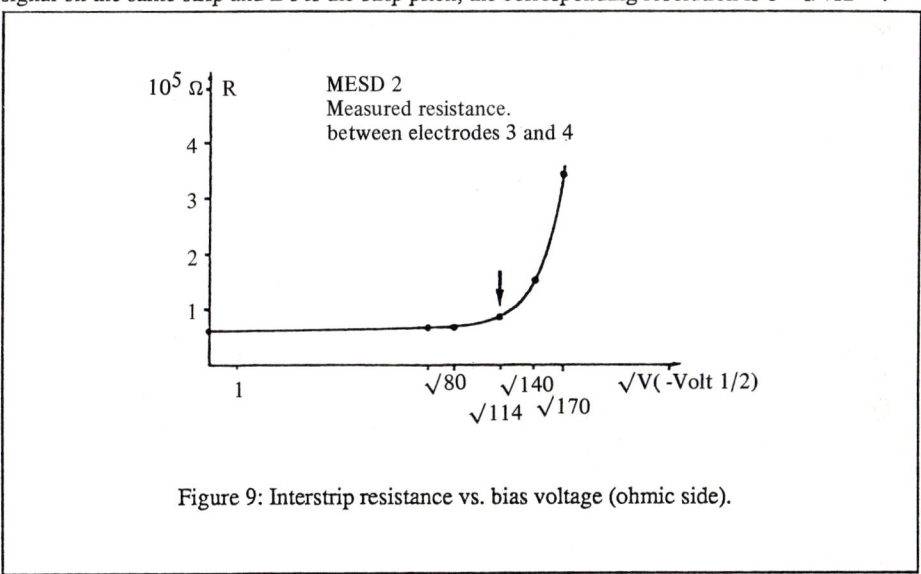

Figure 9: Interstrip resistance vs. bias voltage (ohmic side).

The only way to improve the resolution of such a detector is to decrease the pitch, a procedure that is usually limited by the enormous amount of readout channels. It is possible to increase the resolution without increasing the number of readout channels by operating the crystal in charge partition mode, that is subdividing the total charge produced among different adjacent strips.

This can be done in two ways.

1. Resistive charge division [19]: the crystal is underpolarized so that a partially undepleted zone remains between the strips. This zone acts as a low value resistance (see Fig. 9) that divides the charge proportionally to the impact point of the electron cloud (see Fig.10). As discussed in section 2, the depletion layer begins always near the junction and therefore this mode of operation is possible only in detectors with strips on the ohmic side. A resistive charge division prototype detector is shown in Fig. 11. The method can be sensitive to temperature and impurities density variations.

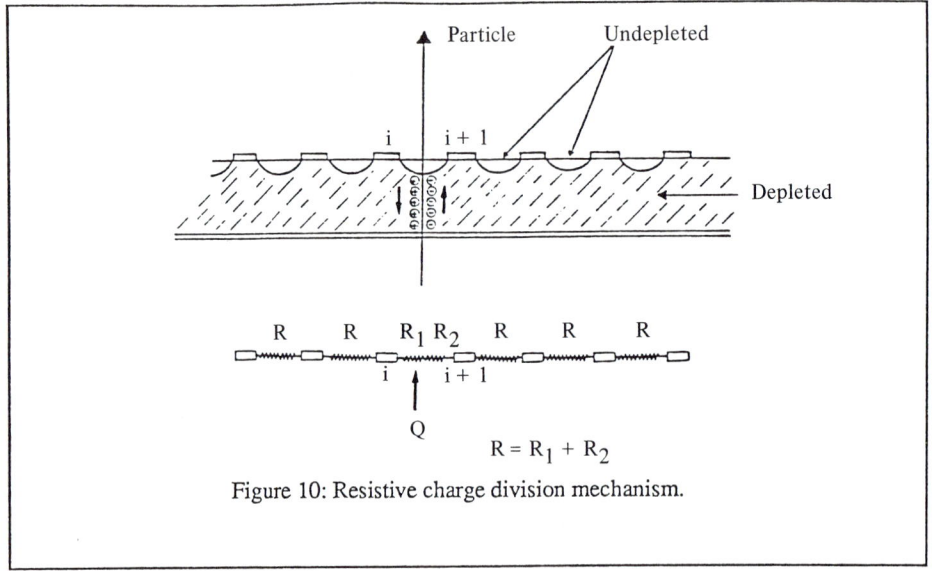

Figure 10: Resistive charge division mechanism.

2. Capacitive charge division [20]: the crystal is overdepleted and the interstrip resistance on the junction side is high ($\approx 10^2$ MΩ). Only one strip out of N is connected, while the remaining N-1 are left floating. The charge collected on intermediate strips capacitively induces a signal on the readout strips. This signal is again linear with the distance of the strip from the impact point. The readout is on the junction strips.

In both charge division methods the position is reconstructed through an interpolation algorithm. If the particle passes between strip i and i+1, in position x with respect to strip i and loses and energy E, the pulse heights will be:

$$PH_i = (1 - x/s) \times E \qquad PH_{i+1} = x/s \times E$$

and the position is reconstructed as

$$x = (PH_i \times x_i + PH_{i+1} \times x_{i+1}) / (PH_i + PH_{i+1})$$

The precision depends on the noise σ_{noise} of preamplifiers, as well as on strip pitch. For resistive charge partition the resolution is:

$$\sigma_{res} = (\sigma_{noise} / <E>) \times s$$

where $<E>$ is the mean energy loss in the detector. For capacitive charge division it becomes:

$$\sigma_{cap} = (s^2 /12N^2 + \sigma_{res}^2)^{1/2}$$

where s is the readout strips pitch and N is the number of intermediate floating strips. Using very low noise amplifiers, and of course a high density of strips, the resolution obtainable with these detectors can be pushed down to 5 µm.

Strip detectors can be operated under vacuum and inside intense magnetic fields, although corrections to position measurement due to Lorentz forces during charge collection must accounted for.

Figure 11: Micro-strip detector prototype for position measurements.

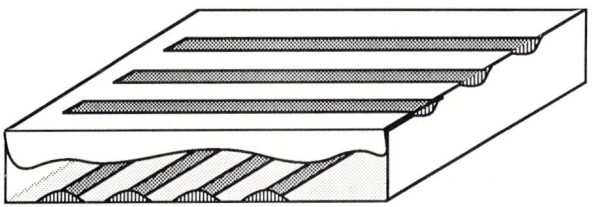

Figure 12: Double side read-out of a silicon detector.

The construction of double face strip detectors is now being investigated in several laboratories and great efforts are currently being devoted to this project in view of its in future physics experiments. This kind of device (see Fig. 12) provides the simultaneous measurement of two orthogonal projections of the particle impact point and offers some appreciable advantages over the more conventional single side detectors.

- Only one silicon wafer is needed to measure both projections and therefore total thickness crossed by particles can be halved.
- The charge collected is exactly the same on each side and correlations can be found to solve ambiguities in the coupling of coordinates. Let a single wafer be crossed by two particles. Referring to the symbols of Fig. 13, the total energy losses of the two particles are E_1 and E_2. The coupling ambiguities can be solved if the difference $|E_1 - E_2|$ is much greater than the average difference due to noise, that is $\sqrt{n} \times \sigma_{noise}$, where n is the number of strips fired by a single particle. This can be achieved by taking the opportunity offered by Landau fluctuations in the energy loss.

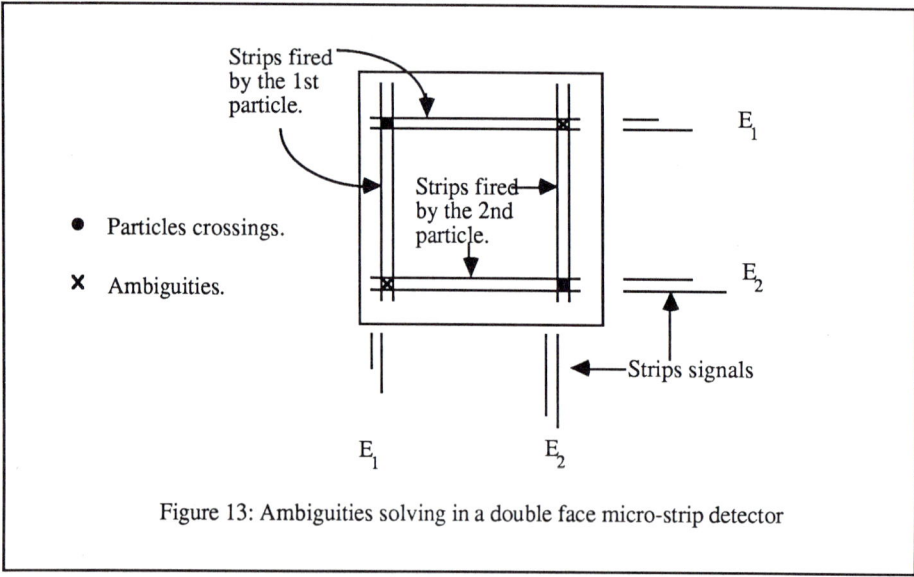

Figure 13: Ambiguities solving in a double face micro-strip detector

Several problems also arise in the construction of such devices.
- Double side wafer processes require non-standard techniques.
- The interstrip resistance on the ohmic side is rather low (see Fig. 9) compared to the junction side. This phenomenon is due to the formation of a negative space charge region near the $Si-SiO_2$ interface that cannot be drained by the insulating oxide. On the junction side the contact between this negative space charge and the p^+ doped strips acts as a reverse biassed diode junction producing a very high interstrip resistance. On the ohmic side this phenomenon dramatically decreases the interstrip resistance, making it difficult to operate the crystal in capacitive charge division mode. Solutions of the problem, employing more complicated doping structures seem possible.

4.3 Charge coupled devices.

CCD's have been developed by the electronic industry (firstly in Bell Laboratory) and produced mainly for television. It is a typical matrix device (pixel structure) with a serial output after a time shifting of analog information. The latter feature has produced for CCD's a wide range of applications in electronics as an analog shift register, although its main use remains in applications as sensor in TV cameras, in the infrared radiation detection in astronomy and more recently as an ionizing particle detector.

The description of CCD's is outside the intentions of this paper and it can be easily found in the specialized literature [21)-22)], nevertheless a short illustration of the working principle can help in understanding the main difference between a CCD and a Silicon position sensitive detector.

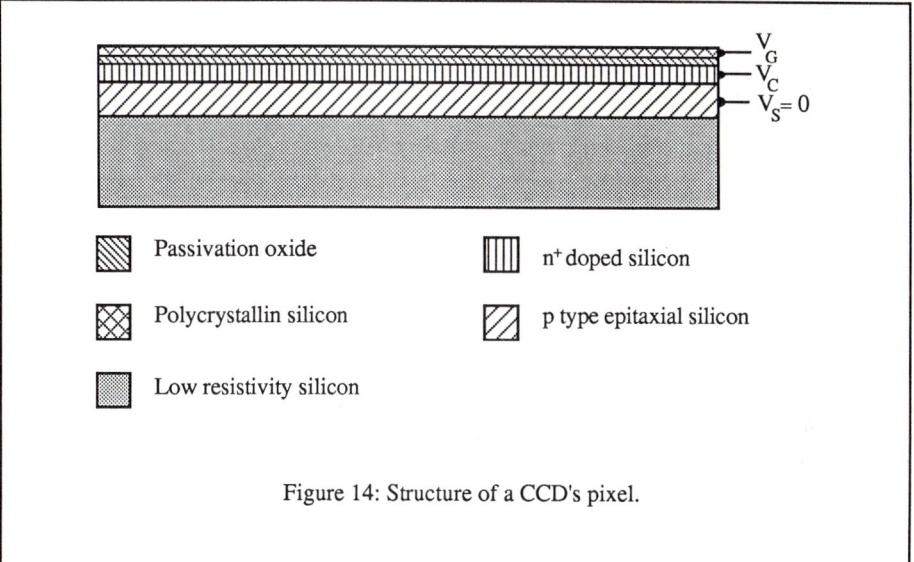

Figure 14: Structure of a CCD's pixel.

The structure of the sensitive element of the CCD (the pixel) can be sketched as follows (see Fig. 14) :
- A substrate of Silicon of elecrtonic grade (low resistivity) is used to support the structure.
- On top of it a hyperpure p type Silicon layer of several tens of microns is grown by epitaxy.
- A layer of ≈ 1 μm in the external surface of the p type support is heavily and uniformly n^+ doped to form an n^+-p junction.
- The n^+ surface is passivated by oxidation.
- A thin conductive polycrystallin Silicon film (≈ 2000 Å) is deposited onto the SiO_2 surface.
- Electrical connections are stated to the p substrate, the n^+ layer (n channel) and to the conductive film deposited over the SiO_2 (gate).

Let the p substrate be grounded, $V_S=0$, the channel and gate be positively biased, $V_C>0$ $V_G>0$. The n^+-p junction is then reverse polarized and a zone around the junction is depleted according to the p Silicon resistivity and the bias value.

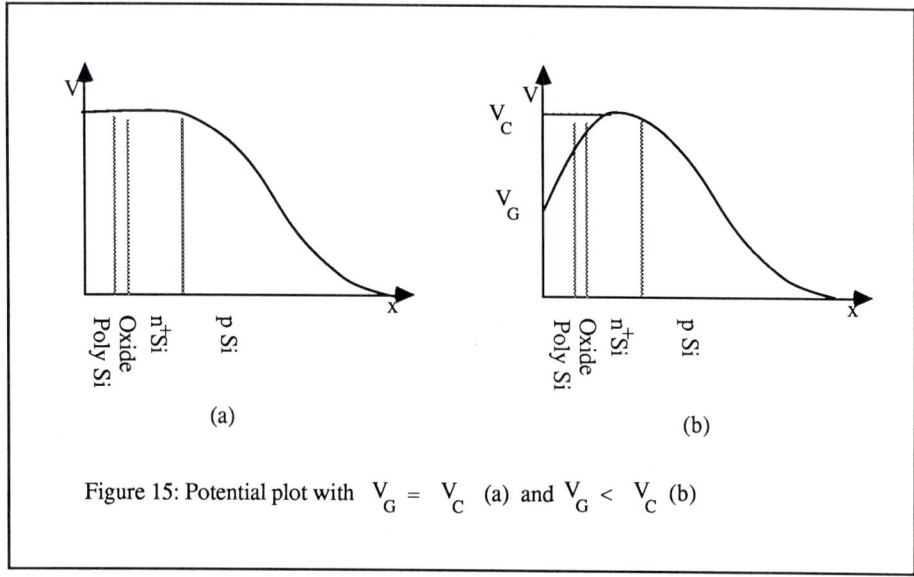

Figure 15: Potential plot with $V_G = V_C$ (a) and $V_G < V_C$ (b)

An example of the voltage dependence as a function of the distance from gate is shown in Fig.15 when $V_G=V_C$. Electrons produced by an ionizing particle inside the depleted region drift towards to n^+-SiO_2 interface, where many traps and metastable states are present. If one decreases the gate voltage, $V_G<V_C$, the potential plot shows a well, because of the typical capacitance of the gate-channel structure (behaving as a flat face capacitor).

Electrons are then captured inside the well (buried channel). By acting on V_G of adjacent pixels, a suitable change in the depth of the well allows the migration of the charge from pixel to pixel throughout the device.

CCD's are beautiful structured position sensitive detectors, being the true 2-dimensional device. In principle they can cope with all the requirements of experimentalists, good two track resolution, optimum precision, a minimum number of cables to connect the device to data acquisition electronics and computer, thanks to the shifting feature. In fact, several problems arise in running CCD's:

1. Due to the structure of detector, the active thickness is rather low, limited by the thickness of the epitaxially grown layer (≤ 20 μm). Therefore the charge produced by an ionizing particle is about one order of magnitude less than in normal Silicon detectors, despite the fact that energy loss and Coulomb multiple scattering are the same.
2. To make such small signals detectable, the thermal noise and trapping effects must be limited, so the device needs to be cooled.
3. The overall readout time is at least microseconds and even milliseconds, due to the shifting that is regulated by the frequency of the clock.
4. Radiation damage hardness is similar to that of elecrtonics. CCD's cannot stand more than 10^5 rad.

In Table 3 some CCD's characteristics are summarized [23)-24)].

TABLE 3
CCD's typical characteristics

Total thickness	300 μm
Active thickness	≈15 μm
Total energy loss for M.I.P.	84 keV
Visible energy	30 keV
Readout noise	≈ 80 e⁻ (ENC)
Two track separation	40 μm
Maximum resolution	5×5 μm²
Readout time	40 ms
Physical dimensions	≈1 cm²
Operating temperature	< 200°K

4.4 Semiconductor drift chambers.

The idea of this semiconductor device was firstly presented at the 2nd Pisa Meeting on Advanced Detectors 1983 by E.Gatti and P.Rehak [25]. Later on it was realized and tested in several versions with differrent designs. The interested reader can find construction details in recent literature [26)-27)]. The basic concepts of charge transport inside bulk are shortly illustrated in Fig.16.

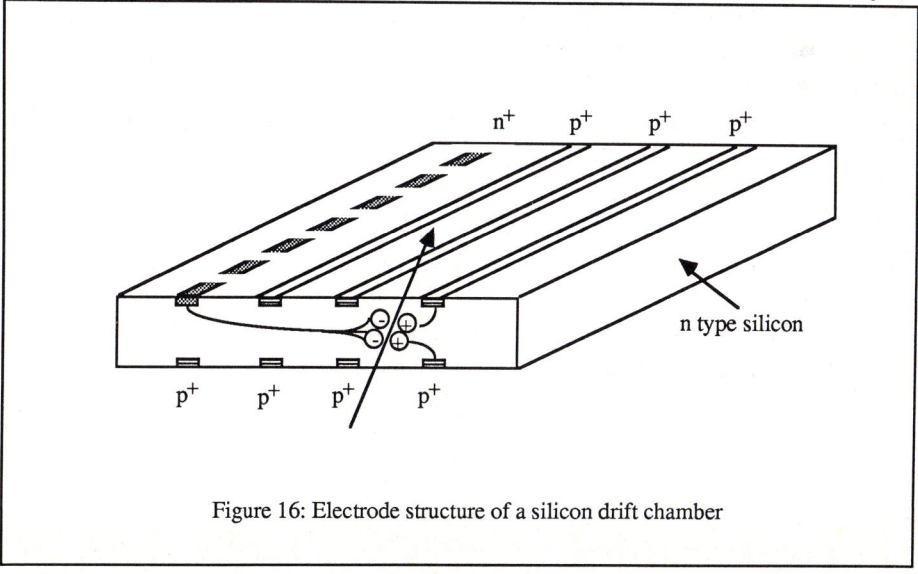

Figure 16: Electrode structure of a silicon drift chamber

The detector is built on a standard n type high resistivity silicon wafer 300 μm thick. A structure of p⁺-n junction parallel strips is built on both surfaces of the detector and a structured n⁺ implanted anode is put on one side. By biasing independently the p⁺ strips one can get a

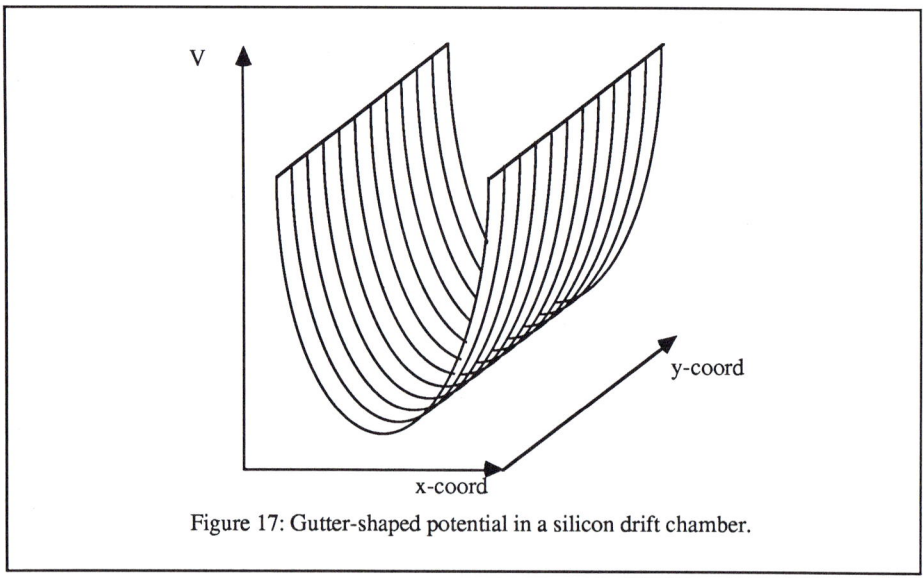

Figure 17: Gutter-shaped potential in a silicon drift chamber.

gutter-shaped potential distribution (see Fig.17). Electrons produced inside the detector by ionization mechanism drift inside the potential gutter toward the minimum and are collected by one element of the anode. If operated in stable conditions, once the drift velocity is known this detector behaves as a 2-dimensional device. One coordinate is obtained very precisely through the electrons drift time measurement and the other is determined by the segmentat ion of the anode. The measured perfomance of Silicon drift chambers are reported in Table 4.

TABLE 4

Silicon drift chambers characteristics

Drift device used as position sensitive detector

Maximum drift length used in tests	16 μm
RMS resolution expected	
for time measured coordinate	4 μm
Measured resolution in test beam at NA31	≈10 μm

Drift device used as large area low capacitance X-ray sensor

Anode diameter	200 μm
Capacitance	0.06 pF
Energy resolution (100 ns integration constant,	
FET preamp. with 12 pF input capacitance)	2.6 keV

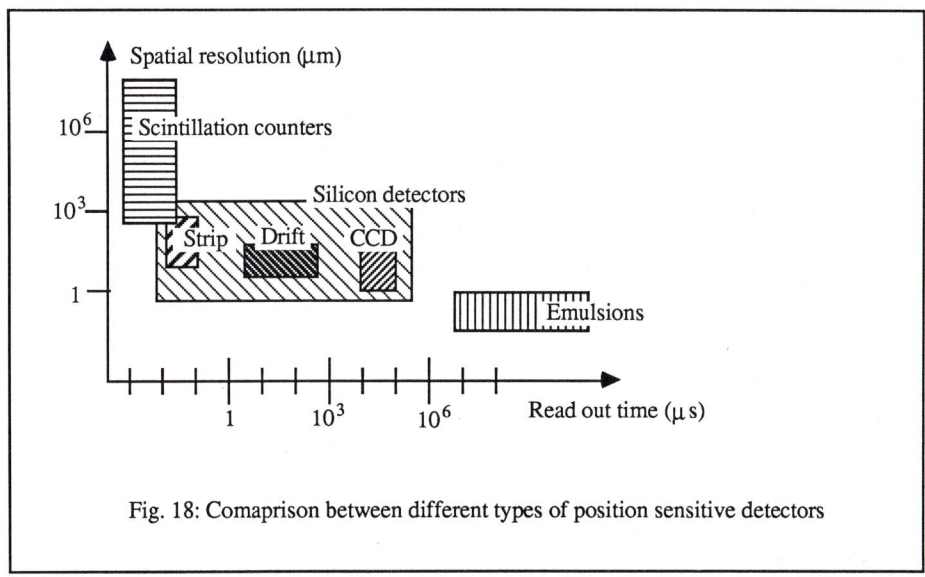

Fig. 18: Comaprison between different types of position sensitive detectors

The Silicon drift chamber appears to be one of the most promising detectors for future, although a few question marks remain on the stability of its working conditions. For example, the effect of operating temperature variation and of crystal and doping inhomogeneities on the drift velocity, which is not constant or saturated as in gas drift chambers operation. Certainly the above described charge transport mechanism will find useful applications in the construction of ultra-low capacitance devices as large photodiodes or radiation sensors.

To conclude: at present the position sensitive semiconductor detectors show a very good precision in tracking measurement and seem to cope at best with the needs of those experiments in which the high precision requirement is coupled with a little physical space. In particular they appear to be suitable for tracking apparatuses of very small dimensions used for vertex reconstruction in colliding beam experiments. In a pictorial way a comparison between the various devices employed in position measurement is made in Fig. 18, where the read out time and the coordinate precision are the axes of the scatter plot.

REFERENCES

1) See for instance, F.S. Goulding, Nucl. Instr. and Meth. *43* (1966) 1,
   or, for a more comprehensive discussion,
2) G. Dearnaley and D.C. Northrop, Semiconductor Counters for Nuclear Radiations (E. & F.N. Spon, London, 1966).
3) For references before 1982: G. Bellini, L. Foa', M.A. Giorgi, Phys. Rep. *83* (1982) 1.

4) M.A.Giorgi, Microdetectors for high energy physics, in: General Meeting on LEP, Villars-sur-Ollon, Switzerland, June 1981, editor M.Bourquin, ECFA 81/54 (Geneva, 1981) pp. 179-210.
5) C.Canali et al., Nucl. Instr. and Meth. *160* (1979) 73.
6) H.F. Wolf, Semiconductors (Wiley-Interscience, New York, 1971).
7) S.R. Amendolia et al., Nucl. Instr. and Meth. *176* (1980) 449.
8) P.F. Manfredi, Electronics for silicon detectors in high energy experiments, in: Miniaturization of High-Energy Physics Detectors, editor A. Stefanini (Plenum Press, London 1983) pp. 77-86.
9) S.R. Amendolia et al., IEEE Trans. on Nucl. Sci. NS31 (1984) 945.
10) H.W. Kraner, Nucl. Instr. and Meth. *A225* (1984) 615,
and
T.Kondo et al., Radiation damage test of silicon microstrip detectors, in: Proc. of 1984 Summer Study on the Design and Utilization of the Superconducting Super Collider, eds. R.Donaldson and J.G. Morfin (Division Particles and Fields of the APS, 1984) pp.612-615,
and
M.G.D. Gilchriese, Radiation damage and rate limitation in tracking devices, ibidem, pp. 607-612.
11) J.T. Walker et al., Nucl. Instr. and Meth. *A226* (1984) 200.
12) J. Kemmer et al., Nucl. Instr. and Meth. *A226* (1984) 89.
13) R. Bailey et al., Nucl. Instr. and Meth. *A226* (1984) 56.
14) G.Bellini et al., Live targets as tool to study short range phenomena in elementary particle physics, in: Miniaturization of High-Energy Physics Detectors, editor A. Stefanini (Plenum Press, London 1983) pp. 41-56.
15) S.R. Amendolia et al., Nucl. Instr. and Meth. *A226* (1984) 117.
16) S.R. Amendolia et al., Nucl. Instr. and Meth. *A226* (1984) 78.
17) S.R. Amendolia et al., Nucl. Instr. and Meth. *176* (1980) 457.
18) E.H.M. Heijne et al., Nucl. Instr. and Meth. *178* (1980) 331.
19) S.R. Amendolia et al., Nucl. Instr. and Meth. *A226* (1984) 82.
20) J.B. England et al., Nucl. Instr. and Meth. *185* (1981) 43.
21) C.J.S. Damerell et al., Charge-coupled devices for particle detection with high spatial resolution, in: Miniaturization of High-Energy Physics Detectors, editor A. Stefanini (Plenum Press, London 1983) pp. 131-150.
22) C.J.S. Damerell, Lectures presented at the SLAC Summer institute (1984) SLAC Report 281, p.43.
23) C.J.S. Damerell, IEEE Trans. on Nucl. Sci. *33* (1984) 51.
24) A.J. Bross and D.B. Clegg, Nucl. Instr. and Meth. *A427* (1986) 309.
25) E. Gatti and P. Rehak, Nucl. Instr. and Meth. *A225* (1984) 608.
26) P.Rehak et al., Nucl. Instr. and Meth. *A235* (1985) 224
27) E. Gatti et al., IEEE Trans. on Nucl. Sci. *NS32* (1985) 1204.

THE RADIOFREQUENCY QUADRUPOLE LINEAR ACCELERATOR*

M. PUGLISI

University of Pavia, Department of Theoretical and Nuclear Physics, Pavia, Italy.

The seminar is aimed to give a comprehensive picture of an RFQ. After a short description of the accelerating structure the T-K expansion is treated and the fundamental formula for the potential is derived. The vane tips shaping, completed to first order is followed by the physics of the machine where the most important parameters are listed and illustrated. Since the RFQ is essentially a cavity resonator this topic has been given particular attention. Design and other technical considerations complete the picture, while in the last paragraph the new ideas are briefly outlined.

1. INTRODUCTION

The RFQ is a linear accelerator for ions that uses electric fields to simultaneously focus, bunch and accelerate a beam of heavy particles.

While, in principle, the RFQ can accept, focus and accelerate to the desidered energy any kind of charged particle this machine is particularly convenient for accepting and accelerating an intense, low velocity beam from a continuous dc injector. In this case the main advantages of the RFQ can be summarized as follows: small size, low voltage dc injection, compatibility with complex ion sources, bunching with high efficiency, high beam current capacity, high output beam quality, easy operation. Viceversa at high energy (2-3 MeV/AMU) most of the above advantages becomes scarcely significant and the standard linacs are preferable.

Before going further we like to remind that the RFQ was invented by Kapchinskii and Tepliakov in 1970 [1], the first russian test was in 1975. Later on the work began at Los Alamos (1978) and subsequently in many other places as for instance Berkeley, Brookhaven, CERN, Chalk-River, GSI, Frankfurt, Saclay, Tokio. The so called Proof of Principle "POP"

---

* Reproduced from CERN Accelerator School. Oxford September 1985 (Proceedings to be Published) was given at Los Alamos in 1980. Since that time many RFQ have been succesfully realized and the interest in this machine is no longer limited only to the high energy physicist because many industries are now planning to use the RFQ for ion implantation, testing of materials, medical purposes. Finally the RFQ can play an important role as a heavy ion accelerator for the inertial fusion.

Fig. 1 - The BNL RFQ

2. THE ACCELERATING STRUCTURE

Basically an RFQ (Radio Frequency Quadrupole) is made by four equal electrodes symmetrically placed around the beam axis, excited by an appropriate radio frequency voltage and contained in a vacuum tank with high conducting walls.

Depending upon their shape the electrodes are named as "vanes" or "rods" and in Fig. 2 a group of four idealized vanes is sketched in order to show the geometry of the arrangement and for giving a rough idea of the vane tips shaping that is needed for creating the appropriate accelerating fields.

Actually the whole beam dynamics of the machine depends upon the shape of the vane tips and it is rather evident that the vane tips shaping of a physical machine will be determined by a compromise among many conflicting requirements.

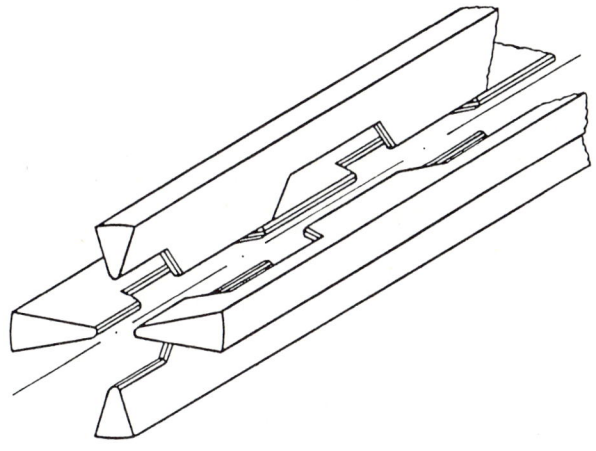

Fig. 2 - Schematic view of the four vanes assembled for creating the longitudinal field.

Nevertheless for sake of clarity the physics of the machine will be discussed on the basis of the sketch already seen.

In Fig. 3 only a horizontal and a vertical vane have been sketched together with the appropriate reference axes. The peaks for each tip are the nearest points to the z axis and the valleys are the points that being on the coordinate planes are the most distant from the z axis.

The distance between two adjacent peaks or valleys is changing along the beam axis according to the beam dynamics and the structure is non periodic.

We consider now the assembly of four vanes and two planes normal to the z axis passing through two adjacent peaks of the horizontal vane. The space limited by the two planes is called an elementary unit and each unit is made up with two cells.

In Fig. 4 again one horizontal and one vertical vane are represented with the vertical vane rotated 90° into the same plane as the horizontal vane. The horizontal vanes are supposed to be held at a dc potential equal to + V/2 while the vertical ones are supposed to be held at a potential equal to - V/2.

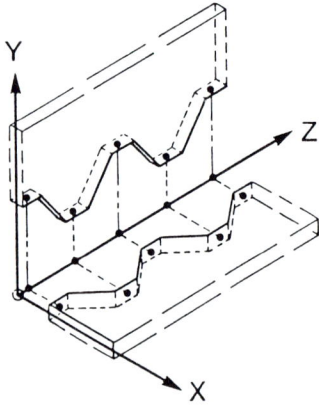

Fig. 3 - Schematic view of two adjacent vanes

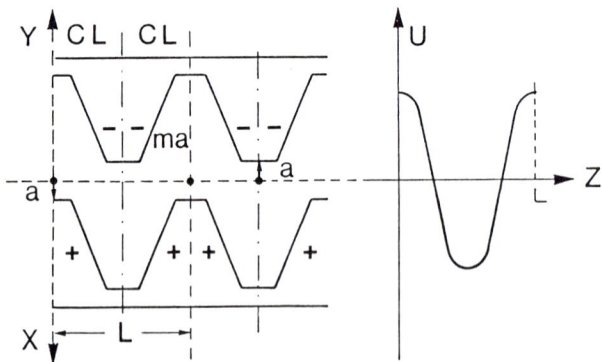

Fig. 4 - Two adjacent vanes are shown on the same horizontal plane. The potential between the vanes is sketched

It is now rather evident that a positively charged particle passing through the first cell (with length equal to l) will gain energy, while the same particle will loose energy in going through the second as it is required by the potential along the beam axis.

If now we suppose to excite the four vanes with an RF voltage then the whole structure can be accelerating if the particle spends half period of the RF voltage for passing through each cell. It should be noted that under the previous hypothesis when one cell is focusing on the vertical plane (and necessarily defocusing on the horizontal) then the following cell will be focusing on the horizontal plane while is defocusing on the vertical.

The net result can be focusing as predicted by the theory for the alternate gradient machines.

Since both the transverse and the longitudinal focusing critically depend on the shape of the vane tips, a detailed knowledge of the fields in the beam region is required.

Unfortunately, employing the geometry already indicated the calculation of the electromagnetic field distribution is very complicated and some simplifying hypotheses are to be sought.

Actually if one attempts to design a physically realizable (and useful) structure then any kind of calculation (even the simplest one) shows that the important part of the RFQ cross section has very small dimensions if compared with the free space wavelength of the accelerating field.

According to that reason M. Kapchinskii and V.A. Tepliakov did the calculations assuming the electrostatic distribution for the field inside the beam region. The subsequent experience proved that this analysis is sufficently accurate.

3. OUTLINE OF THE T-K EXPANSION

The distribution of the electrostatic field due to the vanes is normally known as the T-K expansion [1,2].

This distribution is easily obtained solving the Laplace equation, written in cylindrical coordinates, with the technique of the separation of variables in our cylindrical reference frame.

The z axis is the beam axis and the origin is located as in Fig. 3. The coordinate is equal to zero on the positive section of the x axis.

If $U = U(r,z,\psi)$ is the unknown potential then the Laplace equation is as follows:

$$\frac{\partial^2 U}{\partial r^2} + \frac{1}{r}\frac{\partial U}{\partial r} + \frac{1}{r^2}\frac{\partial^2 U}{\partial \psi^2} + \frac{\partial^2 U}{\partial z^2} = 0 \qquad (1)$$

in order to separate the variables we assume that:

$$U = f(r,\psi) \cdot \phi(z) \qquad (2)$$

Upon substituting (2) into (1) and separating we obtain the system:

$$\begin{cases} \dfrac{\partial^2 f(r,\psi)}{\partial r^2} + \dfrac{1}{r}\dfrac{\partial f(r,\psi)}{\partial r} + \dfrac{1}{r^2}\dfrac{\partial^2 f(r,\psi)}{\partial \psi^2} = h^2 f(r,\psi) & (3) \\[2ex] \dfrac{\partial^2 \phi(z)}{\partial z^2} = -h^2 \phi(z) & (4) \end{cases}$$

where h^2 is an arbitrary constant to be determined using the boundary conditions.

If we assume that h^2 is a positive number and that A and B do not depend upon z then

$$\phi(z) = A\cos(hz) + B\sin(hz) \qquad (5)$$

is a solution of equation (4). Because of the linearity of the previous equation then any linear combination of functions like (5) is a solution of equation (4) that could obey the boundary conditions.

At this point we assume that our structure is periodic. In this case the potential we are looking for should be a periodic even function of z with period equal to L. It follows that:

$$F(z) = \sum_{n=0}^{\infty} A_n \cos\left(\frac{2\pi n z}{L}\right) \qquad n = 1, 2, 3 \qquad (6)$$

is a solution of equation (4) that can fit the actual boundary conditions of the particular problem if the appropriate values are given to $A_n = A_n(r,\psi)$.

From Eq. (6) the values of the separatrix constant are also known because h should be equal to $2\pi \dfrac{n}{L}$.

In this continuation the quantity $2\pi/L$ will be set equal to k (normally known as the phase costant).

Now for each value of h we have a solution of Eq. (3). In order to satisfy the fundamental relationship already assumed, $U = f(r,\psi) \cdot \phi(z)$ each solution of Eq. (3) should be multiplied by the corresponding $\phi(z)$.

Consequently the quantities appearing in Eq. (6) should be those functions which are solutions of Eq. (3).

Each solution A being determined by the eigenvalue nk that pertains to the corresponding $\phi_n(z)$

The method used above can be employed for solving Eq. (3). Again we separate the variables assuming the product solution:

$$f(r,\psi) = R(r) \cdot \Theta(\psi)$$

where R and Θ are respectively functions of r and ψ only. Substituting and separating the variables we obtain:

$$\begin{cases} \Theta''(\psi) = -m^2 \Theta(\psi) & (7) \\ R''(r) + \frac{1}{r} R'(r) - \left[(nk)^2 + \left(\frac{m}{r}\right)^2 \right] R(r) = 0 & (8) \end{cases}$$

where m is a new arbitrary constant to be determined by the boundary conditions. If we assume that m is a positive number a solution of Eq. (7) is as follows:

$$\Theta(\psi) = p \cos(m\psi) + q \sin(m\psi) \qquad (9)$$

where p and q depend neither on z nor on ψ.

The electrical excitation of the structure (the vertical vanes are in parallel as the horizontal ones) requires that the potential U should be a periodic even function of the variable ψ with period equal to π.

This means that for meeting the above requirements we should have:

$$q = 0 \quad , \quad m = 2s \qquad s = 1, 2, 3$$

and, consequently, the Θ functions should have the form:

$$\Theta(\psi) = A_s \cos(2s\psi)$$

Again each value of m must be substituted into Eq.(8) in order to obtain a radial function that depends upon both n and m.

The general solution will be written term by term assuming that m ranges from 0 to ∞ and that for each value of n the index s can assume all the values that fit the boundary conditions.

In general the boundary conditions above mentioned can be summarized as follows:

1) For $r = 0$ the potential should remain finite.

2) For $kz = \pi/2$ and $kz = 3\pi/2$ each unit should exhibit a four pole symmetry. (The potential on the axis is equal to zero).

Taking into account the above conditions for $n = 0$ the contribution U_o to the total potential U is as follows:

$$U_o = \sum_{s,J}^{\infty} A_J r^{2J} \cos(2J\psi) \qquad (10)$$

where for meeting the four pole symmetry (independently of z) we must have:

$$J = 2s + 1 \qquad s = 1, 2, \ldots$$

For $n \neq 0$ Eq.(8) is solved by the modified Bessel functions of order 2s and as a contribution to the total potential we obtain:

$$U_n = \left\{ \sum_s A_s I_{2s}(nkr) \cdot \cos(2s\psi) \right\} \cos(nkz) \qquad (11)$$

where $(n + s)$ should be odd in order to fit the four pole symmetry that the structure periodically exhibits. (The Neumann functions are excluded because, as already said, the potential must remain finite for $r = 0$).

Adding the various contributions we obtain the well known T-K expansion:

$$U(r,z,\psi) = U_o + \sum_n^{\infty} U_n \qquad (12)$$

And the electrostatic problem is virtually solved.

4. THE VANE TIPS SHAPING

As was demonstrated in the previous paragraph the T-K expansion is very complicated and consequently if the vane tips geometry is assigned then a sufficiently accurate description of the electrostatic field in the beam region could be a very difficult problem.

On the other hand an adequate study of the beam dynamics inside the machine can be done only if the field in the beam region is well known.

A sensible procedure for overcoming the trouble can be the following [3]:

"The vane tips are to be shaped in such a way as to coincide with the equipotentials described by few terms of the T-K expansion".

In other words we select a reasonable form for the potential in the beam region.

When the potential is known the field is known, and the beam dynamics associated with the selected potential can be defined completely.

If the calculated beam dynamics is satisfactory then the vane tips must have the form of the equipotentials that limit the beam region.

The simplest function that describes a potential consistent with a boundary condition of the beam region is obtained mantaining only the lowest order terms of the T-K expansion. Accordingly, if we set n = 1 and s = 0 we obtain that the simplest form for the potential in each unit is as follows:

$$F(r,z,\psi) = A_1 r^2 \cos(2\psi) + A_2 I_0(kr)\cos(kz) \tag{13}$$

where A_1 and A_2 are two constants to be defined and $k = 2\pi/L$ is the phase constant. Since the four vanes are powered by an RF voltage with period T (radian frequency ω equal to $2\pi/T$) the complete form of the quasi-static potential is:

$$U(r,z,\psi,t) = F(r,z,\psi) \cdot \sin(\omega t + \phi) \tag{14}$$

where ϕ is the phase of the RF voltage when the charged particle enters the unit.

From the physical picture of the accelerator we know that the "synchronous" particle should pass through the unit exactly in one period of the radio frequency voltage. Consequently if β is the average normalized velocity of the synchronous particle we can write:

$$\frac{k}{\ } = \frac{2\pi}{L} = \frac{2\pi}{\beta cT} = \frac{2\pi}{\beta\lambda} \tag{15}$$

where λ is the free space wavelength of the applied field.

A_1 and A_2 indicate two constants which depend on the geometry of the unit and this means that with this choice we can input only two boundary conditions. It should be noted that those boundary conditions are not as arbitrary as they could be thought because once A_1 and A_2 are given the resulting structure should be physically realizable and electrically compatible.

Accordingly, with the scheme given in Fig. 3 our boundary conditions at $z = 0$ are as follows:

$$\psi = 0 \quad ; \quad r = a \qquad ; \quad U = \frac{V}{2} = \frac{V_0}{2}\sin(\omega t + \phi)$$

$$\psi = \frac{\pi}{2} \quad ; \quad r = ma \qquad ; \quad U = -\frac{V}{2} = -\frac{V_0}{2}\sin(\omega t + \phi)$$

Upon substituting the boundary conditions in Eq. (14) we obtain the system:

$$\begin{cases} A_1 a^2 + A_2 I_0(ka) = \dfrac{V}{2} \\ -A_1(ma)^2 + A_2 I_0(mka) = -\dfrac{V}{2} \end{cases}$$

solving with the Kramer rule we obtain:

$$A_1 = \frac{I_0(mka) + I_0(ka)}{a^2 I_0(mka) + (ma)^2 I_0(ka)} \cdot \frac{V}{2}$$

$$A_2 = \frac{m^2 - 1}{I_0(mka) + m^2 I_0(ka)} \cdot \frac{V}{2}$$

now A_1 and A_2 are dimensional quantities and this may create troubles in the subsequent manipulations. For this reason two new adimensional parameters, A and X, are defined as follows:

$$\begin{cases} X = \dfrac{2}{V} a^2 A_1 = 1 - AI_0(ka) \\ \\ A = \dfrac{2}{V} A_2 = \dfrac{m^2 - 1}{I_0(mka) + m I_0(ka)} \end{cases} \quad (16)$$

Upon substituting in Eq. (14) we obtain:

$$U = \dfrac{V}{2} \left[X \left(\dfrac{r}{a}\right)^2 \cos(2\psi) + A I_0(kr) \cos(kz) \right] \quad (17)$$

and it should be remembered that the intervane voltage V is equal to $V_0 \sin(\omega t + \phi)$.

Now we assume that the potential U is the potential actually existing in one unit and we want to shape the vane tips in such a way as to realize this potential distribution.

Once a, m and k are given then the shape of the unit is uniquely determined and one of the simplest way to arrive at the vane profile is to determine a reasonable number of vane cross sections along the z axis.

Those cross sections can be obtained by solving, numerically, Eq. (17) in which U is set equal to the potential of the considered vane, the z coordinate is an input for each cross section and a series of values is given to ψ. For each value of ψ the corresponding value of r is calculated.

For calculating the profile of the cross section of a horizontal vane the potential must be set equal to + V/2.

For a chosen value of $z = z_n$ we select a series of values for ψ ($0 < \psi < \pi/4$) and find the corresponding values of r solving Eq. (18):

$$1 = X \dfrac{r^2}{a^2} \cos(2\psi) + A I_0(kr) \cos(kz_n) \quad (18)$$

An identical procedure has to be followed to find the cross section of a vertical vane. The potential must be set equal to - V/2 and $\pi/4 < \psi < \pi/2$. Consequently Eq. (19) is the one to be solved.

$$-1 = X \dfrac{r^2}{a^2} \cos(2\psi) + A I_0(kr) \cos(kz_n) \quad (19)$$

The procedure outlined above cannot be followed blindly and some remarks are in order.

A unit entirely generated using Eq. (18) and (19) might lead to a physically unrealizable structure. This means that once the pole tip profiles are found then the remainder of the vanes has to be determined using different criteria. Moreover for $z = 0$ and $z = \beta\lambda$ the unit has identical cross sections while in an accelerating unit the initial and final sections are different.

If the particles are not relativistic their velocity increases during the acceleration. This means that the distance travelled by the sinchronous particle during one first half cycle of the accelerating field should be shorter then the one travelled during the second half of the same cycle.

The distance travelled by the synchronous particle during half cycle of the accelerating voltage is called the unitary cell length and is obviously equal to $\beta\lambda/2$ where β is the average normalized velocity of the transit particle.

A portion of a horizontal vane that contains two adjacent cells is sketched (exagerating for sake of clarity) in Fig. 5. The very nature of the RFQ accelerator requires that if, for instance, the cell "n" begins at a peak of a horizontal vane then the "n + 1" cell begins where a peak on the vertical vane occurs.

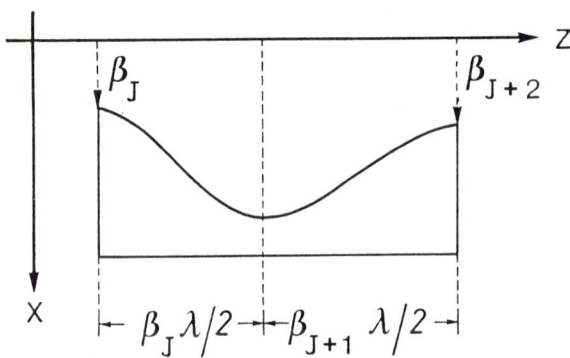

Fig. 5 - Portion of a horizontal vane that is part of two adjacent cells.

This means that we could make use of the symmetrical expansion of a function (the vane profile) pretending that the distance $\beta\lambda$ is just twice the spatial period of a non physically existing cell (with even symmetry properties) that as regards the first half do coincide with the actual one.

Consequently the shape of an actual cell can be determined with the same procedure outlined above but a new cell begins every time $z = \beta\lambda/2$.

In other words once each value is assigned to the four parameters k, a, m, β, then the detailed procedure for finding the profile of the vanes is listed below.

1. The Pth cross section of the horizontal vane of the J cell is obtained by solving Eq. (18) written as follows:

$$1 = X_J \left(\frac{r}{a_J}\right)^2 \cos(2\psi) + A_J I_0(k_J r) \cos(k z_p)$$

where z_p is the p portion of the cell length l_J and the ψ coordinate is varied from zero to $\pi/4$.

2. The Pth cross section of the vertical vane of the same J cell is obtained from the same equation where + 1 is substituted by - 1, and the ψ coordinate varies from $\pi/4$ to $\pi/2$ (again $k_J z$ will vary from 0 to π).

3. The next cell (J + 1) now begins with a radius $a_{J+1} \cdot m_{J+1}$ on the horizontal vane (U = + V/2, $\psi = 0$). On the other hand the cell length is now l_{J+1} and this length should be divided in many intervals as above but the argument of the cosine should vary from π to 2π. This means that the quantity π should be added to the argument of the cosine.

From the above arguments it is evident that there is no continuity between adjacent cells and the previous procedure should be modified in such a way as to obtain continuity between adjacent cells as shown in Fig. 5. For instance a and m can be made linear functions of z.

On the other hand even the modified procedure could generate cells in which the curvatures along z and in the X - Y planes may create serious mechanical and electrical problems.

In Fig. 6 the assembly of four physical vanes is sketched together with some cross sections of the whole machine.

The above procedure can be followed and the vane tip profiles exhibit a very complicated shape that should be machined with a high degree of accuracy. Moreover especially at the low energy end the electric field may result too much enhanced.

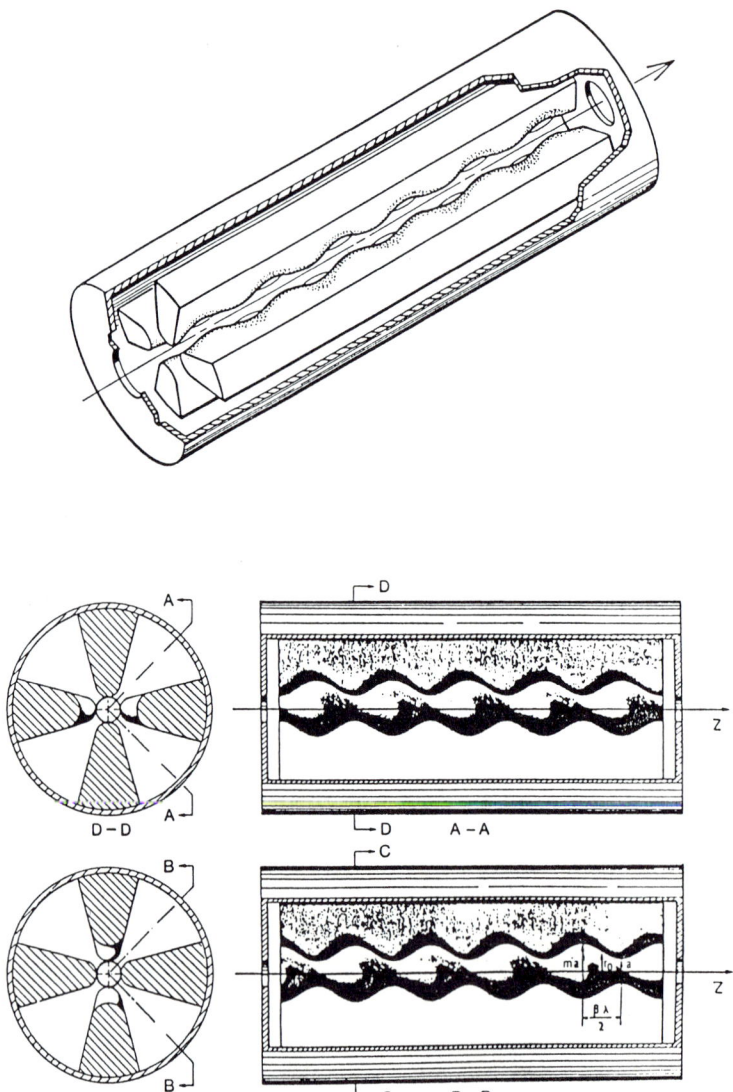

Fig. 6 - Possible shape for the four vanes. Assembly in the container and cross sections.

A better mechanical solution is obtained if the pole tip cross section may have a constant curvature radius. This can be achieved introducing higher order multipoles in the potential function.

Starting from an optimized two term potential structure at every cell one can try a formula with more than two terms in order to minimize the deviation of the pole cross section from a circle of constant radius.

Obviously many other important manipulations can be done to cope with the particular performance required, but the procedure outlined above remains substantially the same.

5. Physical considerations

The lowest order potential function depends upon three parameters: a, m and k and we have seen that each cell is completely determined whenever the value of those parameters is specified.

From the gradient (changed in sign) of the potential function (17) we obtain the fields inside the beam region as follows.

$$\begin{cases} E_r = -\dfrac{XV}{a^2} r\cos(2\psi) - \dfrac{kAV}{2} I_1(kr)\cos(kz) \\ \\ E_\psi = -\dfrac{XV}{a^2} r\sin(2\psi) \\ \\ E_z = \dfrac{kAV}{2} I_0(kr)\sin(kz) \end{cases} \quad (20)$$

where $V = V_0 \sin(\omega t + \phi)$ is the intervane voltage [3].

It is immediately clear, by inspection, that a particle travelling on the z axis does not experience any transverse force while for this particle (charge q and mass m_0) the accelerating force f_z becomes:

$$f_z = q\frac{kAV}{2}\sin(kz) = \frac{\pi q AV_0}{\beta\lambda}\sin\left(\frac{2\pi z}{\beta\lambda}\right)\sin(\omega t + \phi) \quad (21)$$

This means that the quantity X has to do with the transverse focusing force while A is connected with the voltage gain per cell. In fact the potential difference ΔU that exists on the axis between the beginning and the end of each cell is equal to AV as can be easily verified upon substitution.

$$U_o - U\left(\frac{\beta\lambda}{2}\right) = \frac{V}{2}\left\{A - A\left[\cos\left(\frac{2\pi}{\beta\lambda} \cdot \frac{\beta\lambda}{2}\right)\right]\right\} = AV$$

Another important element for the design of the accelerator is the dynamic gain of energy per cell ΔE. Taking into account that in an RFQ the accelerated particles are not relativistic, the motion of a particle travelling on the z axis is as follows:

$$\frac{f_z}{m_o} = \ddot{z} = \frac{\pi A q V_o}{\beta \lambda m_o} \sin(kz)\sin(\omega t + \phi) \qquad (22)$$

where β is the normalized average velocity of the transit particle.

It has been demonstrated (3) that under very broad conditions the above equation can be solved analitically and consequently the quantity ΔE can be calculated with a high degree of accuracy. Nevertheless in a "normal" RFQ the relative variation of velocity per cell $\Delta\beta/\beta$ is always small and this means that the average and the instantaneous velocity of the transit particle are rather close.

In this case we can write $\omega t \cong kz$ and consequently the force f_z becomes:

$$f_z = \frac{\pi q A V_o}{2\beta\lambda}\left\{\cos(\phi) - \cos(2kz + \phi)\right\} \qquad (23)$$

Integrating over the cell length we obtain:

$$\Delta E = \frac{\pi}{4} q A V_o \cos\phi \qquad (24)$$

the numerical computations and the experience on the actually existing RFQ have proved that (24) is sufficently accurate (in the above formula $\pi/4$ is the value of the transit time factor for a longitudinal field with space variations equal to sin kz).

Beside Eq. (22), which describes the motion of a particle travelling on the z axis, we need the equations for the motion in the transverse planes.

While the general theory of the transverse motion is very complicated it is rather easy to define the parameters which determine the stability of the beam on the transverse planes.

For instance using the expression of the gradients (Eq. 20) we can write the differential Eq. of the displacement along x as follows:

$$\ddot{x} = \frac{qE_x}{m_o} = -\frac{q}{m_o}\left[\frac{XV}{a^2} x - \frac{kAV}{2} I_1(kx) \cos(kz)\right] \quad (25)$$

The modified Bessel function $I_1(kx)$ can be expanded and for small values of x and we obtain $I_1(kx) \cong kx/2$. Substituting into Eq. (25) and using the esplicit formula for V and k we obtain:

$$\ddot{x} = -\left[\frac{qXV_o}{a^2 m_o} \sin(\omega t + \phi)\right] x - \left[\pi^2 \frac{AV_o q}{\beta^2 \lambda^2} \cos\left(\frac{2\pi}{\beta\lambda}\right) \sin(\omega t + \phi)\right] x$$

Again we can make the hypothesis $kz = \omega t$ and after little algebra we obtain:

$$\ddot{x} = -\left[\frac{qXV_o}{a^2 m_o} \sin(\omega t + \phi)\right] x +$$

$$-\left[\frac{\pi^2}{2} \frac{AV_o q}{m_o \beta^2 \lambda^2} \sin\phi\right] x + \quad (26)$$

$$-\left[\frac{\pi}{2} \frac{AV_o q}{m_o \beta^2 \lambda^2} \sin(2\omega t + \phi)\right] x$$

now in order to obtain non dimensional coefficients we multiply by T^2/λ both sides of Eq. (26) and change the variables as follows:

$$\lambda x = \xi \quad ; \quad \alpha T = t$$

Upon substituting into Eq. (26) we obtain:

$$\frac{d^2\zeta}{d\alpha^2} = -\left[B\sin(2\pi\alpha + \phi) + \Delta \right] x - \gamma x \cos(4\pi\alpha + \phi) \tag{27}$$

where, following the symbology used at Los Alamos, [3]

$$B = X \frac{qV_0}{m_0 c^2} \left(\frac{\lambda}{a}\right)^2 \quad ; \quad \Delta = \frac{\pi^2}{2} \frac{AVq}{\beta^2 m_0 c^2} \sin\phi \tag{28}$$

and $\gamma = \Delta / \sin\phi$.

It should be noted that B and Δ can be interpreted as normalized forces. B is responsible of the focusing effect while a defocusing effect corresponds to Δ when ϕ is negative, as it should be in a linac.

The solutions of the above equation can be convergent or divergent depending upon the numerical value of the parameters B and Δ. Since Δ is always very small (typically $\cong 0.05$ at the injection) then a good degree of stability is obtained if B is larger than few units and smaller than $\cong 15$.

A more general analysis of the transverse stability will not be undertaken here because it requires the use of techniques too much specialized for a comprehensive seminar on the RFQ. (CERN Accelerator School 1984. Vol. 1 Pag. 176-177).

Nevertheless it should be emphasized that the general theory of the radial stability, valid for a linear accelerator, is applicable to the RFQ.

Before leaving the problem of the radial stability it is important to consider the role of the radius r (r being the distance of the vane tips from the axis) that occurs for any cell when $\cos(kz) = 0$.

In fact if the above condition is fulfilled then Eq. (17) reduces to:

$$U = \frac{V}{2} X \left(\frac{r}{a}\right)^2 \cos(2\psi) \tag{29}$$

It follows that both for $\psi = 0$ and $\psi = \pi/2$ the distance r form the axis of the pole tips is as follows:

$$r = r_0 = \frac{a}{\sqrt{X}} \tag{30}$$

This particular value of the radius is the so-called four-pole radius because on the planes for which $kz = \pi/2$ the vanes show perfect four-polar symmetry with hyperbolic cross sections defined by the equations:

$$r = \frac{r_o}{\sqrt{\cos(2\psi)}} \qquad (0 \le \psi \le \pi/4) \qquad \text{horizontal vane}$$

$$r = \frac{r_o}{\sqrt{-\cos(2\psi)}} \qquad (\pi/4 \le \psi \le \pi/2) \qquad \text{vertical vane}$$

Moreover a very simple calculation can show that when the radius is equal to r_o also the radius of curvature at the pole tips is equal to r_o (on the X-Y plane).

Returning to the transverse focusing we observe that if V is constant, keeping the focusing strength at a fixed value requires that the quantity X/a^2 remains constant along the machine and this means (Eq. 30) that the radius r should remain constant.

Moreover a fixed value of r_o can be expected to minimize variations in the vane-to-vane capacitance and should facilitate the design of an RFQ in which the pole tip voltage distribution is required to be flat over its entire length. For the above reason the quantity r_o can be regarded as a characteristic average radius of the RFQ pole tips that affects all the design of the machine.

6. THE STRUCTURE OF AN RFQ

The accelerating and focusing fields depend upon the voltage applied to the four vanes. This is normally obtained creating a cavity resonator where the vanes are a fundamental part of the whole structure.

More specifically the RFQ cavity (vanes and container) should be designed for resonating, at the working frequency, in such a way as to create the desired voltage on the vanes.

This is a problem that requires some knowledge of the microwave technique. Because the RFQ is essentially a radio frequency device where the microwave techniques play a major role it may be useful to give, in the following, a short outline of this topic.

An electromagnetic field that depends upon the time as a sine wave can exist and propagates inside an hollow cylindrical pipe with perfectly conducting wall if some condition are met.

A short way for arriving at significant results is to assume that both the fields E and H depend upon z and t as follows:

$$F(r,\psi,z,t) = f(r,\psi) e^{j\omega t - \gamma z} \tag{31}$$

where F and f stand both for E and H and ω is the radian frequency of the field.

Moreover assuming that no currents are contained in the bounded volume then the Maxwell Eq. that we need can be written as follows:

$$\begin{vmatrix} 0 & +\gamma E_\psi & +j\omega\mu H_r & 0 \\ -\gamma E_r & 0 & 0 & +j\omega\mu H_\psi \\ -j\omega\varepsilon E_r & 0 & 0 & +\gamma H_\psi \\ 0 & -j\omega\varepsilon E_\psi & -\gamma H_r & 0 \end{vmatrix} = \begin{vmatrix} -\dfrac{1}{r}\dfrac{\partial E_z}{\partial \psi} \\ \dfrac{\partial E_z}{\partial r} \\ -\dfrac{1}{r}\dfrac{\partial H_z}{\partial \psi} \\ \dfrac{\partial H_z}{\partial r} \end{vmatrix}$$

Solving with the Kramer rule and using the normal notation:

$$K_c^2 = \omega^2 \varepsilon \mu + \gamma^2$$

we obtain the transverse components of the field as functions of the longitudinal components E_z and H_z):

$$\begin{aligned}
E_r &= -\frac{1}{K_c^2}\left[\gamma\frac{\partial E_z}{\partial r} + \frac{j\omega\mu}{r}\frac{\partial H_z}{\partial \psi}\right] \\
E_\psi &= -\frac{1}{K_c^2}\left[\frac{\gamma}{r}\frac{\partial E_z}{\partial \psi} - j\omega\mu\frac{\partial H_z}{\partial r}\right] \\
H_r &= -\frac{1}{K_c^2}\left[-\frac{j\omega\varepsilon}{r}\frac{\partial E_z}{\partial \psi} + \gamma\frac{\partial H_z}{\partial r}\right] \\
H_\psi &= -\frac{1}{K_c^2}\left[j\omega\varepsilon\frac{\partial E_z}{\partial r} + \frac{\gamma}{r}\frac{\partial H_z}{\partial \psi}\right]
\end{aligned} \tag{32}$$

This means that if the longitudinal components of the field are known then the transverse ones can be obtained by derivation. Moreover if E_z is always zero then $H_z \neq 0$ (because otherwise the whole field is zero) and we have the family of the so called TE modes, where TE is an abbreviation for transverse electric mode. If $H_z = 0$ and consequently $E_z \neq 0$ we have the family of the TM modes (transverse magnetic).

Since the above system is linear then the superposition principle applies and any synusoidal field can be reduced to a linear combination of TE and TM modes[*]. It is now important to recognize that the structure containing the vanes should be excited with a TE mode. In fact only a TE mode can create the four-polar focusing field that, on the other hand, cannot be accelerating. The modulation on the vane tips introduces the local perturbation that is adequate for creating, locally, the accelerating field.

Consequently we are naturally led to find out a possible solution for H_z in the structure already described.

Under the previous hypotheses the Maxwell Eqs. are as follows:

$$\nabla \cdot E = 0$$
$$\nabla \cdot H = 0$$
$$\nabla \times E = -j\omega\mu H$$
$$\nabla \times H = -j\omega\varepsilon E$$

Taking the curl of the last Eq., substituting $\nabla \times E$ and recalling that:

$$\nabla \times \nabla \times H = \nabla (\nabla \cdot H) - \nabla^2 H$$

we obtain the familiar wave equation for the vector H. Since the same procedure applied to the curl of E gives, formally, the same result we can write:

$$\nabla^2 \left\{ \begin{array}{c} E \\ H \end{array} \right\} + \omega^2 \mu\varepsilon \left\{ \begin{array}{c} E \\ H \end{array} \right\} = 0 \qquad (33)$$

Expanding the above equation and retaining the longitudinal z component of H we obtain:

$$\frac{\partial^2 H_z}{\partial r^2} + \frac{1}{r}\frac{\partial H_z}{\partial r} + \frac{1}{r^2}\frac{\partial^2 H_z}{\partial \psi^2} + K_c^2 H_z = 0 \qquad (34)$$

---

[*] In the litterature concerned with particle accelerators the TM modes are called "accelerating modes" while the TE modes are called "deflecting modes".

Eq. (34) can be solved with the same technique that has been used for the T-K expansion.

We can assume that $H_z = R(r) \cdot \theta(\psi)$ where $R = R(r)$ is function of r and $\theta = \theta(\psi)$ is function of ψ.

Substituting and manipulating we obtain:

$$r^2 \frac{R''}{R} + r \frac{R'}{R} + K_c^2 r^2 = -\frac{\theta''}{\theta} \qquad (35)$$

The left side is function of r alone, the right of ψ alone. Consequently if both sides are to be identical for all values of r and ψ then both sides must be equal to the same constant: for instance ν^2 (assumed positive).

By substitution we obtain:

$$\begin{cases} R'' + \frac{1}{r} R' + \left(K_c^2 - \frac{\nu^2}{r^2} \right) R = 0 \\ \theta'' = -\nu^2 \theta \end{cases}$$

The first equation is solved with the Bessel and the Neumann functions of order ν whereas the second is solved with sinusoids.

For $r = 0$ H_z cannot be infinite and this means that the Neumann function does not fit this boundary condition, on the other hand the field should be the same every time we vary ψ by a multiple of 2π. This means that ν must be an integer. Moreover a proper selection of the origin for the coordinate will allow us to use either the sine or the cosine in the trigonometric part of the solution.

Consequently we obtain:

$$H_z = H_o J_\nu(K_c r) \cdot \cos(\nu \psi) \qquad (36)$$

The field described by Eq. (36) is parallel to the conducting wall and automatically obeys the boundary conditions. Viceversa from the second of the (32), using (36), we obtain E that being parallel to the boundary must be zero on the perfectly conducting walls.

This means that on the boundary (r = a) the derivative of J_ν ($K_c \cdot r$) must be zero and we obtain:

$$J_\nu' (K_c a) = 0 \qquad (37)$$

where a is the inner radius of the cylinder while J_ν' ($k_c \cdot a$) indicates the value of the derivative of the Bessel function of order ν for r = a.

Eq. (37) determines the infinite series of the K_c and for each K_c we have a particular solution indicated as the $TE_{\nu\ell}$ mode.

Specifically ν indicates the number of variations along ψ and l indicates the order of the zero which determines the particular solution.

If we are looking for a transverse field with four-pole symmetry and no variations along z (as required from the assumed form of the potential) we have to set:

$$\gamma = 0 \quad ; \quad \nu = 2$$

and from (37), selecting the first zero, we obtain:

$$K_c a = \omega \sqrt{\varepsilon\mu}\, a = 3.05424 \quad \text{or} \quad f_c = \frac{145.8}{a} \; MH_z \qquad (38)$$

where f_c is the so called cut-off frequency of the selected mode (Fig. 7). (In a waveguide the cut-off frequency is always the one for which $\gamma = 0$).

This means that an infinitely long cylindrical lossless pipe with inner radius equal to a can support the axially uniform four-pole mode. The relationship between frequency and mode being defined by Eq.(38).

An infinitely long wave guide is not a practical device but it is possible to build a physical structure where for a long portion of the axis the field has four-pole symmetry and is adequately uniform (this structure is the RFQ resonant cavity where the cylindrical wall, the vanes and the end sections are fundamental parts of the whole structure).

In order to have some ideas about the cylindrical cavity resonators we suppose to short circuit, with a conducting wall normal to the axis, both ends of our hollow pipe leaving a clearance equal to L between the short circuits.

Now, beside the above conditions on the cylindrical wall (Eq. (37)), the electric field of a TE mode should be zero on the short circuiting surfaces (which are parallel both to E_r and to E_ψ) and we have a third condition that enters in the determination of the cavity resonant frequency.

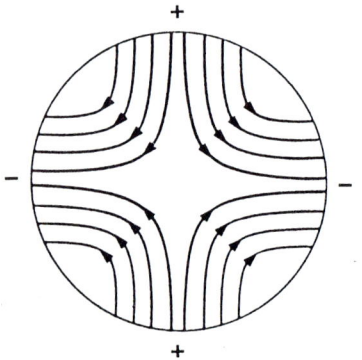

Fig. 7 - Electric lines of force for TE mode in the cross section of a uniform cylindrical waveguide.

It is nearly obvious that this condition is fulfilled if the distance L is an integer multiple of half wavelength of the field as measured inside the pipe.

Let us call $R_{\nu\ell}$ the value of the argument that satisfies Eq. (37). Consequently we have:

$$\gamma^2 + \omega^2 \mu \varepsilon = \left(\frac{R_{\nu\ell}}{a}\right)^2 \qquad (39)$$

and we see that to ω and γ any value consistent with (39) can be given.

In order to build up a stationary field we need propagation in both directions of the z axis.

This means that γ should be imaginary and we put $\gamma = j\delta$. If λ_g is the wavelength inside the pipe (the so called guide wavelength) then it is rather obvious thats $\delta = 2\pi/\lambda_g$.

In fact when we pass through a length equal to λ_g the field has to repeat itself because the argument of $e^{j\delta\lambda_g}$ changes by 2π.

Substituting in Eq. (37) and recalling that $\omega^2 \varepsilon \mu = (\frac{2\pi}{\lambda})^2$, where λ is the free space wavelength of the field, we obtain:

$$\left(\frac{2\pi}{\lambda}\right)^2 = \left(\frac{R_{\nu\ell}}{a}\right)^2 + \delta^2 = \left(\frac{R_{\nu\ell}}{a}\right)^2 + \left(\frac{2\pi}{\lambda_g}\right)^2 \tag{40}$$

Rearranging and introducing the third condition, that is the resonator length L can be equal only to an integer number, say p, of half guide wavelength we obtain:

$$\lambda = \frac{2L}{\sqrt{p^2 + (2L)^2 \cdot (R_{\nu\ell}/2\pi a)^2}} \tag{41}$$

where λ is the free space wavelength of a cylindrical resonator of radius a and length L operating in the $TE_{\nu\ell p}$ mode.

At this point a very short outline of the TM modes for a cylindrical cavity seems in order.

Eq. (33) can be solved for E_z and following step by step the outlined procedure we obtain:

$$E_z = E_o J_\nu(K_c r) \cdot \cos(\nu \psi) \tag{42}$$

The E_z component is, by definition, parallel to the perfectly conducting wall and consequently E_z must be zero for r = a. This condition is verified if:

$$J_\nu(K_c a) = 0 \tag{43}$$

Now it is rather evident that since E_z is always normal to the short circuit at the ends of the cylindrical cavity then an infinite series of modes can exist with no variations along z ($TM_{\nu\ell o}$ modes). As a consequence it happens that a cylindrical cavity can support any TM mode independently of its length.

Beside the above degenerate modes a cylindrical cavity can exhibit a TM resonance if the cavity length L is equal to an integer number of half wavelength measured inside the cavity. Again following the procedure outlined for the TE modes we find that the resonant wavelength of a TM mode is given by the formula (41) where now $R_{\nu\ell}$ is the zero of order 1 of the Bessel function of order ν.

In Fig. 8 examples of resonant modes are illustrated.

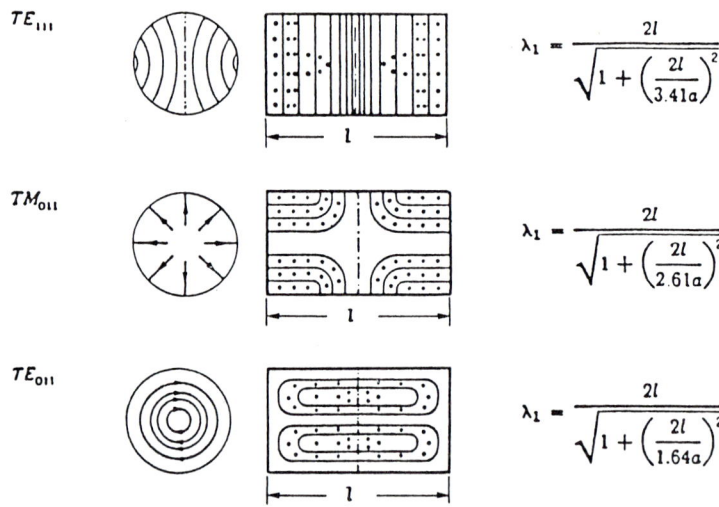

Fig. 8 - Three different modes excited in the same cylindrical resonator.

The degeneracy already seen allows zero value for the last index of a TM mode. This cannot happen for a TE mode because the transverse electric field must be always zero on the short circuiting walls at the end of the cavity. Therefore if no variations are allowed along z (last index equal to zero) then the whole field should go to zero.

Up to this point we considered only the elementary cylindrical resonator where the fields E_z or H_z are completely described with only their eigenfunctions and the resonant frequency is the corresponding eigenvalue.

However the cavity resonators used as accelerators, even mantaining the cylindrical symmetry, are often much more complicated and in order to satisfy the boundary conditions dictated by a technical resonator, the complete set of the cylindrical eigenfunctions is normally required.

Even a simple outline of the general theory would go beyond the purposes of this rather intuitive treatment. Many powerful computer programs are now available for analyzing, with good accuracy, practically any useful cylindrical resonator [6].

The cavity for an RFQ originates from a TE_{211} cylindrical resonator which is loaded with four V-shaped vanes symmetrically connected to the cylindrical wall as shown in Fig. 9.

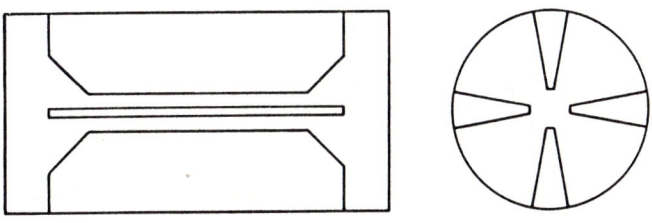

Fig. 9 - Simplified axial and longitudinal cross sections of an RFQ.

The vanes terminate at some distance from the short circuiting wall and consequently the central vane section is symmetrically coupled to the two end sections. (We should observe, in passing, that this resonator is no longer uniform along the abscissa).

The boundary conditions provided by the end sections allow the whole cavity to resonate in a very complicated manner where the fields are nearly uniform inside a large portion of the vane section.

More specifically this condition is obtained if the TE_{21} cut-off frequency of the uniform guide represented by the vane section is slightly below the operating frequency of the whole cavity.

Fig. 10 shows one of the four pole section of the BNL RFQ. Since in this machine the focusing force is held constant then it follows that all the four pole sections should be equal.

As explained above the choice of "constant r_o" minimize the difference of the electrostatic capacity between different portions of the same structure but those differences are non vanishing. This means that some "distributed tuning" along the structure would help in obtaining the good uniformity of the field that is really needed. For this reason each vane has been loaded with two bars tapered along the z axis and in fact a very good uniformity of the field has been achieved after a careful adjustement of the bars.

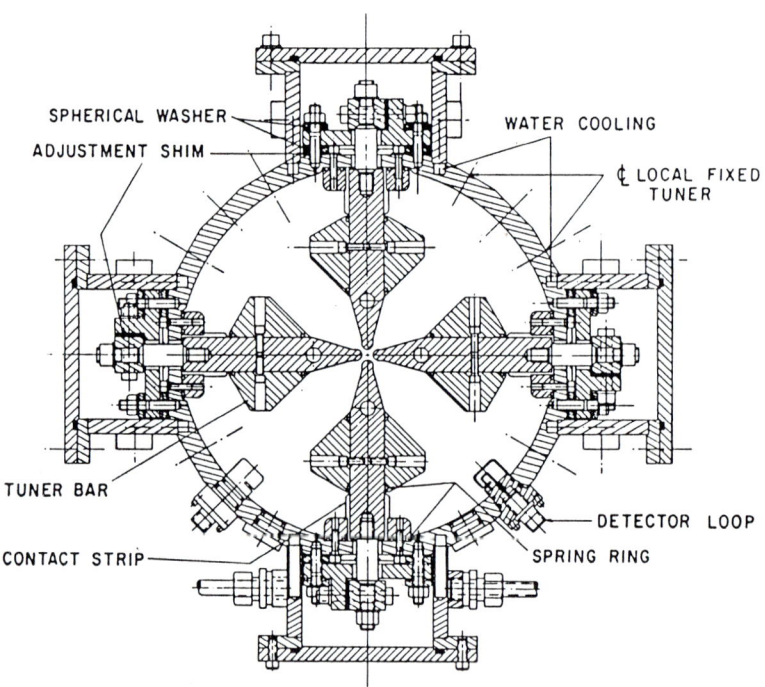

Fig. 10 - Symmetric cross section of the BNL RFQ.

The already mentioned "distributed" tuning eliminates the field distortions that a series of lumped tuners would certainly introduce.

Nevertheless the vane tips modulation, the unavoidable tuners at the end sections, the devices for feeding the power and many other mechanical complications always make very strong the spurious modes which are near the wanted TE_{210} .

Fig. 11 shows the electric lines of force between two adjacent pole tips of an RFQ for the quadrupole (TE_{211}) and dipole (TE_{111}) modes.

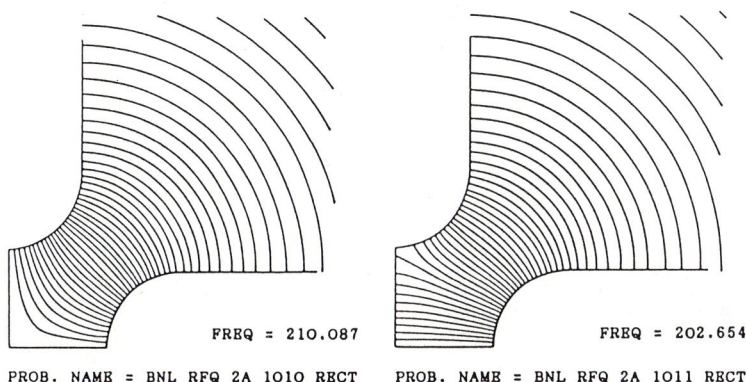

Fig. 11 - Lines of force the electric field between adjacent pole tips for the TE_{211} and TE_{111} mode.

In a uniform cavity the dipole mode is always below the quadrupole mode and the same should happen in a well balanced RFQ cavity. In this case as the whole cavity should be tuned just above the cut-off frequency of the guide corresponding to the vane section of the cavity then it follows that dipole mode results enhanced.

For this reason mode suppressing special techniques are requested.

7. DESIGN AND TECHNICAL CONSIDERATIONS

A full technical description of the machine, together with practical design considerations would go beyond the purposes of this seminar.

Nevertheless some of the problems concerning the whole RFQ will be illustrated in order to improve the general picture of the machine.

7.1 TUNING AND EXCITATION OF THE CAVITY

Fig. 12 shows an idealized section of an RFQ where the horizontal vanes have been removed for simplicity. From the drawing it is evident that the electrodes placed on the end sections load both the ends of each vane with an adjustable capacity towards ground that helps very much in balancing the vanes and tuning the whole cavity.

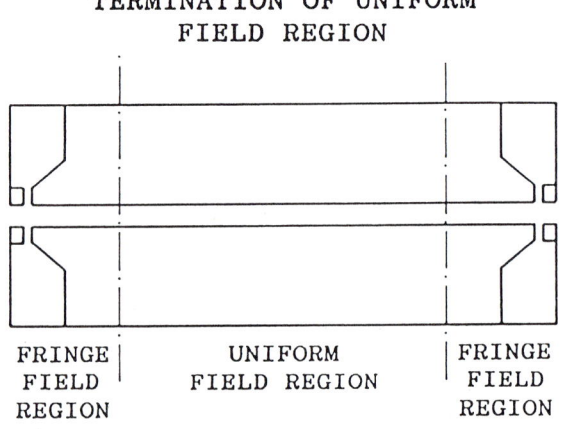

Fig. 12 - Schematic axial section of an RFQ.
The tuners placed on the end section are evidenced

Any cavity resonator can be fed in many ways. Electrodes capacitively coupled with the vanes and connected to the RF generator are not favoured because they tend to arch in case of temporary mismatch (poor vacuum, multipacting, detuning...).

The loop coupling, with one or more exciting loops placed near the end sections and coupled with the magnetic fields which exist between the vanes, is much more used.

Beside the "lumped" devices for coupling to the RF power source, many other "distributed" coupling methods can be used. These methods, which are well known and widely used in microwave techniques, have been used at Los Alamos since the beginning and later on were adopted in many other laboratories.

The solution proposed by Los Alamos can be described as follows.

A large portion of the RFQ cavity can be symmetrically inserted into a shorter cylindrical cavity and the new structure can be considered as a coaxial cable shorted at both ends, where the surface of the inner conductor coincides with the outer boundary of the RFQ cavity.

A coaxial cable shorted at both ends resonates, in a transverse electromagnetic mode, when its length is equal to half the free-space wavelength of the exciting RF field.

If this cavity is made exactly equal to $\lambda/2$ and some coupling slots are opened on the outer wall of the RFQ cavity, then the excitation of the coaxial resonator also excites the RFQ cavity.

This coaxial cavity that matches the RF power generator to the RFQ cavity is known as the coaxial manifold.

Fig. 13 - Example of coaxial manifold (Los. Alamos)

It should be noted that this technique allows to excite the RFQ cavity from many positions uniformly distributed along the outer wall of the machine. Moreover this distributed excitation is obtained without introducing electrodes in the regions between the vanes as shown in Fig. 13.

7.2 SUPPRESSION OF THE SPURIOUS MODES

As it was seen in the previous paragraph, an ideal resonant cavity can oscillate in an infinite number of modes. Actually an RFQ always exhibits a large number of strong resonances that very often are randomly bunched into very small intervals of frequency.

Those modes reduce the amount of power that could excite the wanted one (the TE_{211}) and, by distorting severely the field, impair the calculated beam dynamics. Particularly dangerous is the dipole mode already seen.

For the above reasons many useful devices have been invented in order to eliminate as many spurious modes as possible, at least in the neighbourhood of the working frequency.

Two different methods will be quoted here for giving an idea of the problem.

With the first method [4] the vanes of the same polarity are electrically connected with conducting rings as shown in Fig. 14.

It is interesting to note that the same technique was succesfully used at the dawn of the microwave tubes when the 8 resonant cavities of the magnetron were synchronized by connecting "with a conducting wire" the homologous edges of two adjacent cavities (the so called strapped magnetron).

The second method contemplates the insertion of loops coupled with both modes TE_{211} and TE_{111}. The loops are to be connected in such a way as to short circuit the TE_{111} mode while allowing the existence of the TE_{211} mote.

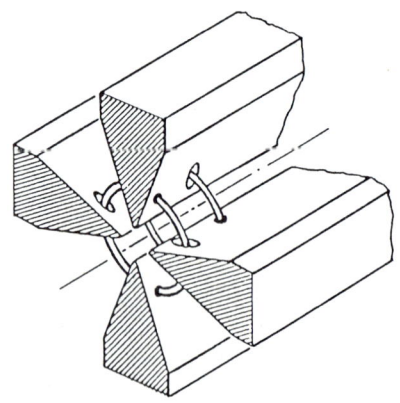

Fig. 14 - Technique for mode suppression (Berkeley)

A practical device based on this criterion was realized at BNL [5] where the RF power is fed to the RFQ through two groups of loops (4 for each group) placed inside the two end sections of the machine. The 8 loops are excited in parallel and are coupled to the H field that exists among the vanes. Selecting the proper orientation for each loop is possible to short circuit the TE_{111} mode.

Fig. 15 shows a picture of the power splitter connected with the 8 coaxial cables.

Fig. 15 - The power splitter used at BNL for feeding the RFQ from 8 places and simultaneously suppressing the TE_{111} mode.

7.3 DESIGN CONSIDERATIONS

The operating frequency is a very important design parameter. Since the cell length is equal to $\beta\lambda/2$, it follows that the higher is the frequency the shorter is the machine. On the other hand for very high frequencies the length of the cells becomes too short at the low energy end of the machine Moreover the working frequency determines the radius of the RFQ cavity and too low frequencies demand a very large diameter.

Another important parameter is the voltage between the adjacent vanes. As a general rule this voltage should be as high as possible obviously avoiding the risk of sparking.

If the ion species with their initial and final energies are specified, and f, the frequency, and intervane voltage are given, then the RFQ design is determined when the three independent functions $a(z)$, $m(z)$, $\phi(z)$ are given, where z is the axial distance along the accelerator. Two different ways for arriving at the above functions are indicated in [3].

The methods used at Los Alamos can be better understood with the aid of Fig. 16, which shows a functional block diagram of an RFQ where beside the acceleration the greatest attention was paid for limiting the growth of the radial emittance of the beam.

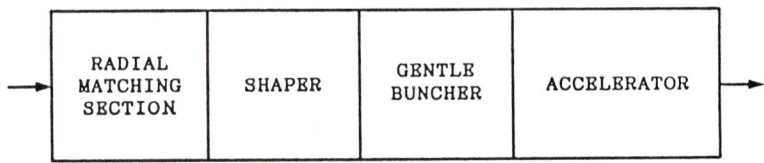

Fig. 16 - Functional diagram of an RFQ.

As it is indicated, the first section accomplishes the transition from a beam having time independent characteristics to one that has the proper variations with time (in this section the profile of the vanes is smooth). In order to obtain high capture efficency the bunching and the energy of the beam should be slowly varying functions of z. This is achieved in two different sections of the machine. Typically the quantity A increases very little in the shaper, while it undergoes a significant change in the gentle buncher. In the last section since the bunching is nearly completed both the synchronous phase and the value of A are held constant.

Fig. 17 shows the suggested variations of the parameters along the machine. It is important to recognize that when λ, ϕ and k are given than a and m are consequences of the assigned values for A and X.

Another important parameter is the maximum value of the electric field E_s that should be kept always below the sparking limit.

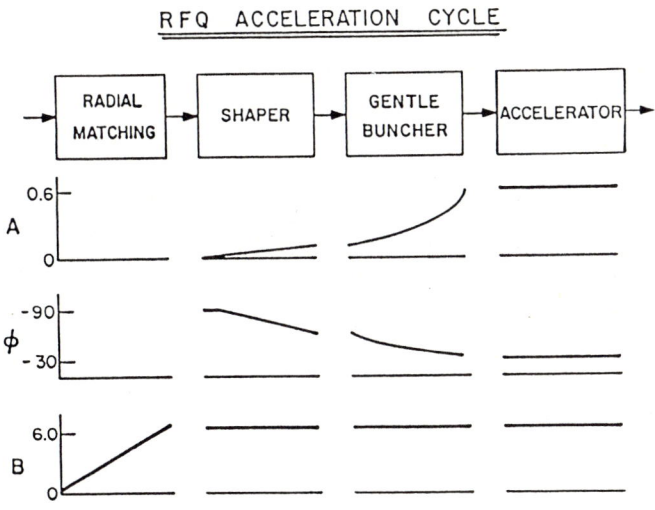

Fig. 17 - A possible choice for the function A(z), $\phi(z)$ and B(z).

Analytical and numerical calculations show that the maximum field E_s occurs around the middle of each separation.

A good approximation for E_s can be as follows

$$E_s = \alpha \frac{V_o}{r_o} \tag{44}$$

where α, the enhancing factor, is near 1.4 and obviously depends on the pole tip shaping.

If the operating wavelength λ, the normalized focusing force B and the maximum field E_s are assigned, then combining the equations (28), (30) and (44) we obtain the value for r_o

and V_o as follows [3]:

$$r_o = \frac{\lambda}{\alpha B} \sqrt{\frac{q\lambda E_s}{m_o c^2}} \quad ; \quad V_o = \frac{E_s \lambda}{\alpha} \sqrt{\frac{q\lambda E_s}{\alpha^2 B m_o c^2}} \quad (45)$$

If we call E_o the average wave of the amplitude of the electric field in each cell we can write:

$$E_o = \frac{AV_o}{\beta\lambda/2} = E_s \sqrt{\frac{2Aq\lambda E_s}{\alpha^2 B \beta m_o c^2}} \quad (46)$$

and this means that once the value of E_o is assigned (accordingly with the selected energy gain per cell) then the value of A is determined.

From Eq. 16 and taking into account that X and r_o are correlated (Eq. 30) we can calculate (numerically) the values for m and a.

Values for m equal to one at the low energy end and near to two at the high energy end normally produce a good compromise between acceleration and focusing efficiency.

8. THE ACTUAL DEVELOPMENTS

The RFQ described is a very complicated radio frequency resonator which has the purpose of creating the special RF fields capable of focussing and accelerating a beam.

Since the first proposal from Kapchinski and Tepliakov it was clear that any device capable to excite four suitably shaped electrodes could be used; the outstanding solution studied and realized at Los Alamos was succesfully adopted in many laboratories and real new developments lasted till the Santa Fe conference on "Particle Accelerators" in 1982.

The leading idea that was very simple can be summarized as follows [6]:

"A non uniform transmission line, made with four bars with circular cross section, can be used for creating the special field needed in an RFQ".

Fig. 18 shows a very simple and effective arrangement.

It is rather evident that each bar can be manufactured with a lathe, while for shaping the vane pole tips the very complicated and expensive tridimensional milling machine was mandatory.

Moreover the reciprocal position of the bars can be easily adjusted without interfering with the container that, on the other hand, can have a cross section independent of the working frequency.

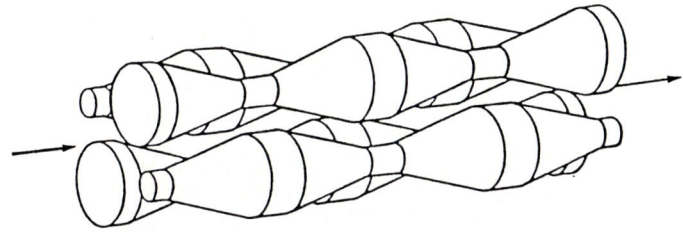

Fig. 18 - Transmission line formed with the four bars.
The indicated shaping produces the focusing and accelerating field (Frankfurt Univ.).

While the mechanical advantages obtained with the four bars are really enormous there are some doubts about the electrical efficency of this structure.

The surface offered by the four bars to the RF currents is always smaller than the one offered by the equivalent vanes and the corresponding four vanes RFQ exhibits a larger shunt impedance.

The choice of the best way for designing an RFQ cannot be decided on theoretical basis. Only the purpose for which each machine is designed can indicate what is more important; the mechanical simplicity or the RF power consumption.

Fig. 19 shows a sketch of the fundamental structure of an RFQ realized with the bars. Only two bars, of opposite polarity are shown for sake of clarity. The U-shaped support can be considered as a piece of uniform transmission line realized with two parallel metallic tapes.

One end of the transmission line (the bottom) is short circuited while the other end is loaded with the four bars that, as a first approximation, behave as a "distributed" capacity.

Let Z_o be the characteristic impedance of the transmission line and c the loading capacity of the bars. Then, if losses and radiation are neglected, the structure will exhibit an infinite impedance at the open end if the length of the support, the radian frequency ω and the loading capacitance obey the well known relation:

$$\frac{1}{\omega C} = Z_o \tan\left(\frac{\omega \ell}{v_f}\right) \qquad (47)$$

where V_f is the phase velocity that, in our case, can be set equal to the speed of light in vacuum.

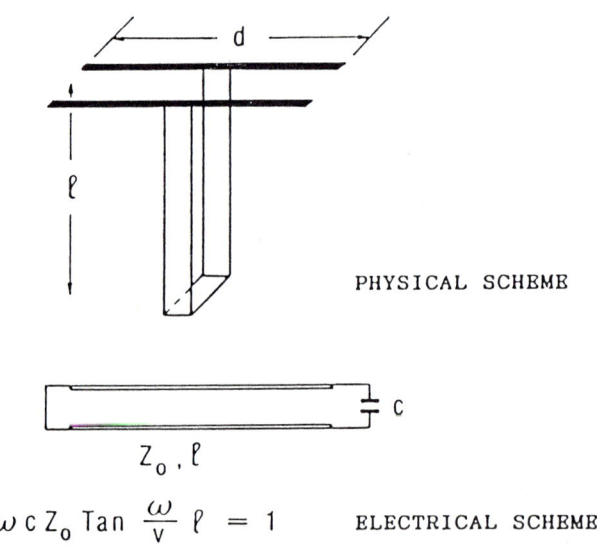

Fig. 19 - The resonant support (foreshortened quarter wavelength support).

If the above condition is verified then the metallic support do not perturb the bars. (the so called λ/4 support).

The whole structure of the machine can be realized by supporting, periodically, the bars with resonating supports.

From the first proposal many different resonating supports have been invented [7,8] and a large variety of devices have been tested.

It is important to note that each resonating support is magnetically coupled, at least, with the neighbouring one. Taking advantage of this situation it is possible to obtain that all the elements of the structure resonate in phase, independently of the physical length of the bars. Consequently the amplitude of the voltage which excites the bars is constant.

The four bars and the supports should be contained into an appropriate metallic tank in order to prevent radiation from the structure but, in this case, the container is not part of the fundamental structure as for the vane RFQ.

AKNOWLEDGEMENTS

The author is indebted with C. Rossi for his cooperation and to C. Guida and G. Bonaschi for editing the present work.

The drawings and the pictures obtained from my friends in Los Alamos and Brookhaven have been essential for this seminar.

REFERENCES

[1] I.M. Kapchinskii and V.A. Tepliakov. Linear ion accelerator with spatially homogeneous strong focusing.
Translated from: Pribory i Tekhnika eksperimenta N.2 pp. 19-22 March-April 1970.
Los Alamos Scientific Laboratory Collection of Papers on the Radiofrequency Quadrupole (RFQ) presented by accelerator technology division personnel March 79 to May 81.

[2] J.L. Laclare and A. Ropert. The Saclay RFQ
LNS. 063 1 June 1982. Laboratoire National Saturne.

[3] K.R. Crandall, R.H. Stokes and T.P. Wangler. RF quadrupole beam dynamics design studies.
1979 Linear Acc. Conference.
Los Alamos collection of papers .....

[4] S. Abbott et al., Lawrence Berkeley Lab. LBL 14624 (1982).

[5] S. Giordano and M. Puglisi. 1 x 8 coaxial manifold. Polarized proton technical note N.18 - BNL 1982.

[6] H. Klein - Development of the different RFQ accelerating structures and operation experience.
1983 Particle Acc. Conf. Santa Fe. IEE Trans. Nucl. Sci. August 1983.

[7] A. Schempp, H. Deitinghoff, M. Ferch, P. Junior and H. Klein. Four-rod $\lambda/2$ RFQ for light ion acceleration.
Eighth Conference on the application of accelerators in research and industry.
Denton, Texas, 12/14 Nov. 1984.

[8] S.O. Schriber. Present status of RFQ.
Los Alamos National Laboratory, At-Do Ms H811 - Los Alamos, NM 87545.

ANTIPROTONS TRAPPING AND COOLING

N. BEVERINI, F. SCURI and G. TORELLI
Dipartimento di Fisica, Universita' di Pisa
INFN, Sezione di Pisa

INTRODUCTION

The possibility to store antiprotons at subthermal energies in a small volume in the laboratory allows an interesting experimental insight in order to compare static properties of p and $\bar{p}$. The measurements on p are usually done in flight or by stopping the antiprotons in the matter; in both cases the antiproton can be observed only during a very short time. Conversely, the stored $\bar{p}$ can be observed for a longer time and, therefore, higher accuracies can be obtained in measuring static properties.

Obviously the possibility to store antiprotons become realizable only in presence of a suitable source of $\bar{p}$. In fact the first proposal for $\bar{p}$ trapping and cooling was presented in the context of the physics program for the the proposed LEAR facility at CERN [1].

1. - Trapping and cooling technique.

1.1 - General formulation

We first expose what we mean by " particles storing at rest in the laboratory " and we will later discuss the procedure proposed to achieve our goal .

A bunch of charged particles is stored in a small volume at rest in the laboratory when these particles are forced to move inside a finite volume. Magnetic or electrostatic forces alone are not able to satisfy these requirements (Maxwell equations), but a suitable combination of both forces can be used in the following way: a strong magnetic field confines the particles in the radial direction and an electric field confines the particles in the axial direction (see fig.1) [2].

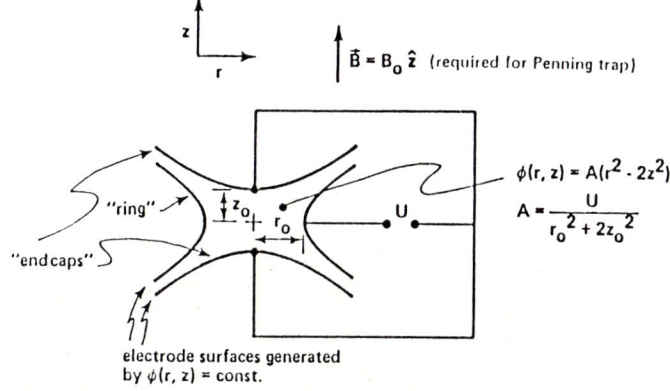

Fig.1

In this field configuration the residual individual energy of the stored particles is divided among three degrees of freedom per particle which are the oscillation along the z-axis due to the electrostatic confinement, the cyclotron motion in the magnetic field, and the magnetron motion in the crossed electric and magnetic fields.

The energy of the trapped particles is related to the value of the magnetic field B and to the potential difference beetwen the electrodes which generate the electric field; the stability of the motion can be obtained only if the cyclotron frequency ω_c is greater $\omega_z\sqrt{2}$, where ω_z is the frequency of the axial oscillation.

In a perfect harmonic trap the three degrees of freedom are completely independent, but unavoidable or artificial defects quickly mix the axial and the cyclotron motion which are thermal motions. The thermalization time constant depends on the particles density and is usually small compared to the mean-life of the particles in the trap. At constant trap conditions, the volume of the trapped bunch decreases as the energy of these thermal motions decreases.

Conversely, the magnetron motion is not thermal and it doesn't mix with the other two; the radius of the magnetron motion increases as its energy decreases.

If electrodes of proper shape (conjugate hyperboloids of revolution around the z-axis) are used, the electrostatic potential along the z-axis is parabolic and the axial oscillation is harmonic; in this case the frequencies of the three motions can be written as [3] :

$$\omega_z = [(4eV_0)/(m\,(r_0^2 + 2z_0^2))]^{1/2} \qquad (1)$$
$$\omega_c' = \omega_c/2 + [(\omega_c/2)^2 - \omega_z^2/2]^{1/2} \qquad (2)$$
$$\omega_m = \omega_c/2 - [(\omega_c/2)^2 - \omega_z^2/2]^{1/2} \qquad (3)$$

where m is the mass of the trapped particles, ω_m is the magnetron frequency, and the other symbols are defined in fig.1.

If $\omega_z \ll \omega_c$, expression (2) and (3) can be rewritten as :

$$\omega_c' \approx \omega_c \qquad (2')$$
$$\omega_m \approx \omega_z^2 / 2\,\omega_c \qquad (3')$$

1.2 Trapping technique

The proposed procedure to obtain a bunch of cooled $\bar{p}$ at rest in the laboratory is the following [4] :

i) A bunch of particles is extracted from LEAR;

ii) The extracted bunch is decelerated to a very low energy and send into the trap;

iii) The particles are captured when are travelling in the trap;

iv) Once trapped, the bunch is cooled at very low temperature.

This procedure is determined by the trapping technique. Since we cannot use a purely static field between the electrodes to capture the particles, the easier trapping method is to send the bunch at very low velocity component along the magnetic field direction into a trap through a small hole in the center of one grounded end-cup; the bunch travels the trap in the repulsive field generated by the second end-cup and, when the bunch is moving inside the trap, the potential of the first end-cup is abruptly rised up.

We can reasonably use a potential of a few kV on the trap electrodes; so we need incoming bunches with c.m. energy lower than a few keV and energy spread of the same order of magnitude in order to achieve a good capture efficiency.

To compare these energy values with the typical ones for a bunch extracted from LEAR we refer to the machine perfomancies after completation of the improvement program for LEAR started in the meatime; the lower extraction energy will be of the order of a few Mev, i.e. a factor 1000 greater than the maximum trapping energy in our case. The deceleration and the driving system of the beam will be improved, so that the energy spread of an extracted bunch and its volume in the longitudinal phase-space, the relevant quantity, will be as small as our trap can accept. In these conditions the bunch can be decelerated in a RF electrostatic device studied to preserve the energy spread, and a good capture efficiency (about 10% of the extracted bunch) can be achieved. On the other hand, if a thin foil absorber is used to decelerate the bunch, the energy and angular spread greatly increase, with a consequent dramatic reduction of the capture efficiency to about 10^{-4}, but the method is much less expensive. In both cases these relatively high value of capture efficiency can be obtained only using traps of length comparable with the bunch length (10-20 cm). The trap to be used is so much longer than those usually employed in high precision experiments.

1.3 Cooling technique

Once the particles are trapped and move up and down in the trap, they must be cooled by reducing the energy of their uncorrelated thermal motions, the cyclotron and the axial motions. Conversely, the magnetron motion energy must be increased to maintain the particles in the trap and to shrink the radius of the bunch.

First we discuss the cooling of the thermal motion. This energy can be transferred to the exterior only via the coupling between the moving particles and the trap electrodes. The energy transfer is efficient only if the particle signal induced on the electrodes can be recovered in the electronic noise, so that the particles motions must be almost harmonic.

There are essentially two cooling technique useful for the antiprotons. The first one, the dissipative cooling, has been already largely used to cool the cyclotron and the axial motions [2]. For the axial motion, the particles moving in the z-axis direction induce an alternate current between the two cup-electrodes, set at the same potential and connected by an high Q value electric circuit tuned to the axial frequency ω_z (see fig. 2). The kinetic energy can be so dissipated by the circuit resistance. For the cyclotron motion the same technique can be applied, cutting the ring electrode in two or four slides connected by a suitable resonant circuit.

Since the individual motions are uncorrelated, the cooling rate λ_c is inversely proportional to the square root of the number of the trapped particles. For standard configuration λ_c is also inversely proportional to the square of the electrodes distance $2z_0$. One has:

$$\lambda_c \approx (e^2 R) / 4 m (z_0^2 + 2 r_0^2) \qquad (4)$$

where e is the particle charge and R is the damping resistor. This dissipative technique seems to be most efficient to reach very low temperature, once the precooled bunch has been transferred into a small trap. According to (4), the cooling rate decreases with the particle mass and in the case of $\bar{p}$ the dissipative cooling technique can be improved by thermalizing the antiprotons bunch with an electron bunch on which the dissipative cooling acts more efficiently.

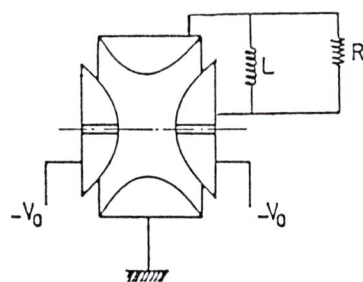 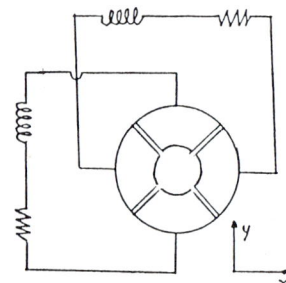

Fig. 2

The maximum cooling rate achievable with this technique is limited by the thermalization rate λ_t of the two particles bunch; in first approximation it seems possible to obtain a good value for λ_t with reasonable values of the electron density [5]. Recent calculations demonstrate that the value of λ_t is strongly reduced by the fact that two bunches of particles with similar energy and very different masses have a poor spatial overlap in a trap [6]. Experimental tests on this problem are in progress.

The second cooling technique, the stochastic cooling, is still under study and has been thought as an active method to reduce the uncoherent energy of the bunch [7].

We briefly remind the principle of the stochastic cooling in an accelerator. A pick-up measures the spread with respect to the mean value of a given variable (momentum, radial position, etc.) of a sample of the beam and a kicker further acts on the same sample to reduce the spread. The efficiency increases with the number of random samples which can be measured. During the process the samples randomization is reduced, so that, in ideal conditions, the cooling stops after a few turns. However, due to the sample mixing, the process goes on with reduced efficiency for a longer time. Other efficiency losses are due to the electronics noise measured by the pick-up to the mixing effect.

In this very simple description the cooling rate λ_c in an accelarator can be written as [8]:

$$\lambda_c = [2N_s / N T] \, g \, (A - g B) \tag{5}$$

where N_s is the samples number, N is the total number of particles in the accelerator, T is the revolution period, g is the gain of the pick-up signal amplifier and A and B are related to the mixing between the pick-up and the kicker and to the signal-to-noise power ratio respectively.

We would like to point out now the strict analogy between the motion of particles in a accelerator and the axial oscillation in a trap. The main difference is the envelope shape of the bunch: a thoroid with transverse dimension almost negligeble with respect to the longitudinal one in the case of a ring and almost spherical elissoid in a standard Penning trap. We will later examine the consequencies of this fact.

A second difference is that in each section of the trap the particles are moving in both direction and consequentely each sample is divided in two sub-samples of particles moving in opposite directions; the original sample is recomposed after a period and in the same position, except in the case of the trap center where the recombination is done after an half period. This last observation suggests to use a device as pick-up and kicker alternatevely; during a period this device is used as a pick-up giving the signal and after a proper delay the amplified signal is envoided to the same devise used now as a kicker. In our case this technique is applied to cool the axial motion of the particles. The r.m.s. value of the current induced by the particles on a special electrode is measured at center of the trap. For bunches of 10^6 particles a cooling time of few seconds has been estimated if a ten sample can be independently measured [7].

Since a pick-up of typical transverse dimension d and leght l measures the average spread of beam sample of lenght l+d , we need to satisfy the condition

$$l + d \ll z_0 \qquad (6)$$

to be able to measure many samples in the trap; this implies that we must use a very long cylindrical trap of small diameter. A cylindrical structure done of ring electrodes set to proper potential can reproduce a roughly quadratic distribution of the potential along the axis (see fig.3a). This trap structure should be efficient also for the dissipative cooling method, if the two central rings are coupled to the dissipative circuit.

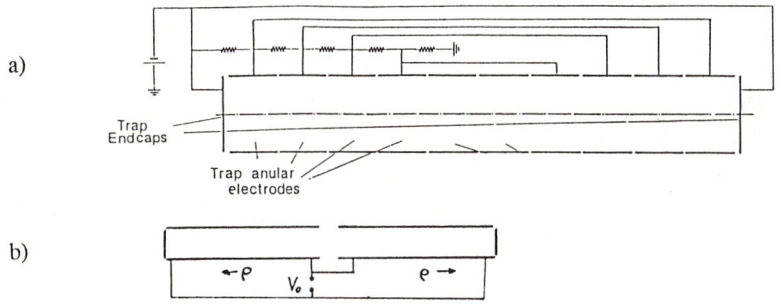

Fig.3

Another interesting structure can be composed by two tubes of insulating material coaxally aligned and closed each other (see fig.3b). If both tubes are internally coated with a layer of conductive material with resistivity linearly increasing from one side to the other, and the tube layers are set to proper voltage, an harmonic potential can be obtained along the axis.

On the basis of this simplified analysis we can conclude that the use of long cylindrical and harmonic traps can strongly increase the cooling rate for both dissipative and stochastic cooling. The technique for building such traps will be developed the next year.

As far as regard the magnetron motion, energy must be transferred at the frequency ω_m in order to reduce the magnetron radius r_m. The sideband excitation method is useful in this case [9]; an inhomogeneous electromagnetic field at frequency $\omega = \omega_z + \omega_m$ is excited in the trap volume;

the induced particles motion at frequency ω_z is damped by the z-motion cooling system, while the component at ω_m is absorbed in the magnetron motion with a consequent reduction of r_m. The efficiency of this technique has been demonstrated also for protons [10].

The above description of the cooling process is however not so encouraging; moreover our model was quite approximate. The space charge effect cannot be neglected; its effect is the shifting of the oscillation frequencies with consequent detuning from external circuit and broadening of the linewidth. More serious problems can arise from the collective effects, because the particle bunch interacts as a charge plasma if the value of the particles density is greater than a given value, let say 10^6-10^8 cm^{-3}; which is lower than the particles density in which we are interested [6].

2. Measurement of the gravitational acceleration of the anti-matter

Several line of research that are based on the use of trapped and cooled p; we list here few examples of these interesting measurements:

- annihilation rates on various channel from different states;
- antihydrogen formation and investigation of properties;
- inertial mass of the $\bar{p}$;
- gravitational acceleration of the $\bar{p}$.

We will limit to discuss the last point. The measurement of the gravitational acceleration of particles and antiparticles is a very sensitive test of some hypothesis of the Grand Unification Theory. This test is not a test of the matter-antimatter symmetry in the framework of CPT invariance as in the case of the other three forces of nature, but it is a direct check on the deviations of the gravity from the Newton's law.

In particular this measurement is sensitive to the presence of a scalar and a vector Yukawa component of the gravitational field, coupled to the barion number; its sensitivity is due to the fact that the vector component is attractive in the case of matter-antimatter interaction and repulsive in the case of matter-matter interaction.

The principle of the proposed experiment, which is a relative measurement, is very simple. A particles bunch is cooled at very low temperature (1-10 K); in these conditions a small fraction of the bunch has an average energy of the order of the gravitational potential energy difference for a drift lenght of 1 meter. At a given time an electrode of the trap is grounded in order to launch the bunch of cooled $\bar{p}$ or H$^-$ alternatively in the same vertical drift-tube. The arrival time distribution depends on the gravitational acceleration on the particles. The comparison of the distributions in the cases of $\bar{p}$ and H$^-$ allows to measure a difference in the gravitational acceleration.

This is experimentally possible if any other acceleration, due to stray forces, is reduced at a level lower than the gravitational one, which is equivalent to that produced on an elementary charge by an electrostatic field of 10^{-7} V/m. For this reason stray electric fields and magnetic gradients must be reduced; moreover the drift-tube surface has to be manufactured in order to reduce the patch-effect [12].

The contact potential between the trap and the drift-tube can be experimentally compensated, but not calculated. Conversely, the charge-image effect of both H$^-$ and $\bar{p}$ on the drift-tube walls is relevant, but can be calculated with high accurancy.

Other experimental difficulties are due to the fact that the particles to be counted have so low energy that, to be detected, they must be accelerated at the end of the drift-tube, and the accelerating fields can produce stray effects.

Finally the most difficult effect to be taken into account is the repulsive force between particles travelling in the drift-tube. The effect of this force is to reduce the number of particles in the low energy range which contribute to the tail of the time of flight distributions, these particles are the most sensitive to the gravitational effect to be measured.

The space-charge effect can be reduced by lowering the density of the trapped particles; this can be done either reducing the number of the trapped particles or increasing the bunch volume. In the first case the statistic is worsened while in the second case the uncertainty in the particles position is increased.

Finally the space-charge effect probably determines an intrinsic limit in the sensitivity of the proposed method and this limit can be lowered only lowering the bunch temperature to the level of few mK.

REFERENCES

1) G.Torelli - An Unusual Approach to the Low Energy Antiproton Physics - presented to the Karlsruhe Workshop 1979
2) H.G.Dehmelt - RF Spectroscopy of Stored Ions Part I: Storage - Advances in Atomic and Molecular Physics $\underline{3}$,53 (1967)
   D.S.Wineland and H.G.Dehmelt - Principles of the Stored Ions Calorimeter - J. of Appl. Phys. $\underline{46}$,919 (1975)
3) J.Birne and P.S.Farago - Proc. Phys. Soc. (London) $\underline{86}$,201 (1973)
4) N.Beverini et al. - CERN Proposal PSCC/83-14/P68
   N.Beverini et al. - CERN Proposal PSCC/86-2/P94
5) W.Kells, G.Gabrielse and K.Helmersin - On Achieving Cold Antiprotons in a Penning Trap - Fermilab-Conf-84/68-E8055.000
6) C.F.Driscoll - Containement of Single-Species Plasmas at Low Energies - presented to the Anti-Matter Facility Workshop, Univ. of Wisconsin - Madison, October 3-5, 1985
7) N.Beverini, L.Bracci, G.Torelli,V.Lagomarsino and G.Manuzio - Stochastic Cooling of Charged Particles in a Penning Trap - Europhys. Lett.,$\underline{1}$,435 (1986)
8) D.Mohl - Stochastic Cooling for Beginners - PS/LEA/Note 84-12-14 Dec.1984¨
9) R.S.VanDyck, P.B.Schwinberg and H.G.Dehmelt in "New Frontiers in High Energy Physics "- B.Kursunoglu,A.Perlmutter,L.F.Scott eds. Plenum Press. New York 1978
   D.J.Wineland - J. Appl. Phys.,$\underline{50}$,2528 (1979)
10) R.S.VanDyck Jr, P.B.Schwinberg and S.H.Bailey in "Atomic Masses and Fondamental Constant 6" - J.A.NolenJr,W.Benenson eds. Plenum Press. New York Pag 173
11) J.B.Jeffries, S.E.Barlov and G.H.Dunn - International J. of Mass Spectrometry and Ion Processes,$\underline{54}$,169 (1983)
12) F.C.Witteborn and W.N.Fairbank,Phys.Rev.Lett.,$\underline{19}$,1049 (1967)
    F.C.Witteborn and W.N.Fairbank.Nature,$\underline{220}$,436 (1968)

AN INTRODUCTION TO BEAM COOLING TECHNIQUES

R. CALABRESE

Dipartimento di Fisica dell'Università - Ferrara, Italy
Istituto Nazionale di Fisica Nucleare - Gruppo di Ferrara, Italy.

1. INTRODUCTION

In the past few years the development of beam cooling techniques allowed the construction of machines more and more sophisticated, and then the possibility to perform a new generation of experiments (it is enough to think at the proton-antiproton colliders). I discuss here several cooling techniques. In order to damp the particle oscillations, the electron machines make use of the energy losses due to synchrotron radiation. For the protons and heavier particles such a process is negligible. In this case we employ other techniques: the stochastic cooling, where one gets damping by means of a feedback system, and the electron cooling in which the damping is obtained in the interaction of the heavy particle beam with an electron beam.

2. GENERAL CONCEPTS

2.1. The aim of cooling

The "beam cooling" and "beam temperature" terms come from the kinetic theory of gases. As well known the temperature of a gas is correlated to the average square velocity of the molecules in the gas. In the same way the longitudinal or transverse beam temperature of a particle beam in a storage ring is correlated to the spread of the average square velocity of the particles in the beam. The cooling aim is to reduce both the spread and the size of the particle beam. In particular, with cooling techniques it is possible:

a) to accumulate rare particles (antiprotons); this has allowed the construction of a new generation of machines ($\bar{p}p$ collider);

b) to obtain high resolution beams to perform precision experiments, as the heavy quark spectroscopy.

Of course, beam cooling must be accomplished without beam losses. This is not a simple task, as the Liouville's theorem gives a constraint for the increase of the particle density.

2.2. Liouville's theorem

The canonical variables $\vec{q}$ (position) and $\vec{p}$ (momentum) play a fundamental role in the classical mechanics. A beam particle can be identified at any moment with a point in the 6-dimensional phase space $(\vec{q}, \vec{p})$. Then we can identify the beam, at a fixed time, with the volume that includes all the points $(\vec{q}, \vec{p})$ of the single particles, as shown in figure 1 for a 2-dimensional space.

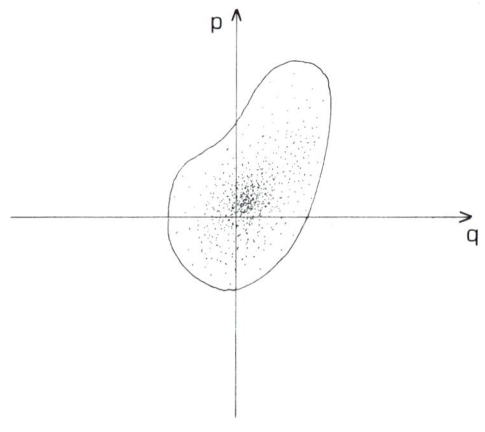

FIGURE 1
Liouville's theorem: the area $\int p dq$ is conserved

The Liouville's theorem[1] states that the volume in the phase space is constant if the particles move in an external magnetic field, or in a general external field whose forces do not depend on the velocity. According to Liouville's theorem, in order to enhance the phase space density one has to introduce dissipative forces in the system. A simple dissipative force is a "friction" force, directed against the velocity of each particle, which increases when the particle velocity goes up. For a particle moving on curved orbits this is the mechanism of emission of synchrotron radiation.

3. COOLING BY RADIATION DAMPING

It is well known that when a charged relativistic particle is accelerated in a force field it will radiate electromagnetic energy at a rate which depends on the angle between the force and the particle velocity. This rate is larger by the factor γ^2 when the force is perpendicular to the velocity than when it is parallel. If we consider a circular accelerator, or a storage ring, the rate of energy losses due to forces perpendicular to velocity is[2]

$$P_\gamma = \frac{c \cdot C_\gamma \cdot E^4}{2\pi\rho^2}$$

where E is the energy of the particle of mass m, ρ the local radius of curvature of the trajectory, and

$$C_\gamma = \frac{4}{3}\pi \frac{e^2}{(mc^2)^4}$$

For the electrons we have $C_\gamma = 8.8 \cdot 10^{-5}$ m/GeV^3, and the energy losses are important (17 KeV/turn for 1 GeV electrons and $\rho = 5$ m). The rate of energy losses increases when the electron velocity increases, and this mechanism causes a compression in the phase space volume. The damping coefficients for the energy and betatron oscillations are[3]

$$\alpha_\epsilon = \frac{\langle P_\gamma \rangle}{2E_0}(2+D), \quad \alpha_z = \frac{\langle P_\gamma \rangle}{2E_0}, \quad \alpha_x = \frac{\langle P_\gamma \rangle}{2E_0}(1-D)$$

where $\langle P_\gamma \rangle$ is the average rate of energy losses, E_0 the nominal energy of an electron circulating on the design orbit, and D is a quantity which depends on the constructive characteristics of the machine.

In the previous considerations we assumed that the energy loss is a continuous process. This is not true in that every phenomenon of electromagnetic radiation occurs by emission of quanta of energy. Each emission of a photon perturbs the electron trajectory, and the cumulative effects of such perturbations give rise to a noise which increases the oscillation amplitudes. The equilibrium condition is reached when the quantum excitation is balanced by the radiation damping. Let me point out that quantum excitation and radiation damping are not two different physics processes, the only physics process being the quantized emission of energy from the electron.

If we look at the equilibrium condition, we observe a Gaussian distribution of the energy and betatron oscillations. The radiation damping mechanism works very well for electron machines, but for the proton ones? One finds C_γ (proton)$= 7.8 \cdot 10^{-18}$ m/GeV^3, and this mechanism is negligible. We must use other techniques.

4. STOCHASTIC COOLING

At the beginning of the seventies S. Van der Meer[4] suggests the possibility to enhance the phase space density with an electronic feedback system which reduces the stochastic fluctuations of the particles. The system is sketched in figure 2. A beam pick-up detects the error signal. This signal is amplified and then applied to a correction system, the fast kicker, which damps the oscillations of the particles.

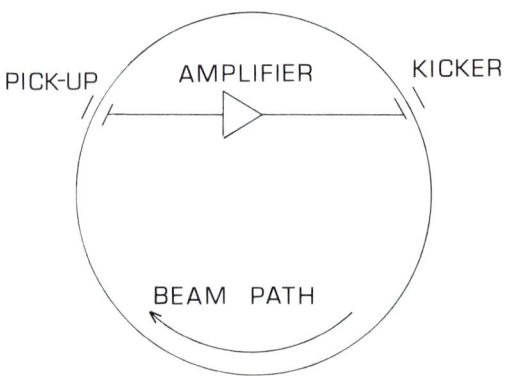

FIGURE 2
A sketch of the stochastic cooling system

But what about the Liouville's theorem? The Liouville's theorem is really applicable only when we consider continuous systems. On the contrary it is possible, with stochastic cooling, to "see" the granular structure of the phase space: the pick-up identifies the particles at the edge of the phase space area and then by means of the kicker we can redirect them to the centre of the area. In this case the Liouville's theorem has no influence.

Now let's turn our attention to the device in figure 2. We can think of the pick-up as constituted of two plates. When a particle passes through the pick-up, the error signal will be proportional to the difference between the two signals induced in the two plates. When the short error signal from the pick-up arrives at the kicker it is enlarged, owing to the bandwidth W of the low pass amplification system. The typical time length of the correction signal is $T_S = 1/2W$. Then a particle passing at the time t_0 will be subject to the kicks due to the particles passing during the time interval between $t_0 - T_S/2$ and $t_0 + T_S/2$. For this reason the particles passing at the time t, being

$$t_0 - T_S/2 < t < t_0 + T_S/2$$

belong to the same sample. If we compute the number of samples for an uniform beam of N particles, we have

$$L_S = T/T_S = 2WT$$

$$N_S = N/L_S = N/2WT$$

where T is the time revolution of a particle, L_s the number of samples per turn, and N_s the number of particles per sample.

Let's look at one particle, in the scheme of figure 2. The error on this particle, x_i, will become x_{ic} after the crossing of the kicker

$$x_{ic} = x_i - \lambda x_i - \sum_{\substack{\text{sample}\\j \neq i}} \lambda x_j$$

λx_i being the correction due to the particle itself (proportional to the error), and $\sum_{\substack{\text{sample}\\j \neq i}} \lambda x_j$ the contribution due to other particles of the sample.

Then there are two effects:
a) the correction signal of the particle on itself (this implies a cooling due to this coherent effect);
b) the signal of other particles of the sample on the test particle (this implies a random heating due to this incoherent effect).

The total system evolves in the competition between the coherent effect and the incoherent one.

By definition, the average sample error $\langle x \rangle_s$ is

$$\langle x \rangle_s = \frac{1}{N_s} \sum_{\text{sample}} x_j$$

Then for the corrected error x_{ic} we can write

$$x_{ic} = x_i - \lambda N_s \cdot \langle x \rangle_s = x_i - g \langle x \rangle_s$$

where we have defined the gain $g = \lambda N_s$, which depends on the electronic amplification and on the number of particles in the sample. We see that $g = (x_i - x_{ic})/\langle x \rangle_s$, i.e. it is the fractional correction per turn: the cooling system measures the error $\langle x \rangle_s$ on the sample and gives a correction $-g\langle x \rangle_s$ to all the particles of the sample. In order to obtain the cooling rate, we compute the square error on the sample, and its average over many turns. Finally, if we substitute all the sample averages by their expectation values for random samples, we obtain for the cooling rate[5]

$$1/\tau = W(2g - g^2)/N$$

In this relation the $2g$ term shows the contribution of the coherent effect, while the g^2 term shows the incoherent heating. Let me point out that the cooling time is linearly dependent on the number of particles, and this could be a limitation for very high intensity machines.

Until now we have considered an ideal system, but we must take into account the electronic noise of the amplification system. This noise is not negligible in that the pick-up signals often are lower than the amplifier noise. We can think of the noise effect as an equivalent pick-up signal x_n which is seen by the amplifier, as shown in figure 3.

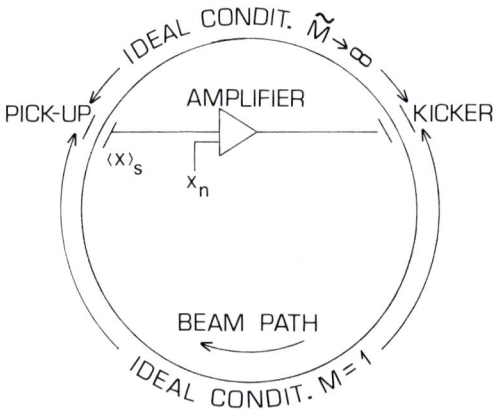

FIGURE 3

Stochastic cooling system including noise and mixing effects

Then the error x_{ic} after a "kick" will be

$$x_{ic} = x_i - g\langle x\rangle_s - gx_n$$

If we consider the noise uncorrelated with the particles, the cooling rate is modified as follows[6]

$$1/\tau = W(2g - g^2 \cdot (1+U))/N$$

where U is the noise to signal ratio.

Another important concept for the stochastic cooling system is the mixing. If we cool the same particles of a sample, for full correction the quantity $\langle x\rangle_s$ will be zero, and then cooling will stop. What really happens is a mixing of the sample population. Owing to the energy spread of the particles, the sample population changes, the sample error is not zero, and the cooling proceeds. If the mixing is so fast that from the kicker to the pick-up there is a complete randomization of the population of the samples, our previous valutation of the cooling rate is correct (disregarding the mixing in the path between the pick-up and the kicker). But if the mixing is not so fast there is an increase of the cooling time. We must take into account another effect, i.e. the mixing between the pick-up and the kicker. Obviously, in this region we want no mixing at all, and what happens in the real situation is another increase of the cooling time. Taking into account both mixing effects, we obtain

$$1/\tau = W(2g(1-\widetilde{M}^{-2})-g^2(M+U))/N$$

where M is related to the mixing between kicker and pick-up (ideal condition, i.e. complete mixing, M=1), while $\widetilde{M}$ is related to the mixing between pick-up and kicker (ideal condition, i.e. no mixing, $\widetilde{M}\to\infty$), as shown in figure 3. Figure 4 shows the coherent and incoherent terms of the cooling rate.

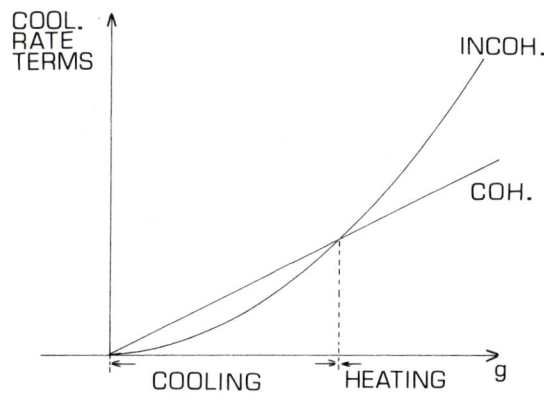

FIGURE 4

Coherent and incoherent effects
in the cooling rate evaluation

We observe that when the cooling proceeds the quantities U and M become greater and greater. Then the cooling can stop when:
a) the energy distribution is so small that the mixing is very slow (M is a great number);
b) the pick-up signal is so small that the noise to signal ratio U is very large.

Let me point out that even in such conditions it is possible to cool the beam, if we suitably decrease g during cooling in order to work in the best conditions, $g=(1-\widetilde{M}^{-2})/(M+U)$. But this implies a long cooling time, $\tau \simeq N\cdot(M+U)/W\cdot(1-\widetilde{M}^{-2})^2$, and the cooling stops definitively when τ is equal to the beam decay constant.

In conclusion, the main characteristics of stochastic cooling are the strong dependence on the number of particles and the weak dependence on the energy.

The typical momentum resolution of a beam with stochastic cooling is $\Delta p/p \approx 10^{-3}$. If we want to perform precision experiments, as the charmonium and bottonium spectroscopy, we need high resolution beams, with $\Delta p/p \approx 10^{-5}$, in order to measure the width of these states with a resolution of about 100 KeV. The stochastic cooling is clearly not sufficient. The Liouville's theorem suggests us that we must search for a system with dissipative forces.

5. ELECTRON COOLING

The electron cooling was suggested by G. Budker[7] in 1966. Let us consider the interaction between an intense "cold" beam of electrons and a beam of protons (or heavier particles) with the same velocity. The electron beam is "cold" in the sense that is has a small momentum spread ($\Delta p/p \approx 10^{-3}$). During the interaction between the two beams the electrons "take away" the proton oscillations, owing to the Coulomb scattering and to the low mass ratio m_e/m_p. The ideal condition of thermodynamic equilibrium is reached when the proton temperature in the rest-frame system of beams becomes equal to the electron temperature, $T_p = T_e$. But this implies for the angular spreads Θ_p (protons) and Θ_e (electrons)

$$\Theta_p = \sqrt{\frac{m_e}{m_p}} \, \Theta_e$$

If $\Theta_e \approx 10^{-3}$ we can obtain $\Theta_p \approx 2 \cdot 10^{-5}$. Hence, as a result of the beam-beam interaction, the phase space volume of the proton beam decreases.

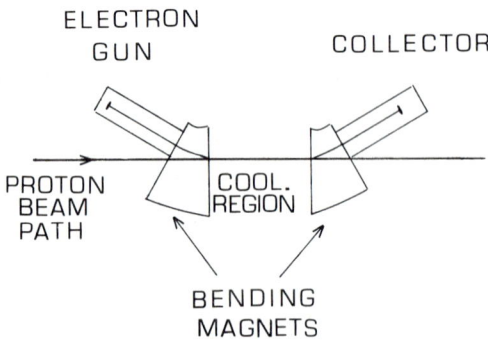

FIGURE 5
A sketch of the electron cooling system

Let's turn our attention to the experimental features of the electron cooling systems. In a right section of a proton storage ring we introduce, by means of a magnetic system, an electron beam with the same velocity of the proton one, as shown in figure 5. In this cooling region the beams are parallel and can interact, as discussed. The electron beam is then deflected, decelerated and collected, in order to achieve the best energy recovery. This recovery is fundamental, as the power involved in the electron beam can be of several Megawatts. It is necessary a particular care both in the design and in the construction of the electron gun, because the electron beam temperature must be as low as possible. Hence a suitable beam optics[8] and a magnetic guide field are needed, in order to compensate the space charge effects.

With the methods of plasma physics it is possible to compute the "friction" force in the electron cooling process. One finds[9]

$$\vec{F} = \frac{4\pi e^4 L_p n_e}{m_e} \int d^3 v_e \cdot \frac{(\vec{v}_p - \vec{v}_e)}{|\vec{v}_p - \vec{v}_e|^3} \cdot f(\vec{v}_e)$$

where n_e is the electron density, L_p the Coulomb logarithm, $\vec{v}_e$, $\vec{v}_p$ the electron and proton velocities and $f(\vec{v}_e)$ the electron velocity distribution. Following the Liouville's theorem, this dissipative force is responsible for the compression of the phase space volume of the proton beam. If we consider an uniform Maxwellian distribution, disregarding the longitudinal magnetic field used to guide the electrons, we obtain the cooling time in the laboratory frame[10]

$$\tau = \frac{3}{2 \cdot \sqrt{2\pi}} \cdot \frac{e}{j_e r_p r_e n L_p} \cdot \begin{cases} \beta^4 \gamma^5 \theta^3_p & \text{if } \theta_p > \theta_e \\ \beta \gamma^2 \left(\frac{T_e}{m_e c^2}\right)^{3/2} & \text{if } \theta_p < \theta_e \end{cases}$$

where j_e is the electron current density, r_e, r_p the classical electron and proton radii, η the ratio between the cooling region lenght and the proton ring circumference. There are two different conditions:
a) the proton beam is hot ($\theta_e < \theta_p$)
b) the proton beam is precooled ($\theta_p < \theta_e$): in this case it is easier to cool because τ is lower than in the case a).

A fundamental characteristic of the electron cooling is the strong dependence on the energy (τ proportional to $\beta \gamma^2$ in the best case). This is a strong limitation for the application of this method to high energy accelerators. The electron cooling can work very well for low and medium energy beams ($E \leq 10$ GeV). A more detailed analysis of an electron cooling system is given in this volume[11]. In the table 1 I show the characteristics for an electron cooling device[12] in construction for the Low Energy Antiproton Ring (LEAR) at CERN, covering the momentum range 0.6-2.0 GeV/c.

Electron max energy	700 KeV
Electron max current	3.5 Amp.
Electron beam diameter	3 cm.
Vacuum in the cooling region	10^{-12} Torr.
Magnetic guide field	3 KG
Power involved	2.5 MW
T_e	2 eV
Rate of energy losses	$\leq$ 10 KW
τ(0.6 GeV/c) without precooling	$\simeq$2 min.
τ(0.6 GeV/c) with precooling	$\simeq$2 sec.
$\Delta p/p$ for the proton beam	$\simeq 2 \cdot 10^{-5}$

TABLE 1

Characteristics for the 0.6-2.0 GeV/c LEAR electron cooling device

6. CONCLUSION

For the hadron physics at intermediate energy it is important the construction of high intensity proton machines and the use of the $\bar{p}p$ collider technique. In fact it is well known that in the e^+e^- machines the particles with $JP \neq 1^-$ are produced by radiative decay of vector meson of higher mass, with the typical resolution ($\simeq$10 MeV) of the detector. If we want to perform a high resolution heavy quark spectroscopy, which is a fundamental test of QCD[13], we must use the $\bar{p}p$ collisions. In this case the states with $JP \neq 1^-$ are directly produced, hence it is possible to exploit the resolution of p, $\bar{p}$ beams to measure the width of these states. For the realization of this physics program it is important to have very high resolution beams ($\Delta p/p \simeq 10^{-5}$). We have previously seen that the stochastic and electron cooling are two complementary techniques. The best scheme for cooling at intermediate energy seems to be the use of the stochastic method as precooling; this will prepare the beam to the final cooling, which is performed by electron cooling. In this scheme, we can reach a beam resolution of 100 KeV. With such a resolution there will be many new possibilities in the field of hadron physics at intermediate energies.

REFERENCES

1) V. Arnold, Methodes Mathematiques de la Mecanique Classique (Ed. MIR, Moscou, 1976).

2) J. Jackson, Classical Electrodynamics (John Wiley, New York, 1962).

3) M. Sands, The Physics of Electron Storage Rings, SLAC-121 UC-28 (1970).

4) S. Van der Meer, Stochastic damping of betatron oscillation in the ISR, CERN/ISR-PO/72-31 (1972).

5) D. Mohl et al., Phys. Rep., 58 (1980) 73.

6) D. Mohl, Stochastic cooling for beginners, in: Cern Accel. School, CERN 84-15 (1984), pp. 97-161.

7) G. Budker, Atomnaya Energiya, 22 (1967) 346.

8) U. Bizzarri et al., Nuovo Cimento 88A (1985) 161.

9) Y. Derbenev, A. Skrinsky, The Kinetics of electron cooling of beams in heavy particle storage devices, CERN 69-18 (1969).

10) VAPP - NAP Group, CERN 77-08 (1977).

11) L. Tecchio, this volume.

12) M. Biagini et al., IEEE Trans. Nucl. Sci., NS-32 (1985) 2409.

13) P. Dalpiaz, Quantitative test of QCD with high intensity cooled $\bar{p}$ beams, in: Nuclear and Particle Physics at Intermediate Energies with Hadrons (Ed. Compositori, Bologna, 1986) pp. 89-96.

ELECTRON COOLING FOR $\bar{p}$-p MACHINES AT INTERMEDIATE ENERGIES

Luigi TECCHIO

Istituto di Fisica Superiore, Università di Torino
Istituto Nazionale di Fisica Nucleare, Sezione di Torino

1. INTRODUCTION

In a project for a particle accelerator or an accumulator ring, experimental physicists requirement's are: a long beam lifetime, an high beam intensity, an high luminosity and a good energy resolution. All these requirements are normally in contrast, and then we need to look for a compromise that allows to reach equilibrium stable conditions optimizing all the required features.

In general, the beam qualities (lifetime, energy resolution,..) are represented in terms of beam temperature. Being $\varepsilon_H, \varepsilon_V$ the horizontal and vertical emittances, β_H, β_V the betatron function values and $\Delta p/p$ the momentum spread we can write the beam horizontal and vertical divergences

$$\theta_H = (\varepsilon_H/\pi\beta_H)^{\frac{1}{2}}$$

$$\theta_V = (\varepsilon_V/\pi\beta_V)^{\frac{1}{2}}$$

which are used to define the longitudinal and trasversal temperatures:

$$T_{//} = m\langle v_{//}^2\rangle = mc^2\beta^2(\Delta p/p)^2$$

$$T_\perp = m\langle v_\perp^2\rangle = mc^2\beta^2\gamma^2\langle \theta_H^2+\theta_V^2\rangle$$

where

$$T = T_{//} + T_\perp$$

represents the total beam temperature.

Furthemore the beam stability condition (Keil-Snell criteriom) gives for the beam current[1]:

$$I < KF' \frac{\beta^2 \gamma |\eta|}{|\frac{Z}{n}|} (\frac{\Delta p}{p})^2$$

where:
$K = 4 \frac{E_o}{e} = 3.753 \times 10^9 \ \{V\}$
$F' = 0.51 \div 1$ form factor depending on particles distribution
$|\frac{Z}{n}|$ machine impedence
$\eta = \frac{1}{\gamma^2} - \frac{1}{\gamma_{tr}^2}$ revolution frequency spread per unit of $\Delta p/p$
γ_{tr} transition energy.

In this way, we can observe a direct contrast between the requirement of an high intensity beam (I) and a good resolution ($\Delta p/p$); in other word that means a contrast between high intensity/luminosity and low beam temperature.

Different methods can be used to obtain a compromise and match at best all parameters in order to reach good machine features.

2. PROTON, ANTIPROTON MACHINE.

Let me now to discuss about the different techniques to be used in a $\bar{p},p$ machine in order to increase the intensity/luminosity and improve the experimental resolution ($\Delta p/p$). These techniques are usually called beam cooling techniques, in fact they act in order to reduce the beam temperature and thus cool the particles. The accelerator physicist know two kind of complementary cooling techniques: stochastic cooling and electron cooling (see ref.2). At the intermediate energies application of both techniques is suggested, in fact they are complementary as one can deduce form table 1.

Since stochastic cooling works better for large beam divergences it can be applied at beam injection; electron cooling can be used in a second time in order to further reduce the beam divergences and obtain as low temperatures as possible. Unfortunately, electron cooling is strongly dependent from the beam energy, in fact the cooling time is proportional to $\beta^4 \gamma^5$. This fact require to inject the beam at low energy, in order to obtain a cooling time shorter as possible. A possible working scheme for a $\bar{p}/p$ machine at intermediate

TABLE 1. Comparison between stochastic and electron cooling.

	Stochastic Cooling	Electron Cooling
Cooling time	proportional to the number of particles $N(10^9/s)$ and to the band width W	Don't depend from N and W
Beam divergences	Work better for large divergences θ	Work better for small divergences θ
Energy	Cooling time Don't depend from energy	Cooling time is proportional to $\beta^4 \gamma^5$

energy is described as follow: the $\bar{p}/p$ beam is injected at relatively low momentum (few GeV/c) in the multiturn mode, a sequence of batches of small emittance will appear, till the total number of $\bar{p}/p$ will be obtained.

Whilst ring filling is taking place, stochastic and electron cooling are being set up. After this operation (precooling), the $\bar{p}/p$ beam can be accelerated/decelerated at the working energy, practically without changing the temperature. In order to compensate beam broadening, due to space charge effect or to beam-beam interaction, electron cooling must be kept on work. Being the beam initially cooled, in this condition the cooling time is quite shorter[2]. This kind of working scheme can be eventually applied to an antiproton storage ring like LEAR[3] or to a $\bar{p}$-p collider machine like Super LEAR[4,5] or EHF facility[6] in order to improve the beam resolution and increase the luminosity, which allow to perform a fine charmonium and bottomonium spectroscopy[7].

I would like to discuss here the advantages of using electron cooling in an intermediate energy machine (2-10 GeV/c) that could operate with internal jet target and in $\bar{p}p$ collider mode.

3. COOLING TIME EVALUATION.

Before beginning the discussion about the electron cooling performances we need to evaluate the cooling time that we can obtain with a typical cooling device. The cooling times are given by[2]:

$$\tau = \begin{cases} \dfrac{1}{2\pi} \dfrac{e\beta^4\gamma^5}{r_e r_p \eta J_e L} \theta_p^3 & \theta_p > \theta_e \\[2ex] \dfrac{1}{2} \dfrac{e\beta\gamma^2}{r_e r_p \eta J_e L} \left(\dfrac{T_e}{M_e C^2}\right)^{3/2} & \theta_p < \theta_e \end{cases}$$

where

r_e, r_p classical radii of electron and proton respectively
J_e electron current density (typical 5×10^3 A/m^2)
$L=20$ Coulombian logarithm
T_e electron temperature
$M_e C^2$ electron mass
η ratio between the cooling region length and the machine circumference

Typical cooling time are ranging from 5 and 60 minutes for beam energies from 2 to 10 GeV/c; as foreseen for the machines above discussed. The cooling time represent the necessary time to reach the thermalization of both beams:

$$T_{\bar{p},p} = T_e$$

or

$$\theta_{\bar{p},p} = \left(\dfrac{m_e}{m_p}\right)^{1/2} \theta_e$$

For high intensity beams the thermalization process is limited by other process that play a destructive role on beam stability.

4. EFFECTS LIMITING THE COOLING.

Theoretically, the electron cooling is a very powerfull method to improve very much the $\bar{p}/p$ beam qualities. Unfortunately, these improvements are in part limited by two effects:
a) the e-$\bar{p}$,p tune shift. The $\bar{p}$,p tune shift due to co-rotating electrons in the cooling region (representing the space charge

effect induced by the electrons on the $\bar{p},p$ beam) is determined by the intensity and size of the electron beam

$$\Delta\nu_{\bar{p},p} = \frac{J_e \ell R r_p}{2Q_v \ell C \beta^3 \gamma^2}$$

where:

- J_e electron current density
- ℓ cooling region length
- R mean radius of the storage ring
- Q_v number of betatron oscillations per revolution in the vertical plane

An evaluation of $\Delta\nu_{\bar{p},p}$ gives values of the order of 10^{-4}-10^{-5} that are negligible compared with other instability effects or with the beam-beam tune shift for the collider case ($\Delta\nu=5\times10^{-3}$);

b) the Laslett space charge tune shift. The application of electron cooling imply a phase space improvement, with a consequentely increase of the beam space charge. The Laslett space charge tune shift of each $\bar{p},p$ beam imposes a limitation on beam density/luminosity and depends on the main parameters of the machine and on the beam cross-section. The value of this tune shift (ΔQ) can be calculated in two ways, depending on the operarion mode of the machine:

- coasting beam

$$\Delta Q = \frac{N_{\bar{p}p} r_p R}{2\pi\beta^2 \gamma Q_v \sigma^2} \left(\frac{1}{\gamma^2} - f\right)$$

- bunched beam

$$\Delta Q = \frac{N_{\bar{p}p} r_p R}{2\pi\beta^2 \gamma Q_v \sigma^2} \left(\frac{1}{\gamma^2} - f\right) \frac{1}{B_F}$$

where

- σ beam radius
- B_F bunching factor (ratio between the bunch length and the machine circumference)

f   neutralisation factor (usually set equal to zero)

These operation modes will be separately analyzed in the next paragraphs.

5. JET TARGET OPERATION.

Let me now to analyse the case in which machine is operated with an internal gas jet target and a coasting beam. Being the electron cooling a method to increase $\bar{p}$-beam density, a direct consequence of this fact is the possibility to use a thin jet as target. From an experimental point of view that means a better vertex reconstruction and an improvement of the vacuum in the target region. Furthemore, the electron cooling is a very efficient method to improve the momentum resolution ($\Delta p/p \sim 2 \times 10^{-5} - 10^{-4}$) and can be also used to counteract the multiple scattering in the target, increasing in such a way the beam lifetime. The Laslett space charge effect limits in part the advantages of using the electron cooling method, but a stable operation mode of the machine can still be obtained with very satisfactory conditions ($N_{\bar{p}}=10^{12}$ particles, $\sigma=1$mm $\Delta Q \sim 10^{-3}$, $L=10^{32}$ cm^{-2}s^{-1}, lifetime $\sim$ 1 day).

6. $\bar{p}p$ COLLIDER OPERATION.

When a machine as those said above operate as collider the electron cooling can be used as a method to improve the resolution and increase the luminosity. I would like to discuss in detail how electron cooling can be used in order to improve a collider luminosity. In a collider machine the tune shift induced over one beam (i.e. $\bar{p}$ beam) by the second beam (i.e. p beam) is determined, at fixed intensity, by the size of the second one, as given by the formula:

$$\Delta \nu_{\bar{p}} \sim r_p \frac{N_p \beta_v}{A_{int}} \left(\frac{1+\beta^{-2}}{\gamma}\right)$$

The interaction area (A_{int}) is given by the effective width and height of the larger of the two beams:

$$A_{int} = \frac{\pi}{4} a_H \cdot a_v$$

The electron cooling permits to reach small transversal beam sizes ($\Delta v_{\bar{p}}$ increases) and, in the same time, avoids a rapid beam decay due to beam-beam interaction. Being the luminosity

$$L = \frac{N_{\bar{p}p} f_{rev} \gamma \beta \Delta v_{\bar{p}}}{(1+\beta^{-2}) r_p \beta_v}$$

directly proportional to $\Delta v_{\bar{p}}$, an increase of the beam-beam tune shift means an increase of the same factor of the luminosity.

This gain factor is strongly depended from the Laslett tune shift of par.4. If the machine is working with a bunched beam the gain factor is relatively low, in fact the space charge effect for high intensity beams is dominant. In order to improve the luminosity it is convenient to operate the collider machine with two separated rings, where a coasting beam solution is possible. In this case the distructive effect of the beam space charge on the electron cooling is rather limited. A gain of till a factor 7 in luminosity ($L > 10^{31} cm^{-2} s^{-1}$) seems possible, and then the momentum resolution ($\Delta p/p$) could range from 2×10^{-5} to 10^{-4}.

7. CONCLUSIONS.

The advantages in using the electron cooling technique as complementary to stochastic cooling in the $\bar{p},p$ machine at intermediate energies have been analyzed. The electron cooling seems the right technique to be used for high precision particle spectroscopy. Unfortunately this technique is quite limited at the low and intermediate energies, but is just in this energy range that more experimental precision is required. At present many machines use this technique and in particular way it is used in the heavy ions machines where the electron cooling allow to store more particles that usual with a long lifetime.

REFERENCES.

1) G. Guignard; Selection of formulae concerning proton storage rings, CERN 77-10, pag.86.

2) R. Calabrese; Lecture published in this book.

3) M. Biagini, U. Bizzarri et al.; Possibility for high energy electron cooling in LEAR, Proceedings on Physics with antiprotons at LEAR in the ACOL era, ed. U. Gastaldi, R. Klapisch, J. M. Richard, J. Tran Thanh Van, Editions Frontières (1985) p.135.

4) D. Möhl et al.; A Superconducting low energy antiproton ring (Super LEAR), Proceedings on Physics with antiprotons at LEAR in the ACOL era, ed. U. Gastaldi, R. Klapisch, J. M. Richard, J. Tran Thanh Van, Editions Frontières (1985), p.83.

5) L. Tecchio; Electron cooling at intermediate energy, Proceedings on Physics with antiprotons at LEAR in the ACOL era, ed. U. Gastaldi, R. Klapisch, J.M. Richard, J.M. Richard, J. Tran Thanh Van, Editions Frontières (1985), p.167.

6) F. Bradamante; Conceptual design of the EHF, INFN/AE-86/7.

7) P. Dalpiaz; lecture published in this book.

CHERENKOV PICK-UPS IN THE MICROWAVE BAND

Giuseppe DI MASSA

Dipartimeno Elettrico, Universita' della Calabria, Arcavata di Rende, 87100 Cosenza, Italy*

and

Vittorio G. VACCARO

Dipartimento di Fisica, Universita' di Napoli, Mostra d'Oltremare Pad. 20, 80125 Napoli, Italy

1. INTRODUCTION

Experimental physics of elementary particles need beams of high density in the phase space. The beams produced from the sources, and in particular the antiproton beams, do not have these prerequisites. Stochastic cooling is a procedure to circumvent these difficulties compacting the particles in the phase space.

The primary devices for the stochastic cooling are the sensor or Pick-Up (PU) which detects the position of the particle to be corrected, the amplifier which amplifies the signal from the PU and feeds the Kicker (k) which is the reciproque of the PU and corrects the particle position.

This system could work instantaneously if the transfer function of the pick-up-amplifier-kicker system were a delta function, namely had an infinite bandwidth.

In general the transfer function has a characteristic time, which can be small but finite and is related to the bandwidth by the following equation:

$$T = \frac{1}{2W} \qquad (1)$$

where W is the bandwidth of the system. As a consequence of this behaviour the system not only corrects the particle in examen, but also disturbs in a certain amount, those which are distant in time T_s. These particles must be corrected and an average correction is obtained over a more or less long period.

In first approximation the spread decays exponentially with a characteristic time, proportional to T_s. As a conclusion, to enlarge the bandwidth means to reduce the cooling time.

It is possible to reach this goal encreasing the central frequency of the pick-up (and of the kicker) making use of the Cherenkov radiation in the microwave range.

---

* Permanent address:
  Dipartimento di Elettronica
  Via Claudio 21, 80125 Napoli. Italy

2. THE CHERENKOV RADIATION IN THE MICROWAVE RANGE

The phenomenon of Cherenkov radiation is a well known one, but we want to treat it in the microwave range, in a qualitative way.

A charged particle of velocity β is a potential source of electromagnetic waves which has an infinite and uniform spectrum. It is well known that if it is travelling in an infinite dielectric medium of index of rifraction n(f), function of the frequency f, we get an electromagnetic radiation for all those frequencies for which the following relation is satisfied:

$$n(f) \beta > 1 \qquad (2)$$

So we have conical waves which have an angle θ with respect to the axis of the motion for which

$$\cos\theta = \frac{1}{n(f)\beta} \qquad (3)$$

We wonder now what would be the behaviour of the electromagnetic radiation in the case that the dielectric has a cylindrical indefinite hole of radius d parallel to the motion. This hole is necessary for the passage of intense beams. As a first guess, if the hole is sufficiently small it is an ininfluent perturbation for the Cherenkov Radiation. Now we must refine the concept of "small or large hole". The particle "sees" the dielectronic by means of its harmonic contents for those wavelengths such that:

$$\lambda \geq d \qquad (4)$$

So that Cherenkov radiation occurs only below certain frequencies (low pass filter behaviour). So that, if the radius d is of the order of centimeters, we expect the radiation at frequencies of GHz.

We understand also that radiation in the optical range is possible only if the dimension of the hole is of the order of microns.

If in addition the dielectric is limited and surrounded by an infinite metallic cylinder (see fig.1), becoming a partially loaded waveguide, we get a cut-off frequency, so the whole system behaves as a pass-band filter.

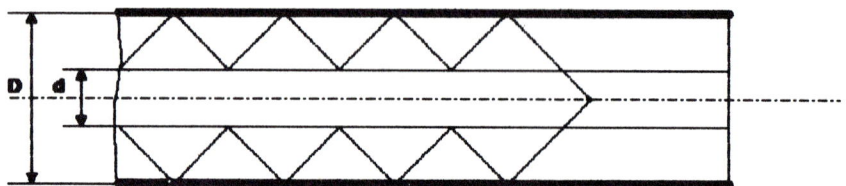

FIGURE 1
Cherenkov radiation in partially filled waveguide

A quantitative analysis can be done making use of the mode expansion of the electromagnetic field in the waveguide matched to the Fourier expansion of the source current.

The configuration of this device gives an additional phenomenon which can be exploited in the microwave range in order to enhance the intensity of radiation. Infact the Cherenkov wave is "trapped" in the dielectric and therefore it is possible to pump power in it in order to encrease the signal to be detected.

We can understand this trapping by considering the wave numbers in the dielectric and in the vacuo. They must satisfy the following triangular relation:

$$k_x^2 + k_y^2 + k_z^2 = k_o^2 \tag{5}$$

$$(k_x^d)^2 + k_y^2 + k_z^2 = n^2 k_o^2 \tag{6}$$

From equation (5) and (6), taking into account that $k_x^d = nk_o \sin\theta$,

$$k_x^2 = (1 - n^2\cos^2\theta)k_o^2 \tag{7}$$

If the relation (2) of Cherenkov Radiation holds, k_x is imaginary. The wave is evanescent towards the vacuum. Not only, but there is no power flux in the direction orthogonal to the direction of the particle. In other words the wave in the dielectric is in condition of total reflection. This means that the wave is trapped, because of the multiple reflections on the metallic walls and on the interface between the dielectric and vacuum. As a result there is the possibility of increasing this power flux if we are able to produce other waves which have an adeguate phase relation among them when the particle beam has a sinusoidal density modulation.

This can be achieved over a bandwidth of frequencies in which we get the maximum sensitivity of the device.

3. PICK-UP DESCRIPTION

A synchronous PU is essentially a slow wave structure in which the phase velocity of the wave propagating in the PU coincides with the beams velocity. In an ordinary rectangular waveguide, with smooth metallic walls, the phase

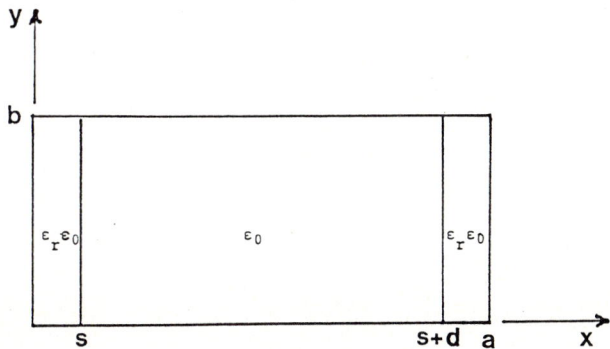

FIGURE 2
Inhomogeneously filled waveguide section

velocity is always larger than the velocity of light. However, if we change the "surface impedance" of two opposite walls of the guide (see fig.2), there can also be slow waves propagating in this inhomogeneous waveguide. Two of the possible solutions to change the surface impedance have been examinated: dielectric slabs and metallic corrugations.

The inhomogeneous waveguide is coupled on both sides with transitions (see fig.3) to ordinary rectangular waveguides. In this transition region the coupling is weak because there is not longer synchronism between the beam and the inducted wave. The problem here is to minimize the reflection such that most of the induced field in the slow wave structure is transformed into power flowing into the output waveguides.

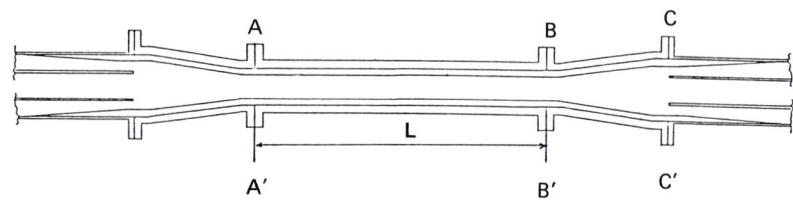

FIGURE 3
Pick-up section

The PU is divided in two parts (fig.3)

- synchronous part AA'-BB' where the phase velocity of the electromagnetic field is approximately the same as that of the particles;
- transition part BB'-CC' where the electromagnetic field excited in AA'-BB' is transformed into a wave propagating in the output waveguide.

The normal modes propagating in this inhomogeneous waveguide are hybrid modes (combination of TE and TM modes). They are classified [1] as Longitudinal Section Electric (LSE) and Longitudinal Section Magnetic (LSM). Fields can be derived from scalar Hertyian potentials: the electric type Π_E for LSM modes and the magnetic type Π_H for LSE modes.

The general solution of the potential equation is a superposition of an even and an odd solution with respect to the transverse dimension. The odd solution for Π involves an even dependence of E_z and suggests to use the inhomogeneous waveguide as a longitudinal pick-up where the hybrid mode is excited by the beam in the centre of the structure. On the other hand, the even solution for π (odd for E_z) is excited where the beam is off-center in the waveguide and this suggests to use the solution for a transverse pick-up.

In order to know the pick-up response we define the pick-up sensivity as the coupling impedance, i.e. the ratio of the complex voltage V on the output termination for a given beam current J_0 travelling through the pick-up.

$$S(x_0) = \frac{V(x_0)}{I_0} \qquad (8)$$

where x_0 is the beam position.

For a transverse pick-up the differential sensitivity S_Δ, for the transverse direction x, is defined as

$$S_\Delta = \frac{1}{I_o} \frac{\delta V}{\delta x}\bigg|_{x=x_o} \qquad (9)$$

4. DESIGN OF A TRANSVERSE DIELECTRIC 10 GHz PU

Selecting the even solution of the potential equation, the fields can be calculated and consequently the power P_T flowing in the waveguide. The following expression for S_Δ results [2]:

$$S = \frac{R_o}{2b\,\omega\varepsilon_o\varepsilon_r k_z(k_y^2+k_z^2)\left[\dfrac{k_x s + \sin k_x s \cos k_x s}{k_x} + \varepsilon_r \dfrac{\cos^2 k_x s}{\cosh^2 q\frac{d}{2}} \dfrac{k_x^d d + \sinh k_x^d d}{2k_x^d}\right]}^{\frac{1}{2}}$$

(10)

$$k_z 2(k_x^d)^2\, \varepsilon_r \frac{\cos k_x s}{\cosh k_x^d \frac{d}{2}}\, \cosh k_x^d (x - \frac{a}{2})\, \sin(\frac{m\pi}{b} y)\, L\, \text{sinc}\left[k_o(\frac{1}{\beta_p} - \frac{1}{\beta_w})\frac{L}{2}\right]$$

where R_o is the PU load, $k_z = k_o/\beta_w$ is the wave propagation constant and $\text{sinc}(x) = \sin(x)/x$.

The design consists in adjusting the dimensions of the PU for a fixed

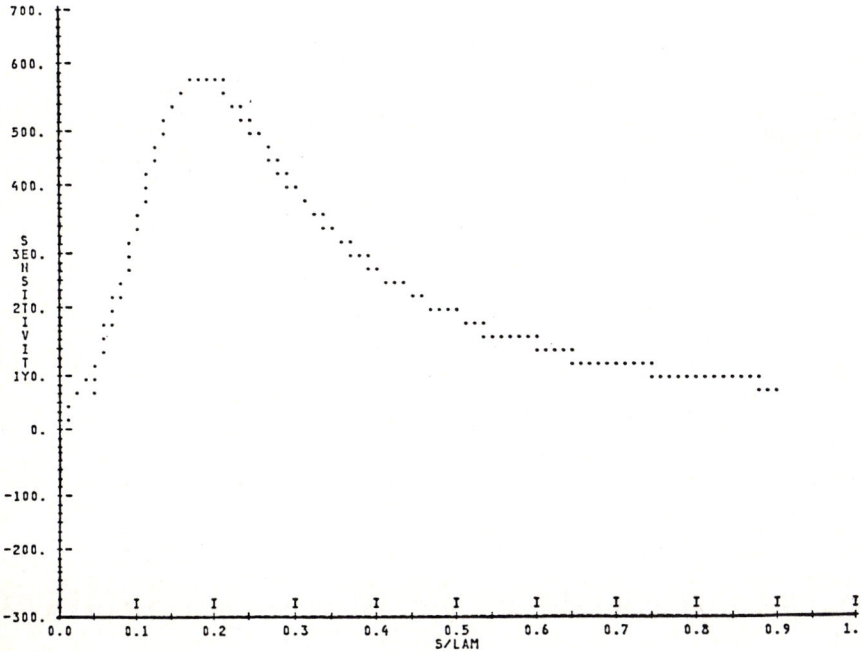

FIGURE 4
Dielectric pick-up. Differential sensitivity as a function of the dielectric thickness.

bandwidth to obtain the maximum sensitivity. The minimum value of d, corresponding to the maximum S_Δ, is selected according to the largest beam dimensions (in our case d=16 mm). For a given length L of the PU and therefore a given bandwidth [3], we compute, for every value of the slab thickness s, the maximum value of S_Δ, in the center of the PU (x=a/2) and at synchronism ($\beta_p=\beta_w$).

In this calculation, we selected the first transverse mode along y (n=1); we can therefore compute b for every value of s, using equations (5) and (6) in which we make $\beta_p=\beta_w=1$ (ultrarelativistic beam), and the dispersion relation. For $\varepsilon_r=2.56$, the result displayed on fig.4, shows a maximum for s=0.19λ, corresponding to b=0.603λ at the frequency of 10 GHz.

For the optimum dimensions obtained previously, the differential sensitivity S_Δ as a function of frequency is displayed on fig.5 for L=400 mm. The

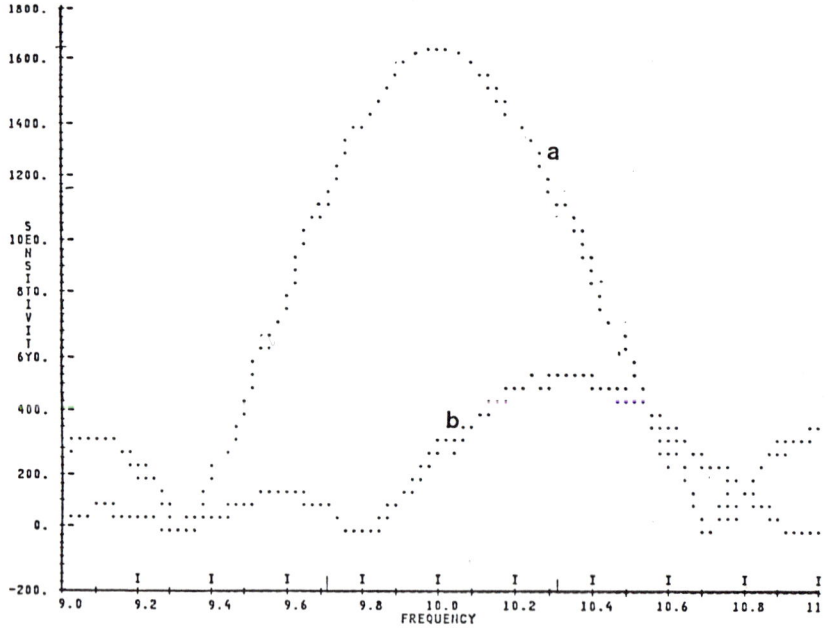

FIGURE 5

Dielectric PU. sensitivity s=.19 λ, b=.603 λ, d=16 mm, L=400 mm.
a) transversal sensitivity. b) longitudinal sensitivity.

longitudinal sensitivity of the same structure (which has been optimized as a transverse PU) is given in fig.6. This is an interesting parameter as it corresponds to the common mode rejection properties of the PU structure.

With another set of parameters: s=0.11λ and b=0.9λ, the curves of fig.6 are obtained. This is certainly not the optimum PU, as far as sensitivity is concerned, but it does have a bandwidth twice as large as the previous one.

Using the previous procedure a dielectric PU model has been specified and built (fig.7). The transition sections were designed empirically, with small angles and tapered dielectric junctions.

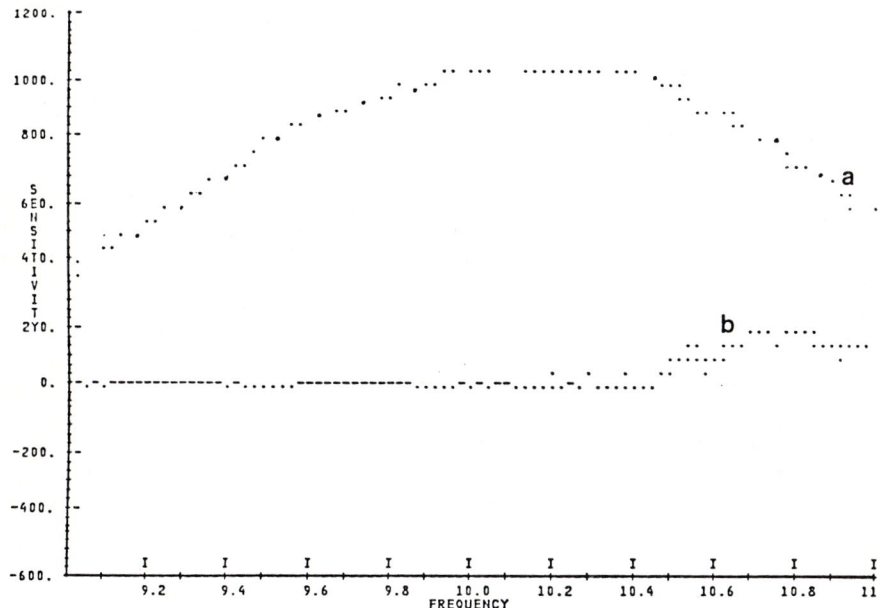

FIGURE 6
Dielectric PU sensitivity. s=.11 λ, b=.9 λ.
a) transverse sensitivity. b) longitudinal sensitivity.

The PU model designed has been measured to determine experimentally its transverse sensitivity. In the chosen method a bifilar line is stretched in the middle of the PU: it represents the beam at two transverse positions separated by the distance x_0 and will directly excite the transverse PU mode.

Fig.8 shows the results obtained for the dielectric wall PU. The top trace represents the line transmission and the bottom line the PU response.

The maximum sensitivity, in the center of the band can be computed obtaining:

$$S_\Delta = 135 \ \Omega/mm$$

Compared with the theoretical values of 200 Ω/mm (eq.10), for a perfect transmission in the transition region this result is in reasonable agreement.

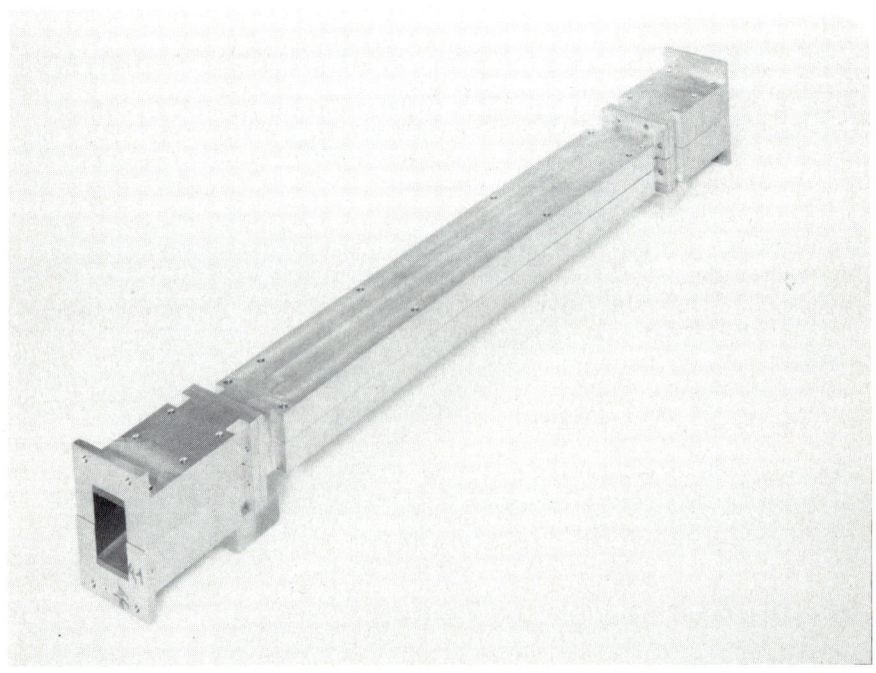

FIGURE 7
Dielectric pick-up

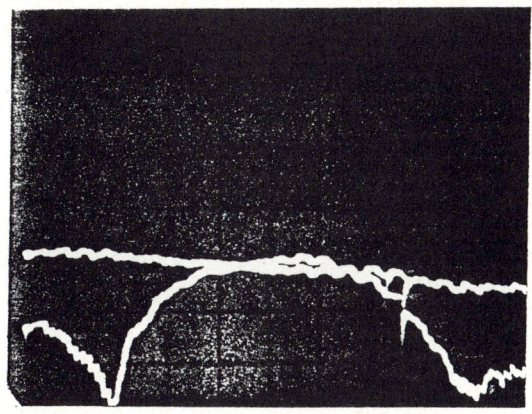

FIGURE 8
Transmission coefficient. Frequency range 8-12 GHz.
Amplification scale 10 dB/div

5. CONCLUSION

Large sensitivities and large bandwidths can be obtained by means of this device which can be used for stochastic cooling and not only for cooling. This sensor can be used for those cases which need sensors in the high frequency range and for large bandwidths.

As an example we already succeeded in deriving a pick-up for measuring the transverse electron temperature of the electron gun for the electron cooling in the LEAR. This device works in the bandwidth 3.5÷7 GHz when the velocity of electrons varies in the range

$$0.91 \geq \beta \geq 0.55$$

The flexibility in the use of this device is due to the large number of parameters, which can be optimized for any specific application.

REFERENCES

1) R.E. Collin, Field Theory of Guided Waves (McGraw-Hill, New York, 1960).

2) D. Boussard and G. Di Massa, CERN Report SPS/86-4(ARF) (1986)

3) G. Di Massa and V.G. Vaccaro, Proc. 5th National Meeting on Applied Electromagnetics, Saint Vincent, Italy, October 1984.

4) G. Di Massa, CERN Report SPS/ARF/note/85-8, 1985.

5) G. Di Massa, CERN Report SPS/ARF/note/85-11, 1985.

LIST OF PARTICIPANTS

A. ATALMI
Dip. di Fisica dell'Università
Via Paradiso 12
44100 Ferrara
Italy

A.M. BADALÀ
Sezione INFN di Catania
Corso Italia 57
95129 Catania
Italy

G.P. BELLINI
Dip. di Fisica dell'Università
Via Celoria 16
20133 Milano
Italy

A. BERTIN
Dip. di Fisica dell'Università
Via Irnerio 46
40126 Bologna
Italy

P. BOCCACCIO
Laboratori Nazionali di Legnaro
Via Romea 4
35020 Legnaro (PD)
Italy

F. BRADAMANTE
Dip. di Fisica dell'Università
Via Valerio 2
34127 Trieste
Italy

T. BRESSANI
Ist. Fisica Superiore dell'Università
Corso Massimo d'Azeglio 46
10125 Torino
Italy

M.P. BUSSA
Sezione INFN di Torino
Corso Massimo d'Azeglio 46
10125 Torino
ITALY

R. CALABRESE
Dip. di Fisica dell'Università
Via Paradiso 12
44100 Ferrara
Italy

F. CANNATA
Sezione INFN di Bologna
Via Irnerio 46
40126 Bologna
Italy

R. CARLIN
Dip. di Fisica dell'Università
Via Marzolo 8
35131 Padova
Italy

A. CENTRO
Dip. di Fisica dell'Università
Via Marzolo 8
35131 Padova
Italy

R. CESTER
Sezione INFN di Torino
Corso Massimo d'Azeglio 46
10125 Torino
Italy

R. CHERUBINI
Laboratori Nazionali di Legnaro
Via Romea 4
35020 Legnaro (PD)
Italy

P.A. CHESSA
Dip. di Scienze Fisiche Università
Via Ospedale 72
09100 Cagliari
Italy

E. CHIAVASSA
Ist. Fisica Superiore dell'Università
Corso Massimo d'Azeglio 46
10125 Torino
Italy

G. COSTA
Ist. Fisica Superiore dell'Università
Corso Massimo d'Azeglio 46
10125 Torino
Italy

P. DALPIAZ
Dip. di Fisica dell'Università
Via Paradiso 12
44100 Ferrara
Italy

F. DE MARCO
Sezione INFN di Torino
Corso Massimo d'Azeglio 46
10125 Torino
Italy

L. DICK
CERN
EP Division
1211 Geneva 23
Switzerland

M. FRISONI
Dip. di Fisica dell'Università
Via Irnerio 46
40126 Bologna
Italy

S. GALASSINI
Facoltà Medicina e Chirurgia
 dell'Università di Verona
37100 Verona
Italy

R. GARFAGNINI
Ist. di Fisica dell'Università
Via Larga 36
33100 Udine
Italy

U. GASTALDI
CERN
EP Division
1211 Geneva 23
Switzerland

M. GIORGI
Dip. di Fisica dell'Università
Piazza Torricelli 2
56100 Pisa
Italy

F. GRAMEGNA
Laboratori Nazionali di Legnaro
Via Romea 4
35020 Legnaro (PD)
Italy

M. GUIDETTI
Dip. di Fisica Politecnico di Torino
Corso Duca degli Abruzzi 24
10129 Torino
Italy

E. LUPPI
Dip. di Fisica dell'Università
Via Paradiso 12
44100 Ferrara
Italy

A. MAGGIORA
Laboratori Nazionali di Frascati
Via E. Fermi 40 - C.P. 13
00044 Frascati (Roma)
Italy

C. MANFREDOTTI
Ist. Fisica Superiore dell'Università
Corso Massimo d'Azeglio 46
10125 Torino
Italy

S. MARCELLO
Dip. Scienze Fisiche Università
Via Ospedale 72
09100 Cagliari
Italy

A. MASONI
Dip. Scienze Fisiche Università
Via Ospedale 72
09100 Cagliari
Italy

G. MEZZORANI
Dip. Scienze Fisiche Università
Via Ospedale 72
09100 Cagliari
Italy

B. MINETTI
Dip. Fisica Politecnico di Torino
Corso Duca degli Abruzzi 24
10129 Torino
Italy

M. MORANDIN
Dip. di Fisica dell'Università
Via Marzolo 8
35131 Padova
Italy

G. MOSCHINI
Dip. di Fisica dell'Università
Via Marzolo 8
35131 Padova
Italy

F. MURGIA
Dip. Scienze Fisiche Università
Via Ospedale 72
09100 Cagliari
Italy

D. PANZIERI
Ist. Fisica Generale dell'Università
Corso Massimo d'Azeglio 46
10125 Torino
Italy

N. PASTRONE
Sezione INFN di Torino
Corso Massimo d'Azeglio 46
10125 Torino
Italy

N. PAVER
Dip. di Fisica dell'Università
Via A. Valerio 2
34127 Trieste
Italy

A. PICCOTTI
Sezione INFN di Torino
Corso Massimo d'Azeglio 46
10125 Torino
ITALY

G. PIRAGINO
Ist. Fisica Generale dell'Università
Corso Massimo d'Azeglio 46
10125 Torino
Italy

P. PISTILLI
Dip. Fisica Università "La Sapienza"
Piazzale Aldo Moro 2
00185 Roma
Italy

E. PREDAZZI
Dip. Fisica Teorica dell'Università
Corso Massimo d'Azeglio 46
10125 Torino
Italy

G. PREPARATA
Laboratori Nazionali di Frascati
Via E. Fermi 40 - C.P. 13
00044 Frascati (Roma)
Italy

G.A. PUDDU
Dip. Scienze Fisiche Università
Via Ospedale 72
09100 Cagliari
Italy

M. PUGLISI
Dip. Fisica Nucl. e Teorica Università
Via Bassi 6
27100 Pavia
Italy

M. RADICI
Dip. Fisica Nucl. e Teorica Università
Via Bassi 6
27100 Pavia
Italy

R.A. RICCI
Dip. di Fisica dell'Università
Via Marzolo 8
35131 Padova
Italy

P. SALVINI
Dip. Fisica Nucl. e Teorica Università
Via Bassi 6
27100 Pavia
Italy

L. SANTI
Ist. di Fisica dell'Università
Via Larga 36
33100 Udine
Italy

F. SAULI
CERN
EP Division
1211 Geneva 23
Switzerland

P. SCHIAVON
Dip. di Fisica dell'Università
Via A. Valerio 2
34127 Trieste
Italy

J. SOFFER
Centre Physique Theor. du CNRS Luminy
Case Postale 907
13288 Marseille Cedex
France

L. TECCHIO
Ist. Fisica Superiore dell'Università
Corso Massimo d'Azeglio 46
10125 Torino
Italy

F. TESSAROTTO
Sezione INFN di Trieste
Via A. Valerio 2
34127 Trieste
Italy

G. TORELLI
Dip. di Fisica dell'Università
Piazza Torricelli 2
56100 Pisa
Italy

F. TOSELLO
Sezione INFN di Torino
Corso Massimo d'Azeglio 46
10125 Torino
Italy

V. VACCARO
Sezione INFN di Napoli
Mostra d'Oltremare - Pad. 20
80125 Napoli
Italy

L. VANNUCCI
Laboratori Nazionali di Legnaro
Via Romea 4
35020 Legnaro (PD)
Italy

E. VERCELLIN
Ist. Fisica Superiore dell'Università
Corso Massimo d'Azeglio 46
10125 Torino
Italy

A. VITALE
Dip. di Fisica dell'Università
Via Irnerio 46
40126 Bologna
Italy

C. VOCI
Dip. di Fisica dell'Università
Via Marzolo 8
35131 Padova
Italy

AUTHOR INDEX

BELLINI, G., 163
BERTIN, A., 189, 201
BEVERINI, N., 427
BRADMANTE, F., 341
BRESSANI, T., 223, 259

CALABRESE, R., 435
CANNATA, F., 249
CESTER, R., 129
CHIAVASSA, E., 311
COSTA, S., 279

DALPIAZ, P., 119
DE VINCENZI, M., 181
DI MASSA, G., 455
DICK, L., 209

FORTI, F., 363

GIORGI, M.A., 363

KUBISCHTA, W., 209

MORONI, L., 163

PAVER, N., 65
PIRAGINO, G., 293
PISTILLI, P., 141, 181
PREDAZZI, E., 49, 103
PREPARATA, G., 3
PUGLISI, M., 387

RICCI, R.A., 235, 321

SCHIAVON, P., 149
SCURI, F., 427
SOFFER, J., 85

TECCHIO, L., 447
TORELLI, G., 427
TRIGGIANI, G., 363

VACCARO, V.G., 455
VITALE, A., 189, 201

SUBJECT INDEX

abrasion, 246
additive quark model, 271
alternating gradient
  synchrotrons, 343
anisotropic chromo dynamics, 16
anisotropic tensor, 42
antimatter abundance in
  early universe, 301
antiproton cooling, 429
antiproton trapping, 428
antiproton-nucleus
- annihilation, 298
- optical potential, 297
- scattering, 295
antiproton-proton annihilation, 119
axial-vector coupling constant, 189

B mean life, 148
baryon-baryon scattering, 87
baryon spectrum, 59
beam
- cooling techniques, 435
- kicker, 437
- losses, 348
- polarization in synchrotrons, 351
beauty physics, 141
beauty spectroscopy, 124
booster synchrotron, 354
bottonium spectroscopy, 123
Buridan's ass, 5

Cabibbo angle, 135
charged coupled devices, 381
charmonium, 50
- spectroscopy in p-p, 123
charm physics, 141
charm spectroscopy, 124
Cherenkov pick-ups, 455
Cherenkov radiation in the
  microwave range, 456
cluster beam target, 211
coherent production, 168
collection time for semiconductor
  detectors, 365
color confinement, 40
color degrees of freedom, 227; 249
conserved vector current, 189
cooling limiting effects, 450
cooling time 452
CP violation in $K^°$-$\bar{K}^°$
  system, 74; 136

$D^{\pm}D^°$ lifetimes, 146
$D^°$, $\bar{D}^°$ mixing, 147
deck effect, 176
deep-inelastic collisions in
  heavy-ion reactions, 240
delta degrees of freedom, 225; 251
deposition experiments, 230
dinuclear system, 242
diquarks, 49
dissipative phenomena in heavy-ion
  reactions, 238
doped semiconductors, 369
double charge exchange
  reactions, 224
double hypernuclei, 272
double spin longitudinal
  asymmetry, 149
doubly composite structrue
  of nuclei, 110
Drell-Yan process, 314

Eikonal approximation, 173
electron cooling
- for $\bar{p}$-p machines, 447
- principles, 442
electroweak theory, 8
EMC effect, 103
eta-nucleus interaction, 318
European Hadron Facility (E.H.F.)
  project, 224; 345
exclusive reactions at large
  angles, 86
experimental areas of E.H.F., 356
extreme conditions of
  nuclear matter, 324

Fermi motion in the EMC effect, 107
fire-string, 19

gauge invariance, 11, 12
gauge theories, 9
generation of quarks and leptons, 65
Glauber model, 165
glueballs, 121
gluon degrees of freedom, 252; 318
gluonia, 253
Goldberger-Treiman relation, 195
Gran Sasso Laboratory, 183
gravitational acceleration, 432

H-particle, 274; 308

hadron scattering experiments, 265
hadron spectroscopy, 49
Hagedorn limiting temperature, 335
heavy flavour
- decay, 145
- hadroproduction, 141
- tagging, 363
helicity conservation rule, 89
helicity formalism, 153
Higgs boson, 33; 126

incoherent production, 164
insertion experiments, 228
intensity frontier, 342

jet targets, 209
junction detectors, 373

$K^+ \to \pi^+$ + anything decay, 132
$K^\circ \to e^\pm + \mu^\mp$ decay, 134
$K^+ \to \mu^- + e^+ + \pi^+$ decay, 134
K_L-K_S mass difference, 135
kaon scattering by nuclei, 287
Kobayashi-Maskawa matrix, 66
Kolbig-Margolis formula, 172
Krolikowski equation, 52-53

lambda-hypernuclei
- decay modes, 269
- excited states, 263
- ground states, 261
- lifetime, 268
- production reaction, 265
- substitutional states, 264
lambda production by
  antiprotons, 306
Landau fluctuations, 371, 375
Laslett space charge
  tune shift, 451
leading baryon, 24
lifetime technique for muons, 203
Liouville's theorem, 436
luminosity, 210
Lund model, 96

magnetic spectrometer for
- hypernuclei, 267
- OBELIX, 230
- SPES II, 294
- Streamer chamber, 298
matter effect in ν oscillation, 185
mean free path of hadrons, 284
meson-baryon scattering, 93
meson spectrum, 28
microstrip detectors, 377
milliweak theories, 76
multifragmentation, 244
multiplicity in p-nucleus
  interactions, 303

muon capture in gaseous
- deuterium, 203
- hydrogen, 202
muon capture in liquid
- deuterium, 204
- hydrogen, 204

neutrino masses, 68
neutrino oscillations, 69; 181
nucleur equation of state, 332
nucleon-nucleon elastic scattering
  matrix, 151
nucleus-nucleus collisions
- at high energy, 321
- at low energy, 235

optical potential, 281

partially conserved axial
  current, 195
participant-spectator model, 326
penguin diagram, 78
percolation of partons, 114
perturbative QCD, 34
phase transition of nuclear
  matter, 322
pion scattering by nuclei, 284
pionic fusion, 225
Pizero spectrometer, 317
polarization in the Λ° decay, 160
polarized atomic beam target, 214
polarized targets, 157
proton scattering by nuclei, 286
proton synchrotron, 342

quantum chromodynamics:
- general, 15
- quantitative tests of, 123
quark degrees of freedom, 311
quark geometrodynamics, 16
quark-gluon plasma, 235; 326
quark-lepton flavour mixing, 65
quasi fusion, 246

radiation damage, 372
rare decays of K mesons
  (general), 129
recombination model, 97
relativistic scheme for
  interacting quarks, 52
RFQ accelerator
- electrostatic field, 391
- general, 387
- structure, 405
- vane tips shaping, 395
Rosenbluth formula, 103

Savvidy states, 36
sea quarks, 112